Springer Texts in Education

Springer Texts in Education delivers high-quality instructional content for graduates and advanced graduates in all areas of Education and Educational Research. The textbook series is comprised of self-contained books with a broad and comprehensive coverage that are suitable for class as well as for individual self-study. All texts are authored by established experts in their fields and offer a solid methodological background, accompanied by pedagogical materials to serve students such as practical examples, exercises, case studies etc. Textbooks published in the Springer Texts in Education series are addressed to graduate and advanced graduate students, but also to researchers as important resources for their education, knowledge and teaching. Please contact Yoka Janssen at Yoka.Janssen@springer.com or your regular editorial contact person for queries or to submit your book proposal.

Sophia Jeong · Lynn A. Bryan ·
Deborah J. Tippins · Chelsea M. Sexton
Editors

Navigating Elementary Science Teaching and Learning

Cases of Classroom Practices and Dilemmas

 Springer

Editors
Sophia Jeong (ID)
Department of Teaching and Learning
The Ohio State University
Columbus, OH, USA

Deborah J. Tippins
Mary Frances Early College of Education
University of Georgia
Athens, GA, USA

Lynn A. Bryan (ID)
Center for Advancing the Teaching
and Learning of STEM
Purdue University
West Lafayette, IN, USA

Chelsea M. Sexton
Mary Frances Early College of Education
University of Georgia
Athens, GA, USA

ISSN 2366-7672   ISSN 2366-7680   (electronic)
Springer Texts in Education
ISBN 978-3-031-33417-7   ISBN 978-3-031-33418-4   (eBook)
https://doi.org/10.1007/978-3-031-33418-4

This Springer imprint is published by the registered company Springer Nature Switzerland AG
The registered company address is: Gewerbestrasse 11, 6330 Cham, Switzerland

This book is dedicated to all of the current and future elementary teachers of science who are committed to making science a meaningful part of children's lives.

Foreword

As a scholar and teacher educator in the early childhood science education community, I have experienced that the reality of science teaching in classrooms involves unpredictable, uncertain, and dilemmatic moments with competing ideologies. High-quality science teaching is not a straightforward and neutral process in which teachers simply apply concepts and methods to classroom situations. Considering the context-specific complexities of science teaching, how can we prepare elementary teachers for meaningful, responsive, equitable, and multi-dimensional science teaching? By marrying theoretical perspectives and contemporary scholarship on elementary science teaching and learning with authentic case narratives, this book is one way to answer this question.

Dialogue with practices through case narratives is at the heart of this book, *Navigating Elementary Science Teaching and Learning: Cases of Classroom Practices and Dilemmas*. The cases in each part present problems and dilemmas that elementary teachers of science have faced in their classrooms. The cases end with reflective questions, not a standardized resolution, and are followed by commentaries from educators in the field. It is notable that the cases were not written for modeling high-quality, idealistic science instruction, but for challenging beliefs, assumptions, and taken-for-granted ways of thinking about elementary science teaching and learning. Various authors' vivid voices and realistic depictions of their day-to-day work serve as a practical resource by stimulating our reflection on the complexities of teacher decision making and readying us for potential science teaching scenarios. More importantly, however, the cases and commentaries in this book trigger a moment of pause and perhaps discomfort about the problems and dilemmas of science teaching, ultimately engaging us in a transformative process to counter our implicit assumptions about children, their ways of learning science, and the sociocultural contexts of the science classroom. The cases and commentaries promote critical awareness of our practices in accordance with the contemporary issues of science teaching.

The topics of nine parts of this book serve as essential elements of a blueprint for improving the quality of science teaching practices and preparation of elementary teachers of science. While considering the challenges of the reconciliation between theory and practice and the rapidly changing needs for context-specific support for both children and teachers, the editors of this book, Sophia Jeong, Lynn A. Bryan, Deborah J. Tippins, and Chelsea M. Sexton, saliently identify the ongoing and timely issues of elementary science education and insightfully update the nine parts. It is to the credit of the editors of this book that they and the diverse authors and commentators produce such an impressive collective work. This book is truly an accessible and informative pedagogical resource for preservice and in-service elementary teachers of science, science teacher educators, and researchers who are committed to creating high-quality science experiences for all young learners.

Sung Eun Jung
School or Department of Teaching
Learning, and Sociocultural Studies
University of Arizona
Tucson, AR, USA
sungeun@arizona.edu

Preface

This textbook, *Navigating Elementary Science Teaching and Learning: Cases of Classroom Practices and Dilemmas*, brings attention to cases that present dilemmas through rich narratives characterizing the lives of elementary science teachers and their students. The cases in this book present authentic problems of elementary science teaching and the experiences of preservice and in-service teachers, science teacher educators, informal science educators, and scientists. Their diverse experiences illustrate the situation-specific and context-dependent nature of science teaching and learning at the elementary level.

The cases are written for, with, and by elementary teachers of science. Each case focuses on contemporary issues of practice relevant to elementary science teaching and learning, followed by one or more commentaries with questions for reflection and discussion. The commentaries build on and extend the case discussions, providing diverse perspectives from multiple stakeholders. To navigate the complex landscape of science teaching, elementary teachers of science require a repertoire of pedagogical approaches to make informed, professional judgments. These judgments require a theoretical basis to guide decision making as teachers navigate the unpredictability and personalized nature of science teaching. The cases and their commentaries serve as the centerpiece for deliberation, assisting elementary teachers of science to foster their professional judgment and develop an understanding of current instructional practices in science at the intersection of science education reform efforts.

We hope that our readers will benefit from these cases as they explore the challenges and dilemmas of elementary science teaching and learning. Though there are no "right answers" or "perfect solutions" to the dilemmas featured in each case, we encourage our readers to use our textbook to engage with well-crafted narratives and elicit meaningful discussion of responses and discourses that lead to a thorough understanding of multiple perspectives, including that of our elementary students. We thank all our contributors who shared their real classroom experiences to help improve elementary science teaching and learning. We are indebted to our teachers and appreciate this opportunity to widely share their knowledge, experiences, and wisdom.

Columbus, OH, USA   Sophia Jeong
                                                                jeong.387@osu.edu

Acknowledgements We would like to acknowledge Mu-Yin (Echo) Lin, whose ideas and technical assistance were invaluable in keeping us on track, organizing our ideas and files, and ensuring the production of a book that will ultimately enhance science teacher preparation. We would also like to thank all our contributing authors whose experiences and advice can enrich the learning of elementary teachers of science for years to come.

Contents

Editors and Contributors

About the Editors

Sophia Jeong is an Assistant Professor of Science Education in the Department of Teaching and Learning at The Ohio State University. Her work draws on theories of new materialisms to explore ontological complexities of subjectivities by examining sociomaterial relations in the science classrooms. Her research interests focus on equity issues through the lens of rhizomatic analysis of K-16 science classrooms. She is passionate about fostering creativity, encouraging inquisitive minds, and developing sociopolitical consciousness through science education.

Lynn A. Bryan is the inaugural Director for the Center for Advancing the Teaching and Learning of STEM (CATALYST) at Purdue University. She is professor of science education and holds a joint appointment in the Department of Curriculum and Instruction and Department of Physics and Astronomy. Her research focuses on science teacher education, particularly teachers' development and enhancement of knowledge and skills for teaching science through the integration of STEM disciplines and modeling-based inquiry approaches.

Deborah J. Tippins is a professor emeritus and distinguished research scholar at the Department of Mathematics and Science Education at the University of Georgia. Her scholarly work focuses on encouraging meaningful discourse around environmental justice in science education. She draws on ecojustice philosophy and the anthropology of science education to investigate questions related to citizen science, sustainability, culturally relevant science, and science teacher preparation.

Chelsea M. Sexton is a former high school environmental science and research methods teacher with a background in marine ecology. She currently teaches elementary preservice teachers while working on her doctoral degree in science education at the University of Georgia. Her research interests center on education for sustainability, justice-centered and case-based pedagogies, and preservice science teacher preparation.

Contributors

Jennifer D. Adams Department of Chemistry, Canada Research Chair in Creativity and STEM, University of Calgary, Calgary, AB, Canada

Rouhollah Aghasaleh School of Education, Humboldt State University, Arcata, CA, USA

Sevil Akaygun Department of Mathematics and Science Education, Bogazici University, Istanbul, Turkey

Valarie L. Akerson Department of Curriculum and Instruction, Indiana University, Bloomington, IN, USA

Caleb Amerman University of Georgia, Athens, Georgia

Holly Amerman Darlington School, Rome, Georgia

Kathryn A. Baldwin School of Education, Eastern Washington University, Cheney, WA, USA

Kelly Baldwin Theiss Elementary School, Klein ISD, Klein, TX, USA

Kathryn M. Bateman Department of Psychology, Temple University, Philadelphia, PA, USA

Ian C. Binns Department of Reading and Elementary Education, Cato College of Education, University of North Carolina at Charlotte, Charlotte, NC, USA

Dearing Blankmann Stout School of Education, High Point Univertsity, High Point, NC, USA

Mark A. Bloom College of Natural Sciences and Mathematics, Dallas Baptist University, Dallas, TX, USA

Alec Bodzin Department Education and Human Services, Lehigh University, Bethlehem, PA, USA

Valarie Bogan Curriculum Development, National Radio Astronomy Observatory, Charlottesville, VA, USA

Stacey Britton Department of Early Childhood Through Secondary Education, University of West Georgia, Carrolton, GA, USA

Katie L. Brkich Department of Elementary and Special Education, Georgia Southern University, Statesboro, GA, USA

Lynn A. Bryan Center for Advancing the Teaching and Learning of STEM, Purdue University, West Lafayette, IN, USA

Terrance Burgess Department of Teacher Education, Michigan State University, East Lansing, MI, USA

Cory Buxton College of Education, Oregon State University, Corvallis, OR, USA

Rodger W. Bybee Biological Sciences Curriculum Study (BSCS), Colorado Springs, CO, USA

Lautaro Cabrera University of Maryland, College Park, MD, USA

Cynthia Canan College of Pharmacy, The Ohio State University, Columbus, Ohio, USA

Heidi Carlone Katherine Johnson Chair in Science Education, Department of Teaching and Learning, Peabody College of Education and Human Development, Vanderbilt University, Nashville, TN, USA

Sarah J. Carrier Department of Teacher Education and Learning Sciences, North Carolina State University, Raleigh, NC, USA

Ingrid S. Carter Department of Elementary Education and Literacy, Metropolitan State University of Denver, Denver, CO, USA

Catherine Citta Department of Communication Sciences and Special Education, University of Georgia, Athens, GA, USA

Jennifer L. Cody State College Area School District, State College, PA, USA; Curriculum and Instruction, College of Education, University Park, PA, USA

Ellie Cowen Nashoba Regional School District, Bolton, USA

Sally Creel STEM and Innovation Supervisor for Cobb Schools, Georgia, United States

Diana M. Crespo-Camacho College of Education, Oregon State University, Corvallis, OR, USA

Kimberly Dinsdale Saratoga Union School District, Saratoga, USA

Helen Douglass Department of Education, The University of Tulsa, Tulsa, OK, USA

Stephanie Eldridge Department of Mathematics, Science, and Social Studies Education, Mary Frances Early College of Education, University of Georgia, Athens, GA, USA

Valery Erickson Fairfax County Public Schools, Fairfax, VA, USA

Ayça K. Fackler Department of Learning, Teaching, & Curriculum, University of Missouri, Columbia, MO, USA

Adronisha T. Frazier Natural Sciences Department, Northshore Technical Community College, Lacombe, LA, USA

Elizabeth A. French Jefferson City Schools, Jefferson, GA, USA

Nick Fuhrman Department of Agricultural Leadership, Education and Communication, University of Georgia, Athens, GA, USA

Thomas Gaudin Sudekum Planetarium, Department of Physics, New Mexico Institute of Mining and Technology, Socorro, New Mexico, USA

Andrew Gilbert College of Education and Human Development, George Mason University, Fairfax, VA, USA

Donna Governor Department of Middle, Secondary, and Science Education, University of North Georgia, Dahlonega, GA, USA

Tamieka M. Grizzle Kimberly Elementary School, Atlanta, USA

S. Selcen Guzey Department of Curriculum and Instruction and Department of Biological Sciences, Purdue University, West Lafayette, IN, USA

Karla Hale College of Education, Western Oregon University, Monmouth, Oregon, USA

Deborah Hanuscin Science, Math and Technology Education, Western Washington University, Bellingham, WA, USA

Kimberly Haverkos Thomas More University, Crestview Hills, Kentucky, USA

Austin David Heil Department of Mathematics and Science Education, University of Georgia, Athens, GA, USA

Danielle Herro College of Education, Clemson University, Clemson, SC, USA

LeeAnna C. Hooper Department of Curriculum and Instruction, The Pennsylvania State University, State College, PA, USA

Sophia Jeong Department of Teaching and Learning, The Ohio State University, Columbus, OH, USA

Natasha Hillsman Johnson Department of Teacher Education, University of Toledo, Toledo, OH, USA

Adam Johnston Center for Science and Mathematics Education, Weber State University, Ogden, UT, USA

M. Gail Jones Department of STEM Education, North Carolina State University, Raleigh, NC, USA

Melanie Kinskey School of Teaching and Learning, Sam Houston State University, Huntsville, TX, USA

Rachel A. Larimore Samara Early Learning, Midland, MI, USA

Heather F. Lavender School of Education, Louisiana State University, Baton Rouge, LA, USA

May Lee College of Education and Human Services, University of North Florida, Jacksonville, FL, USA;
Curriculum and Instruction, College of Education, University Park, PA, USA

Mu-Yin Lin Dual Language Program, Guy B. Phillips Middle School, Chapel Hill, NC, USA

Megan E. Lynch College of Education and Human Services, University of North Florida, Jacksonville, FL, USA

Stefanie L. Marshall Department of Teacher Education, Michigan State University, Twin Cities, MN, USA

Alejandro Gallard Martínez Middle and Secondary Education, Georgia Southern University, Savannah, GA, USA

Lisa Mekia McDonald Teachers College, Columbia University, Science Education, New York, NY, USA

Thomas Meagher Owatonna Public Schools, Owatonna, MN, USA;
Department of Curriculum and Instruction, University of Minnesota, Minneapolis, MN, USA

Felicia Moore Mensah Teachers College, Columbia University, Science Education, New York, NY, USA;
Science Education, New York University-Steinhardt, New York, NY, USA

Alison K. Mercier School of Teacher Education, College of Education, University of Wyoming, Laramie, WY, USA

Olayinka Mohorn Curriculum and Instruction, College of Education, University of Illinois, Chicago, IL, USA

Jaclyn Kuspiel Murray College of Education, Mercer University, Macon, GA, USA

Ryan S. Nixon Department of Teacher Education, Brigham Young University, Provo, UT, USA

Bailey Ondricek Davis School District, Layton, UT, USA

José M. Pavez School of Education, Western Illinois University, Moline, IL, USA

Jeremy Peacock Director of Secondary Education, Jackson County School System, Jefferson, GA, USA

Michelle J. Petersen Georgia, USA

Ashlyn E. Pierson Department of Teaching and Learning, The Ohio State University, Columbus, OH, USA

Julia D. Plummer Department of Curriculum and Instruction, The Pennsylvania State University, University Park, PA, USA

Kate Popejoy Popejoy STEM LLC, Whitehall, PA, USA

Afif Alhariri Pratama Karang Makmur 2 Elementary School, Musi Banyuasin District, South Sumatra, Indonesia

K. Renae Pullen Department of Teaching and Learning, Caddo Public Schools, Shreveport, LA, USA

Cassie F. Quigley School of Education, University of Pittsburgh, Pittsburgh, PA, USA

Sara Raven Texas A&M, College Station, TX, USA

Michael J. Reiss Institute of Education, University College London, London, UK

Alexis D. Riley Science Education, New York University-Steinhardt, New York, NY, USA;
Teachers College, Columbia University, Science Education, New York, United States

Rashida Robinson Teachers College, Columbia University, New York, NY, USA

Maria A. Rodriguez Department of Curriculum and Instruction, University of Texas at Rio Grande Valley, Brownsville, TX, USA

Gillian Roehrig McKinley Elementary STEAM School, Owatonna, MN, USA;
Department of Curriculum and Instruction, University of Minnesota, St. Paul, MN, USA

Troy D. Sadler School of Education, University of North Carolina at Chapel Hill, Chapel Hill, NC, USA

Wahyu Setioko Department of Teaching and Learning, The Ohio State University, Columbus, OH, USA

Elsun Seung Center for Science Education, Indiana State University, Terre Haute, IN, USA

Chelsea M. Sexton Department of Mathematics, Science, and Social Studies Education, Mary Frances Early College of Education, University of Georgia, Athens, GA, USA

Meenakshi Sharma Tift College of Education, Mercer University, Macon, GA, USA

Ji Shen Department of Teaching and Learning, University of Miami, Coral Gables, USA

Teresa Shume School of Education, North Dakota State University, Fargo, ND, USA

Michelle Simon McKinley Elementary STEAM School, Owatonna, MN, USA;
Department of Curriculum and Instruction, University of Minnesota, St. Paul, MN, USA

Mandy McCormick Smith The PAST Foundation, Columbus, OH, USA

Joseph W. Spurlock Department of Teaching and Learning, The Ohio State University, Columbus, OH, USA

Jennifer C. Stark School of Education, University of West Florida, Pensacola, FL, USA

David Steele Alder Graduate School of Education, STEM Education, Redwood City, CA, USA

Michael Svec Department of Education, Furman University, Greenville, SC, USA

Mutiara Syifa Inclusive Science, Technology, Engineering, Art, and Mathematics Education, Department Teaching and Learning, College of Education and Human Ecology, The Ohio State University, Columbus, Ohio, USA

Akarat Tanak Division of Education, Department of Education, Kasetsart University, Bangkok, Thailand

Kristina M. Tank School of Education, Iowa State University, Ames, IA, USA

Joseph A. Taylor Department of Leadership, Research, and Foundations, University of Colorado Colorado Springs, Colorado Springs, CO, USA

Anthony B. Thompson Technology Advancement and Commercialization, RTI International, Research Triangle Park, NC, USA

Deborah J. Tippins Department of Mathematics, Science, and Social Studies Education, Mary Frances College of Education, University of Georgia, Athens, GA, USA

Bhaskar Upadhyay Comparative and International Development Education, Department of Organizational Leadership, Policy and Development, University of Minnesota, Twin Cities, MN, USA

Delinda van Garderen Departement of Special Education, University of Missouri, Columbia, MO, USA

Maria Varelas Department of Curriculum and Instruction, University of Illinois Chicago, Chicago, IL, USA

Geeta Verma School of Education, University of Colorado Denver, Denver, CO, USA

Brianna Wallace SC, USA

Lisa M. Kreklow Weatherbee Westside Elementary School, Sun Prairie School District, Sun Prairie, USA

Julianne A. Wenner Boise State University, Boise, ID, USA

Allison Wilson Early Childhood Education, Phyllis J. Washington, College of Education, University of Montana, Missoula, MT, USA

Randy K. Yerrick Kremen School for Education and Human Development, Fresno, CA, USA

Aarum Youn-Heil Grady School of Mass Communication and Journalism, University of Georgia, Athens, GA, USA

Laura Zangori Department of Learning, Teaching and Curriculum, University of Missouri, Columbia, MO, USA

Introduction: Cases Written for, with, and by Elementary Teachers of Science

1

Meenakshi Sharma and Sophia Jeong

1.1 A New Vision for Elementary Science Education

"Where did the puddles go? What's the moon like around the world? What's hiding in the woodpile? Is the Earth getting heavier?" (Konicek-Moran, 2013; NSTA, 2018). These are just a few examples of authentic questions with an interesting storyline that can be used to engage young children in the investigation of scientific phenomena. Children are innately curious, enjoy exploring and discovering, and instinctively ask questions about the world around them (Trundle, 2015). Elementary teachers of science hold the responsibility to nurture this inherent disposition toward inquiry in children by incorporating phenomena-based practices. Leveraging sensemaking questions and story-lining of scientific phenomena, an elementary teacher of science can anchor science instruction that draws from science principles across disciplines and can provide students with stimulating and engaging spaces for inquiry.

What should science education in K–12 classrooms look like? A benchmark report published by the National Academies of Sciences, Engineering, and Medicine (NASEM, 2021) known as *The Framework*, identified the Next Generation Science Standards (NGSS) and conceptualized a multi-dimensional and integrated view of science teaching. The Next Generation Science Standards (NGSS) and the Framework for K–12 Science Education proposed a

M. Sharma
Tift College of Education, Mercer University, Macon, GA, USA
e-mail: sharma_m@mercer.edu

S. Jeong (✉)
Department of Teaching and Learning, The Ohio State University, Columbus, OH, USA
e-mail: jeong.387@osu.edu

© The Author(s), under exclusive license to Springer Nature Switzerland AG 2023   1
S. Jeong et al. (eds.), *Navigating Elementary Science Teaching and Learning*,
Springer Texts in Education, https://doi.org/10.1007/978-3-031-33418-4_1

multi-dimensional way of science teaching (NRC, 2012). In this document, science teachers are called to align their instruction toward engaging students in authentic science and engineering practices (SEPs), integrating crosscutting concepts (CCCs), and fostering students' understanding of disciplinary core ideas (DCIs). This multi-dimensional view of science teaching and learning suggests that children in science classrooms should not be passively learning, for example, memorizing scientific facts or a list of vocabulary words. By contrast, students should be constructing scientific knowledge by making sense of how and why events, processes, and phenomena happen in the natural world and designing solutions for real-world problems using science and engineering practices. Sensemaking underpins the epistemic goal of science education, which is understanding how we know what we know, and why we believe it (Duschl & Osborne, 2002). The National Academies have since created a consensus document as a "Call to Action for Science Education" which emphasizes the need to prioritize and democratize science education in ways that reach beyond the STEM community while providing rigorous and equitable learning experiences beginning in elementary school (NASEM, 2021).

Children are naturally curious about the world around them, with a proclivity toward exploration and sensemaking (Pullen & Jeong, in this book). Sensemaking involves "being active, self-conscious, motivated and purposeful in the world" (Fitzgerald & Palincsar, 2019, p. 227). Sensemaking sheds light on the idea of "active engagement" by emphasizing what being involved in the act of science can be (Krajcik et al., 2014; Reiser, 2014). In this vein, elementary teachers of science should help students build toward progressively more sophisticated explanations of natural phenomena; understand the core ideas of and engage in scientific and engineering practices; connect science classroom learning experiences with students' interests; and create an equitable and inclusive learning environment for all students to learn science (e.g., DCIs, SEPs). In doing so, elementary teachers of science can nurture a sense of wonder, exploration, inquiry, and critical thinking in young children.

A sociocultural stance toward science teaching and learning recognizes that learning science is a socially and culturally constructed process. Many science educators are of the opinion that sociocultural perspectives are crucial for realizing the "science for all" aspiration advocated by various science reforms, including the NGSS (Hodson, 1999; Lemke, 2001; Rodriguez, 2015). To support equitable science learning discourse in elementary classrooms, teachers must recognize and embrace local, personal, and cultural experiences and diverse ways of knowing and explaining science. Taking a sociocultural stance opens a space for diverse perspectives about providing access to meaningful science in K-12 classrooms for all learners.

1.2   What Should the Next Generation of Elementary Teachers of Science Know and Be Able to Teach?

To enact phenomenon-based teaching effectively, elementary teachers must learn to engage young children in productive, sensemaking conversations that not only empower students to try unpacking and explaining the phenomena, but help students articulate their reasoning and the explanations underpinning these occurrences. To this end, teaching and learning science through sensemaking necessitates a shift in the pedagogy of elementary teachers toward seeing students' ideas and thinking as assets (Pierson et al., 2021; Pierson et al., 2022). This shift requires that teachers recognize and create learning opportunities for young children to show their thinking and articulate their diverse ways of knowing and learning. To effectively build on students' thinking, teachers must develop an ability to "see student learning: to discern, differentiate, and describe the elements of that learning, to analyze the learning and to respond" (Rodgers, 2002).

Elementary teachers of science need preparation to implement science and engineering practices in the classroom (NRC, 2012). The 2022 report by the National Academies of Sciences, Engineering, and Medicine, "Science and engineering in preschool through elementary grades: The brilliance of children and the strengths of educators," recommends that teachers engage young children in their sensemaking and help them construct an understanding about the world. Furthermore, the report suggests that preparing elementary teachers of science to understand and enact science and engineering practices in the classroom is of utmost importance. Accordingly, teachers should receive instruction on creating plans and enacting science teaching and learning that builds on young children's natural curiosity. Teachers should also be prepared to instruct and assess using interesting and relevant phenomena. Such instruction affords collaborative sensemaking and makes science accessible to all learners.

Furthermore, elementary teachers of science need to acquire a knowledge base and ability to enact high-leverage practices (HLPs) in the classroom. HLPs are defined here as research-based teaching practices that, when implemented in the classroom, hold high value for students' intellectual learning (Ball et al., 2009). These practices may look different in different disciplines. For instance, Windschitl et al. (2012) propose the following core set of high-leverage practices (HLPs) for science classrooms: planning for engagement with important science ideas; eliciting students' ideas; supporting ongoing changes in thinking; and pressing for evidence-based explanations. Other research-based HLPs that have the potential to promote science learning entail leading productive classroom discussions.

To advance the equity and epistemic goals of science instruction, teachers need to understand that students may engage through socially and culturally relevant ways of sensemaking; it is important not to conceive their ideas as mere misconceptions (Nasir et al., 2006). Teachers also must develop a critical orientation toward science teaching and learning in order to (a) question the current discourse of teaching and learning science in schools and in their own classrooms; (b) reevaluate what is meant by learning science; and (c) consider who gets to participate in science classrooms and in which ways (Lemke, 2001). Professional standards for science teacher preparation echo aforementioned ideas. For instance, the *National Science Teachers Association* (NSTA), in its position statement, has articulated what science teaching should look like in elementary classrooms (NSTA, 2018). In 2020, NSTA, in collaboration with ASTE, published the latest standards for science teacher preparation. These standards define professional expectations for elementary teachers of science and recognize that, in addition to teaching content, teachers need to engage students in the practices of science and engineering. Most importantly, the standards emphasize equitable science learning; science teachers ought to teach science in a manner that makes learning meaningful and accessible to all learners (Morrell et al., 2020).

1.3   What Are the Systemic Challenges to Preparing Elementary Teachers of Science?

Currently, elementary science educators faces various complexities and challenges. Some studies show that elementary teachers often perceive themselves as having low self-efficacy, low confidence, and high anxiety levels when it comes to teaching science (McClure et al., 2017; Posnanski, 2002). This perception of low confidence can be attributed to factors such as elementary teachers' self-perception of limited content knowledge and conceptual understanding, negative science learning experiences in the past, and limited amount of field experience working in schools (Luera et al., 2005; Smith, 2020a). Latest findings by Horizon Research, reported in the 2018 *The National Survey of Science and Mathematics Education* (NSSME), revealed that only 31% of elementary science teachers feel prepared to teach science, as compared to 73% and 77% for mathematics and English Language Arts (ELA), respectively (Smith, 2020b). In many cases, elementary teachers get extremely limited time to teach science, often rushing through facts without getting a chance to meaningfully engage in in-depth instruction (Banilower, 2019). Only 18 min per day are spent on science instruction in K-3 grades across the nation, as compared to 89 min spent on ELA each day.

According to the 2019 *Nations Report Card*, only 24% of all fourth-grade students in the nation had teachers who spent more than two hours per week on science instruction. Only 30% of the fourth graders got a chance to engage in inquiry-related science activities, up from never to once or twice a year (NAEP, 2019). Findings from these national studies confirm that we continue to face systemic challenges in elementary science education. Additionally, the current status of elementary science education clearly points to inequities and learning gaps across science classrooms in the nation. Elementary teachers must face and overcome these issues to teach science effectively in the elementary grades.

Elementary teachers of science need to engage in "unlearning" to understand and adopt the current vision of science education. They need professional learning that supports implementation of multi-dimensional science teaching and learning along the continuum of their professional career, from preservice to induction teaching and through their time as a veteran educator. We know that traditional, didactic science instruction snuffs children's sense of excitement, wonder, and curiosity. Too many standards and the challenges of day-to-day aspects of teaching science can lead to ineffective teachers and dire consequences for students (Saçkes et al., 2013). Science instruction which is too intense, too early, or too formal, may contribute to a loss of interest, development of a negative attitude, and low academic performance in the classroom (Saçkes et al., 2013). Researchers have found that these problems persist and have long-term implications for students' future academic and career choices. However, paying attention to students' ideas and engaging them in sophisticated and cognitively demanding thinking remains a challenge for many teachers (Davis, 2003; Lehrer & Schauble, 2006; Metz, 1995). Specifically, in relation to elementary science teaching, many teachers need to learn to assist young children with expressing scientific reasoning, reflecting a deficit perspective they may hold regarding student capabilities. The majority of teachers typically emphasize learning science by valuing canonical information (Anderson, 2002; Brown & Melear, 2006; Davis, 2003), while providing students with little to no learning opportunities to engage in scientific sensemaking (Crawford, 2007); Forbes et al., 2013; García-Carmona et al., 2017). Evidence suggests that when provided with rich context, young children are able to articulate their reasoning and use evidence to construct scientific explanations (Tobin et al., 2018; Wonderlin et al., 2022). Therefore, it is imperative to improve learning opportunities for elementary teachers so that they can teach science effectively.

1.4   What Are the Barriers and Affordances for Supporting the Professional Growth of the Next Generation of Elementary Teachers of Science?

How can we prepare elementary teachers of science to implement meaningful, equitable, and multi-dimensional science learning in elementary settings? The landscape of elementary science teaching has become more complex with new challenges. Learning to teach is a crucial phase in the continuum of teachers' professional growth, during which teachers can develop foundational knowledge and practices for teaching. They can continue to build on their teaching repertoire with further professional learning throughout their careers (Feiman-Nemser, 2001). However, various institutional and programmatic barriers may not support teacher learning in a desired manner (Olson et al., 2015). In the context specifically of elementary science teacher preparation, many certification programs offer only one science methods course, which is also sometimes combined with other content areas. Such a slim approach does not afford enough opportunities for preservice teachers to learn to teach science. Field experiences with limited opportunities to teach science further impede preservice teachers' learning by not allowing ample opportunities to hone their practice and receive timely feedback (Abell, 2006). With the new vision of multi-dimensional science teaching and learning as per the NRC (2012)'s Framework, methods courses and field experiences are needed to support elementary teachers' learning and implementation of phenomenon-based instruction, science and engineering practices, and high-leverage practices that can facilitate young children's scientific sensemaking (Bybee, 2014).

Another challenging factor is the formation of science teachers' identities, which are shaped by the teachers' own experiences as science students, opportunities to learn within the university setting, and their current teaching contexts. Teachers also struggle to translate their science content knowledge into effective instruction for young learners (Carrier et al., 2017). Without a doubt, experiences in elementary science teacher preparation courses and in elementary classrooms impact the development of a teacher candidate's identity as a teacher (Kittleson & Tippins, 2012). To this end, elementary science preparation courses can provide a crucial context for realizing a new vision which supports the enactment of meaningful and equitable science teaching. As such, science teacher educators can use various instructional frameworks to model and engage preservice teachers in developing core practices of the profession. For example, a science teacher preparation course centered on developing NGSS aligned HLPs would provide learning opportunities for preservice teachers, weaving cycles of rehearsals and reflection into their practices. To this end, the cases in this book can support course instructors in enhancing their curriculum by providing illustrative dilemmas from real and imagined classroom situations.

1.5 Philosophical and Historical Underpinnings for Using Case-Based Approach in Elementary Teacher Preparation Programs

Various theoretical frameworks have evolved within teacher education that offer unique perspectives regarding how teachers learn best, challenges and strengths to teacher learning, and considerations and stances that matter when supporting teacher learning in various contexts. Decades ago, Lortie (1975) proposed the idea *Apprenticeship of Observation*, according to which preservice teachers come to teacher preparation with tacit insights and preconceived notions about teaching and learning (Grossman et al., 2009; Pajares, 1993). According to Lortie (1975), an average student spends 13,000 h in school before graduating from high school. These hours amount to an informal teaching apprenticeship during which students observe multiple ideas about teaching and learning before entering teacher preparation programs themselves. The idea of apprenticeship generated a huge research base on teacher beliefs, revealing that preservice teachers' tacit theories on teaching and learning deeply affect their classroom practices and learning during teacher preparation. Therefore, it is crucial to pay attention to preservice teachers' prior ideas and beliefs about teaching and learning while developing a teacher education course (Feiman-Nemser, 2001). Learning within a community of peers can help preservice teachers to develop consciousness about their own beliefs and become oriented toward desired forms of dispositions for reform-oriented teaching (Darling-Hammond & Bransford, 2007). In this sense, case-based pedagogy can provide a discussion-based learning environment that elicits and attends to preservice teachers' background knowledge about teaching. In so doing, preservice teachers can have the opportunity to share their ideas as they try to unpack complex teaching situations centered around dilemmas and challenges of teaching highlighted within the cases (Helleve et al., 2021).

More recently, a turn toward practice-based teacher education has shifted attention away from teacher beliefs. Such a shift emphasizes that professional learning of preservice teachers should be situated in the core practices of teaching (Grossman et al., 2009; Windschitl et al., 2012; McDonald et al., 2013; Forzani, 2014). These core practices are being called "ambitious" or high-leverage practices (HLPs). HLPs describe science teaching practices that are research-based and illustrate highly intellectual forms of teaching. These core teaching practices are aligned with reform-oriented approaches to teaching and learning, in contrast to the commonly used traditional, didactic, or "non-ambitious" forms of teaching (Windschitl et al., 2012; Teaching Works, 2020).

Cases and the discussions that accompany them position preservice teachers as decision-makers, critical thinkers, and reflective practitioners who become empowered through improving and changing their own practices (Akbulut & Hill, 2020; Helleve et al., 2021). As such, case-based pedagogy can support ambitious forms of teaching by allowing preservice teachers to engage in classroom-based cases that provide an opportunity to parse, reflect, and critique teachers' practices. Most importantly, case-based pedagogy supports a sociocultural and asset-based viewpoint toward teacher learning (Danish & Gresalfi, 2018). Sociocultural approaches focus on the social activity of learning. Such an approach considers that preservice teachers' learning and their ability to apply their knowledge is intimately integrated with their teaching contexts. A deficit perspective of preservice teachers' learning views their experiences and knowledge as barriers to future learning. An asset-based viewpoint elicits and accepts preservice teachers' ideas as rich resources for learning (Gray et al., 2022), just as young children's ideas and experiences are valued as an asset (Pierson et al., 2021). As such, case-based pedagogy can support learning as a social activity that occurs within a community of teacher educators, peers, and preservice teachers (Arellano et al., 2001). The use of these cases, therefore, can help teachers to be reflective practitioners and consider cultural, social, historical, political, and institutional factors that influence teaching and learning (Heitzmann, 2008).

1.6 What is Meant by Cases and How They Can Serve as a Pedagogical Tool to Promote Science Teacher Learning?

Cases can be defined as narratives based on authentic situations in which teachers or students may encounter unexpected dilemmas. These cases can serve as an impetus for preservice and practicing teachers to build their problem-solving and decision-making capacities (Rosenstand, 2012). Cases are not mere stories or anecdotes but rich representations of situations that are relevant and practical for learning professionals. Historically, educators across professions such as law, business, and medicine have used case-based learning to prepare their student candidates for professional duties (Falkenberg & Woiceshyn, 2008; McLean, 2016). Instructors have also used them as pedagogical tools to scaffold students' to examination of common and authentic problems of the field, generation of solutions, and application of theoretical and practical knowledge. The terms *case-based pedagogy* and *case-based learning* may be used interchangeably, as both terms indicate a fundamental conceptualization of cases as a way to support and facilitate professional learning.

Within the context of science teacher preparation, various studies show the increasing use of case-based pedagogy (Akbulut & Hill, 2020; Darling-Hammond & Hammerness, 2002; Gravett et al., 2017; Heitzmann, 2008; Helleve et al., 2021; Hill, 2020; Papatraianou, 2016). Case-based instruction can be a tool for assisting teachers in analyzing situations and making professional judgments about potential actions in the messy, ambiguous world of teaching. Cases focused on examples of teaching–learning scenarios and situations with challenges and dilemmas can be used to compel preservice teachers to ask questions and offer multiple perspectives on the problem at the center of the case (Cherubini, 2009; Gravett et al., 2017; Helleve et al., 2021). To that end, teachers can study the realistic classroom environments in the cases, through which they can begin to develop effective pedagogical approaches to resolve the emergent dilemmas. Often open ended, through deliberation and reflection on cases, preservice teachers develop insights into the complexities of actual classroom teaching (Kavanagh, 2022). Teachers experiencing dilemmas in the moment may have difficulties seeing the situation through other perspectives; using case-based approaches in teacher education can allow teachers to develop a broader range of perspectives, including that of their very own students. In this way, a case-based pedagogy elicits reflection and perspective-taking through discussions which enable teachers to put themselves into the shoes of different actors within the stories and analyze classroom situations more holistically.

We noted earlier that preservice and in-service elementary teachers have a perception of themselves as having low self-confidence and low self-efficacy when it comes to science teaching (Dennewitz, 2020; Menon et al., 2016). Case-based pedagogy can provide a safe learning environment in which teachers can learn from each other, gaining valuable feedback and developing confidence along the way. This pedagogy provides preservice teachers with an opportunity to learn from complex situations without necessarily having to experience the challenges or difficulties themselves. At the same time, they can engage in examining their profession as elementary teachers more holistically, considering diverse perspectives, variables, and situations. Case deliberations can be discussion-based or involve a written interpretation of issues at the core of the case. Challenges and dilemmas presented in cases intentionally involve preservice teachers in making theory-to-practice and practice-to-theory connections and constructing knowledge in a collaborative manner (Britton & Tippins, 2014; Cherubini, 2009; Koç, 2012). Because case discussions are often implemented among peers, such as during a science teacher preparation course, preservice teachers can engage in co-constructing knowledge about teaching. Accordingly, cases are constructed to compel preservice teachers to not only ask questions, but also offer and critique multiple perspectives (Cherubini, 2009; Helleve et al., 2021).

Researchers are still in the process of gathering robust empirical evidence that connects case-based pedagogy with professional learning of elementary teachers of science, and some studies are showing promising results that, in elementary teacher preparation programs, this pedagogy helps teachers prepare to face and navigate a diversity of issues related to teaching (Arellano et al., 2001; Bhattacharya, 2022; Ching, 2014; Harvard University, 2009; Yadav et al., 2007). Furthermore, science teacher educators can take case-based discussions as an opportunity to assess and evaluate the learning process and ideas central to teacher preparation course design (Bhattacharya, 2022). The use of cases can facilitate science teacher educators' understanding of preservice teachers and their experiences, which they can use as a resource in the learning process. Science teacher educators can also employ cases for the purpose of formative assessment. With this in mind, case-based pedagogy can prompt preservice teachers to apply course themes and knowledge, which instructors can use to assess preservice teachers' growth in their understanding and the rationale behind their decision-making processes (Cherubini, 2009; Helleve et al., 2021).

1.7 Case-Based Pedagogy in Elementary Teacher Preparation

Science is an integral and fundamental part of national and state standards for elementary classrooms and needs to be taught in culturally sustaining ways. Elementary teachers of science need to provide children with opportunities to wonder, discover, think critically, practice methods of inquiry, and develop ideas about how the world works (Tippins et al., 2002). To this end, case-based pedagogy can play a crucial role in elementary science teacher preparation. Science teacher preparation courses can provide learning opportunities to support elementary teachers' learning of science pedagogy per NGSS (2013) and NSTA professional standards. As such, the use of cases in teacher preparation programs can allow preservice and inservice teachers to take a critical stance and ask relevant questions, such as how they can create a meaningful science learning environment for all students.

Science teaching is complex, and while cases are heavily context dependent and unique, they can be used to simulate, model, and illustrate attributes that all effective science teachers should share. Multifactorial aspects of science teaching come into play in elementary teachers' day-to-day practices. Elementary teachers of science can learn to navigate their decision-making processes regarding classroom management, safety issues, students' motivation, standardized testing, and more, and recognize that there is no single perfect recipe for how an elementary science teacher should react, respond, and manage their classrooms (Lewis, 2019). As such, teacher preparation courses and professional development contexts might use a case-based approach to provide an opportunity for preservice and inservice teachers to learn from context-dependent cases of teaching dilemmas and challenges.

Cases do not have to be in a written format. Videocases, for instance, can provide insight into a teacher's practices and students' learning through video clips (Abell & Cennamo, 2003; Goeze et al., 2014). Video clips can be used to model science-specific, high-leverage practices (Tekkumru-Kisa et al., 2018). Similar to written cases, videocases can offer preservice and inservice teachers a risk-free environment to reflect on instruction while adding a layer to developing teachers' abilities to develop a professional vision or teacher noticing (Goodwin, 1994; Sherin, 2007). K-12 science teachers are expected to notice and respond to students' thinking in science classrooms. Thus, videocases can be used to support preservice teachers' noticing and responsiveness to students' scientific and engineering thinking (Clark et al., 2020; Dalvi & Wendell, 2017).

Use of virtual simulations is another example of using case-based pedagogy for teacher preparation, and also for science teachers (Bautista & Boone, 2015; McGarr, 2021; Peterson-Ahmad et al., 2018). These simulations are based on realistic classroom scenarios. Simulated virtual learning environments based on a particular scenario provide preservice teachers real-time interaction with student avatars. These avatars interact with preservice teachers, provide responses, and ask questions based on the scenarios. Science teacher educators can use simulated learning scenarios to help preservice teachers view, analyze, and learn from examples of NGSS (2013) aligned instruction.

1.8 Structure of This Book

This book includes eight sections, each of which begins with an introductory discussion. Each section is organized around a group of cases with diverse, timely topics and will be of particular interest to prospective and practicing elementary teachers of science. Each case is accompanied by commentaries whose purpose is to provide diverse perspectives on the dilemma central to the case. The sections and topics featured in this book are chosen to represent contemporary approaches to teaching elementary science, ranging from new roles for technology to creating inclusive learning environments for all students in elementary science. Other representative topics include assessing children's science understandings, the tensions of standards-based instruction, engaging children in socioscientific issues and reasoning, the role of animals in the science classroom, informal science learning environments, issues of diversity in science teaching, developing mindful science teachers and learners, and preparing children to be lifelong learners of science.

References

Abell, S. K. (2006). Challenges and opportunities for field experiences in elementary science teacher preparation. In *Elementary science teacher education: International perspectives on contemporary issues and practice* (pp. 73–89).

Abell, S. K., & Cennamo, K. S. (2003). Videocases in elementary science teacher preparation. In J. Brophy (Ed.), *Using video in teacher education (Advances in Research on Teaching)*, (Vol. 10, pp. 103–129). Emerald Group Publishing Limited. https://doi.org/10.1016/S1479-3687(03)10005-3

Akbulut, M. S., & Hill, J. R. (2020). Case-based pedagogy for teacher education: An instructional model. *Contemporary Educational Technology, 12*(2).

Anderson, R. D. (2002). Reforming science teaching: What research says about inquiry. *Journal of Science Teacher Education, 13*(1), 1–12.

Arellano, E. L., Barcenal, T. L., Bilbao, P. P., Castellano, M. A., Nichols, S., & Tippins, D. J. (2001). Case-based pedagogy as a context for collaborative inquiry in the Philippines. *Journal of Research in Science Teaching: The Official Journal of the National Association for Research in Science Teaching, 38*(5), 502–528.

Ball, D. L., Sleep, L., Boerst, T. A., & Bass, H. (2009). Combining the development of practice and the practice of development in teacher education. *The Elementary School Journal, 109*(5), 458–474.

Banilower, E. R. (2019). Understanding the big picture for science teacher education: The 2018 NSSME+. *Journal of Science Teacher Education, 30*(3), 201–208. https://doi.org/10.1080/1046560X.2019.1591920

Bautista, N. U., & Boone, W. J. (2015). Exploring the impact of TeachME™ lab virtual classroom teaching simulation on early childhood education majors' self-efficacy beliefs. *Journal of Science Teacher Education, 26*(3), 237–262.

Bhattacharya, A. (2022). Preparing teacher candidates to teach in secondary schools through Socratic case-based approaches. In *Enhancing teaching and learning with Socratic educational strategies: Emerging research and opportunities* (pp. 74–104).

Britton, S. A., & Tippins, D. J. (2014). Practice or theory: Situating science teacher preparation within a context of ecojustice philosophy. *Research in Science Education, 45*(3), 425–443. https://doi.org/10.1007/s11165-014-9430-1

Brown, S. L., & Melear, C. T. (2006). Investigation of secondary science teachers' beliefs and practices after authentic inquiry-based experiences. *Journal of Research in Science Teaching: The Official Journal of the National Association for Research in Science Teaching, 43*(9), 938–962.

Bybee, R. W. (2014). NGSS and the next generation of science teachers. *Journal of Science Teacher Education, 25*(2), 211–221.

Carrier, S. J., Whitehead, A. N., Walkowiak, T. A., Luginbuhl, S. C., & Thomson, M. M. (2017). The development of elementary teacher identities as teachers of science. *International Journal of Science Education, 39*(13), 1733–1754.

Cherubini, L. (2009). Exploring prospective teachers' critical thinking: Case-based pedagogy and the standards of professional practice. *Teaching and Teacher Education, 25*(2), 228–234.

Ching, C. P. (2014). Linking theory to practice: A case-based approach in teacher education. *Procedia-Social and Behavioral Sciences, 123*, 280–288.

Clark, H. F., Sandoval, W. A., & Kawasaki, J. N. (2020). Teachers' uptake of problematic assumptions of climate change in the NGSS. *Environmental Education Research, 26*(8), 1177–1192.

Crawford, B. A. (2007). Learning to teach science as inquiry in the rough and tumble of practice. *Journal of Research in Science Teaching, 44*(4), 613–642.

Dalvi, T., & Wendell, K. (2017). Using student video cases to assess pre-service elementary teachers' engineering teaching responsiveness. *Research in Science Education, 47*(5), 1101–1125.

Danish, J. A., & Gresalfi, M. (2018). Cognitive and sociocultural perspectives on learning: Tensions and synergy in the learning sciences. In F. Fischer, C. Hmelo-Silver, S. Goldman & P.

Reimann (Eds.), *International handbook of the learning sciences* (pp. 34–43). Routledge. https://doi.org/10.4324/9781315617572-4

Darling-Hammond, L., & Bransford, J. (Eds.). (2007). *Preparing teachers for a changing world: What teachers should learn and be able to do.* Wiley.

Darling-Hammond, L., & Hammerness, K. (2002). Toward a pedagogy of cases in teacher education. *Teaching Education, 13*(2), 125–135.

Davis, K. S. (2003). "Change is hard": What science teachers are telling us about reform and teacher learning of innovative practices. *Science Education, 87*(1), 3–30.

Dennewitz, B. J. (2020). *A case study of science teachers' perceptions of self-efficacy in teaching the science and engineering practices* (Doctoral dissertation, Southeastern University).

Duschl, R. A., & Osborne, J. (2002). Supporting and promoting argumentation discourse in science education. *Studies in Science Education, 38*(1), 39–72. https://doi.org/10.1080/03057260208560187

Falkenberg, L., & Woiceshyn, J. (2008). Enhancing business ethics: Using cases to teach moral reasoning. *Journal of Business Ethics, 79*(3), 213–217.

Feiman-Nemser, S. (2001). From preparation to practice: Designing a continuum to strengthen and sustain teaching. *Teachers College Record, 103*, 1013–1055.

Fitzgerald, M. S., & Palincsar, A. S. (2019). Teaching practices that support student sensemaking across grades and disciplines: A conceptual review. *Review of Research in Education, 43*(1), 227–248. https://doi.org/10.3102/0091732X18821115

Forbes, C. T., Biggers, M., & Zangori, L. (2013). Investigating essential characteristics of scientific practices in elementary science learning environments: The Practices of Science Observation Protocol (P-SOP). *School Science and Mathematics, 113*(4), 180–190.

Forzani, F. M. (2014). Understanding "core practices" and "practice-based" teacher education: Learning from the past. *Journal of Teacher Education, 65*(4), 357–368.

García-Carmona, A., Criado, A. M., & Cruz-Guzmán, M. (2017). Primary pre-service teachers' skills in planning a guided scientific inquiry. *Research in Science Education, 47*(5), 989–1010.

Goeze, A., Zottmann, J. M., Vogel, F., Fischer, F., & Schrader, J. (2014). Getting immersed in teacher and student perspectives? Facilitating analytical competence using video cases in teacher education. *Instructional Science, 42*(1), 91–114.

Goodwin, C. (1994). Professional vision. *American Anthropologist, 96*(3), 606–633.

Gravett, S., de Beer, J., Odendaal-Kroon, R., & Merseth, K. K. (2017). The affordances of case-based teaching for the professional learning of student-teachers. *Journal of Curriculum Studies, 49*(3), 369–390.

Gray, R., McDonald, S., & Stroupe, D. (2022). What you find depends on how you see: Examining asset and deficit perspectives of preservice science teachers' knowledge and learning. *Studies in Science Education, 58*(1), 49–80.

Grossman, P., Compton, C., Igra, D., Ronfeldt, M., Shahan, E., & Williamson, P. W. (2009). Teaching practice: A cross-professional perspective. *Teachers College Record, 111*(9), 2055–2100.

Harvard University. (2009). *HBS: How the case based method works.* http://www.hbs.edu/mba/academics/howthecasemethodworks.html

Heitzmann, R. (2008). Case study instruction in teacher education: Opportunity to develop students' critical thinking, school smarts and decision making. *Education, 128*(4), 523–542.

Helleve, I., Eide, L., & Ulvik, M. (2021). Case-based teacher education preparing for diagnostic judgement. *European Journal of Teacher Education, 1–17.*

Hodson, D. (1999). Going beyond cultural pluralism: Science education for sociopolitical action. *Science Education, 83*(6), 775–796. https://www.nsta.org/nstas-official-positions/elementary-school-science

Kavanagh, K. M. (2022). Bridging social justice-oriented theories to practice in teacher education utilizing ethical reasoning in action and case-based teaching. In C. Clausen & S. Logan (Eds.), *Integrating social justice education in teacher preparation programs* (pp. 185–207). IGI Global. https://doi.org/10.4018/978-1-7998-5098-4.ch009

Kittleson, J. M., & Tippins, D. J. (2012). Water can be messy, but that's OK: Reflections on preparing elementary teachers to teach science. *Cultural Studies of Science Education, 7*(1), 41–47.

Koç, K. (2012). Using a dilemma case in early childhood teacher education: Does it promote theory and practice connection? *Educational Sciences: Theory and Practice, 12*(4), 3153–3163.

Konicek-Moran, R. (2013). *Everyday physical science mysteries: Stories for inquiry-based science teaching* (Vol. 1). NSTA Press.

Krajcik, J., Codere, S., Dahsah, C., Bayer, R., & Mun, K. (2014). Planning instruction to meet the intent of the next generation science standards. *Journal of Science Teacher Education, 25*(2), 157–175.

Lehrer, R., & Schauble, L. (2006). Scientific thinking and scientific literacy. In K. A. Renninger & I. E. Siegel (Eds.), *Handbook of child psychology* (6th ed., Vol. 4, pp. 153–196). Wiley.

Lemke, J. L. (2001). *Articulating communities: Sociocultural perspectives on science.*

Lewis, A. D. (2019). Practice what you teach: How experiencing elementary school science teaching practices helps prepare teacher candidates. *Teaching & Teacher Education, 86.*

Lortie, D. C. (1975). *Schoolteacher: A sociological study.* University of Chicago Press.

Luera, G. R., Moyer, R. H., & Everett, S. A. (2005). What type and level of science content knowledge of elementary education students affect their ability to construct an inquiry-based science lesson? *Journal of Elementary Science Education, 17*(1), 12–25.

McClure, E. R., Guernsey, L., Clements, D. H., Bales, S. N., Nichols, J., Kendall-Taylor, N., & Levine, M. H. (2017). *STEM starts early: Grounding science, technology, engineering, and math education in early childhood.* The Joan Ganz Cooney Center at Sesame Workshop.

McGarr, O. (2021). The use of virtual simulations in teacher education to develop pre-service teachers' behaviour and classroom management skills: Implications for reflective practice. *Journal of Education for Teaching, 47*(2), 274–286.

McLean, S. F. (2016). Case-based learning and its application in medical and health-care fields: A review of worldwide literature. *Journal of Medical Education and Curricular Development, 3*, JMECD–S20377.

Menon, D., & Sadler, T. D. (2016). Preservice elementary teachers' science self-efficacy beliefs and science content knowledge. *Journal of Science Teacher Education, 27*(6), 649–673.

Metz, K. E. (1995). Reassessment of developmental constraints on children's science instruction. *Review of Educational Research, 65*(2), 93–127.

Morrell, P. D., Park Rogers, M. A., Pyle, E. J., Roehrig, G., & Veal, W. R. (2020). Preparing teachers of science for 2020 and beyond: Highlighting changes to the NSTA/ASTE standards for science teacher preparation. *Journal of Science Teacher Education, 31*(1), 1–7.

Nasir, N. I. S., Rosebery, A. S., Warren, B., & Lee, C. D. (2006). *Learning as a cultural process: Achieving equity through diversity.*

National Academies of Science, Engineering, and Medicine. (2021). *Call to action for science education: Building opportunity for the future.* The National Academies Press. https://doi.org/10. 17226/26152

National Academies of Sciences, Engineering, and Medicine. (2022). *Science and engineering in preschool through elementary grades: The brilliance of children and the strengths of educators.* The National Academies Press. https://doi.org/10.17226/26215

National Assessment of Educational Progress. (2019). *NAEP report card: 2019 NAEP science assessment.* Retrieve from https://www.nationsreportcard.gov/highlights/science/2019/

National Research Council. (2012). *A framework for K-12 science education: Practices.*

NGSS Lead States. (2013). *Next generation science standards: For states, by states.*

NSTA. (2018). *NSTA position statement: Elementary science education.* https://static.nsta.org/pdfs/ PositionStatement_Elementary.pdf

Olson, J. K., Tippett, C. D., Milford, T. M., Ohana, C., & Clough, M. P. (2015). Science teacher preparation in a North American context. *Journal of Science Teacher Education, 26*(1), 7–28.

Pajares, F. (1993). Preservice teachers' beliefs: A focus for teacher education. *Action in Teacher Education, 15*(2), 45–54.

Papatraianou, L. H. (2016). Case-based learning for classroom ready teachers: Addressing the theory practice disjunction through narrative pedagogy. *Australian Journal of Teacher Education (online), 41*(9), 117–134.

Peterson-Ahmad, M. B., Pemberton, J., & Hovey, K. A. (2018). Virtual learning environments for teacher preparation. *Kappa Delta Pi Record, 54*(4), 165–169.

Pierson, A. E., Keifert, D. T., Lee, S. J., Henrie, A., Johnson, H. J., & Enyedy, N. (2022). Multiple representations in elementary science: Building shared understanding while leveraging students' diverse ideas and practices. *Journal of Science Teacher Education,* 1–25.

Pierson, A. E, Keifert, D., Lee, S., Henrie, A., Johnson, H., & Enyedy, N. (2021, June). Elementary science teachers' use of representations to build shared understanding from students' diverse ideas and practices. In E. de Vries, Y. Hod, & J. Ahn (Eds.), *Proceedings of the 15th International Conference of the Learning Sciences* (pp. 505–508). International Society of the Learning Sciences.

Posnanski, T. J. (2002). Professional development programs for elementary science teachers: An analysis of teacher self-efficacy beliefs and a professional development model. *Journal of Science Teacher Education, 13*(3), 189–220.

Reiser, B. J. (2014, April). Designing coherent storylines aligned with NGSS for the K-12 classroom. In *National Science Education Leadership Association Meeting.* Boston, MA.

Rodgers, C. R. (2002). Seeing student learning: Teacher change and the role of reflection. *Harvard Educational Review, 72*(2), 230.

Rodriguez, A. J. (2015). What about a dimension of engagement, equity, and diversity practices? A critique of the next generation science standards. *Journal of Research in Science Teaching, 52*(7), 1031–1051.

Rosenstand, C. A. F. (2012). Case-based learning. In N. M. Seel (Ed.), *Encyclopedia of the sciences of learning.* Springer. https://doi.org/10.1007/978-1-4419-1428-6_812

Saçkes, M., Trundle, K. C., & Bell, R. (2013). Science learning experiences in kindergarten and children's growth in science performance in elementary grades. *Education and Science, 38,* 114–127.

Sherin, M. G. (2007). The development of teachers' professional vision in video clubs. In R. Goldman, R. Pea, B. Barron, & S. Derry (Eds.), *Video research in the learning sciences* (pp. 383–395). Lawrence Erlbaum.

Smith, P. S. (2020a). 2018 NSSME+: Trends in US science education from 2012 to 2018. *Horizon Research, Inc.*

Smith, P. S. (2020b). What does a national survey tell us about progress toward the vision of the NGSS? *Journal of Science Teacher Education, 31*(6), 601–609.

Teaching Works. (2020). *Eliciting and interpreting student thinking.* Teaching Works Resource Library. https://library.teachingworks.org/curriculum-resources/teaching-practices/eliciting-and-interpreting/

Tekkumru-Kisa, M., Stein, M. K., & Coker, R. (2018). Teachers' learning to facilitate high-level student thinking: Impact of a video-based professional development. *Journal of Research in Science Teaching, 55*(4), 479–502.

Tippins, D., Koballa, T., & Payne, B. (Eds.). (2002). *Learning from cases: Unraveling the complexities of elementary science teaching.* Allyn & Bacon.

Tobin, R. G., Lacy, S. J., Crissman, S., & Haddad, N. (2018). Model-based reasoning about energy: A fourth-grade case study. *Journal of Research in Science Teaching, 55*(8), 1134–1161.

Trundle, K. C. (2015). The inclusion of science in early childhood classrooms. In K. Trundle & M. Saçkes (Eds.), *Research in early childhood science education* (pp. 1–6). Springer.

Windschitl, M., Thompson, J., Braaten, M., & Stroupe, D. (2012). Proposing a core set of instructional practices and tools for teachers of science. *Science Education, 96*(5), 878–903.

Wonderlin, N. E., Lorenz-Reaves, A. R., & White, P. J. (2022). Habitats of urban moths: Engaging elementary school students in the scientific process. *The American Biology Teacher, 84*(5), 284–289.

Yadav, A., Lundeberg, M., DeSchryver, M., Dirkin, K., Schiller, N. A., Maier, K., & Herreid, C. F. (2007). Teaching science with case studies: A national survey of faculty perceptions of the benefits and challenges of using cases. *Journal of College Science Teaching, 37*(1), 34.

Meenakshi Sharma is an assistant professor of Science Education in Tift College of Education at Mercer University. Her work centers around science teacher education and students' learning of science in K-12 classrooms. She specifically focuses on inquiry-oriented science teaching as it aligns with the Next Generation Science Standards (NGSS).

Sophia Jeong is an Assistant Professor of Science Education in the Department of Teaching and Learning at The Ohio State University. Her work draws on theories of new materialisms to explore ontological complexities of subjectivities by examining socio-material relations in the science classrooms. Her research interests focus on equity issues through the lens of rhizomatic analysis of K-16 science classrooms. She is passionate about fostering creativity, encouraging inquisitive minds, and developing socio-political consciousness through science education.

Part I

Considering Diversity Equity and Inclusion

David Steele and Sophia Jeong

"Contemporary science education suffers from a history of stereotyping, linguistic prejudice, and cultural conflict that undermine a school's capacity to provide effective science education for all. The science that is taught today reflects generations of science that was taught with a single audience in mind. This lineage now shapes the way students experience contemporary science classrooms." (Brown, 2019, p. 4).

Brown's (2019) passage situates current discriminatory and oppressive practices found in science classrooms through the lens of the sociopolitical history of science education, where the public perception of a scientist is still envisioned as White, male, heterosexual, and middle class (Yoder & Matheis, 2015). When combined with the narrative that science is an objective field where an individual's identity has no influence on who becomes a scientist, nor on the science produced by those scientists, this perception has the power to create a culture of marginalization that silences individuals with minoritized identities. This powerful combination has created barriers to who is expected to do science and who is welcomed in science spaces, reproducing these hegemonic inequities in STEM classrooms and career fields.

Thus, science teachers need to foster a sense of belonging and inclusion for several important reasons. First, such fostering repositions science teachers to enact anti-deficit and anti-racist frameworks and recognize student identities as assets for learning. Secondly, when teachers recognize the need for a classroom community of belonging and inclusion, they are prompted to develop curriculum that intentionally operationalizes teaching strategies that allow minoritized students to see themselves in the curriculum, while at the same time, positioning students to

D. Steele
STEM Education, Alder Graduate School of Education, Redwood City, CA, USA
e-mail: dsteele@aldergse.edu

S. Jeong
Department of Teaching and Learning, The Ohio State University, Columbus, OH, USA
e-mail: jeong.387@osu.edu

utilize their identities as an epistemic tool for knowledge production in science learning environments.

This part includes nine cases within which elementary teachers of science share their experiences navigating issues of diversity, equity, and inclusion. The reports from the National Academies of Sciences, Engineering, and Medicine (NASEM) highlighted one of the major diversity, equity, and inclusion issues in science education as access to opportunities to learn. To that end, the case written by Cory Buxton and Karla Hale, *No podemos abandonar las ciencias: Supporting Science Home Learning with Multilingual Families*, illustrates how inequitable access to learning opportunities disproportionately affects minoritized students, while Catherine Citta's case, *The Day the Dirt Flew*, shows the importance of accounting for students' differing abilities when teachers consider equitable access and adopt an asset-based perspective in the classroom. In *Can We Start With Science Today?*, Brianna Wallace and Michael Svec illustrate a classroom teacher's efforts to navigate and dispel a deficit perspective of students, while still providing meaningful science learning opportunities.

The 2022 NASEM's report also emphasized the importance of students' science identities and agency. In *I'm Not Stupid in Your Classroom!*, Deborah Hanuscin's case explores this emphasis through demonstrating the importance of including students' self-perceptions as capable learners of science. In *When the Sun Doesn't Shine*, Wahyu Setioko and his teacher-collaborator, Afif Pratama, consider students' diverse ways of knowing, while Ryan Nixon in *Clarissa Says God Doesn't Exist*, illustrates a teacher's affirmation of students' religious backgrounds within the context of science teaching. In *Andrea Draws a Scientist*, Katie Brkich and colleagues highlight a teacher navigating worldviews of students that are different from their own.

The last recommendation by NASEM included an approach to equitable science instruction through seeing science teaching as part of a larger justice movement. The case written by Alexis Riley and Felicia Moore Mensah, *Joseph Has No Money for Groceries!*, examines the intersection of food deserts and students' experience with limited access to healthy foods and questions of race and colorblind curriculum. In *Did I Inherit My Curly Hair from My Mom..., or from My Ma?*, David Steele and Sophia Jeong illustrate the importance of teachers being aware of, recognizing, and rupturing the hidden gender and heteronormative messaging found in science classrooms and instructional practices.

Each case in this part demonstrates how science teachers can rupture and dismantle persistent inequities in science education by engaging in humanizing science teaching practices that both value and affirm the multiplicities of diverse student identities, both visible and invisible. For science to truly be *science for all*, science teachers have to recognize the integral role that students' identity categories—such as race, ethnicity, language, sexuality, gender, socioeconomic status, religion, family structure, and culture—have on student learning and sense of belonging.

References

Brown, B.A. (2019). Science in the city: Culturally relevant STEM education. Harvard Education Press.

National Academies of Sciences, Engineering, and Medicine. (2022). Science and engineering in preschool through elementary grades: The brilliance of children and the strengths of educators. The National Academies Press. https://doi.org/10.17226/26215.

Yoder, J.B. & Matheis, A. (2015). Queer in STEM: Workplace experiences reported in a national survey of LGBTQA individuals in science, technology, engineering, and mathematics careers. *Journal of Homosexuality*, *63*, 1–27.

Case: No podemos abandonar las ciencias: Supporting Science Home Learning with Multilingual Families

2

Cory Buxton and Karla Hale

Abstract

In this speculative case, we imagine the life of an elementary science teacher, Sonia, several years after the COVID-19 pandemic has receded, and schooling has returned to "normal." However, the new normal for students and teachers is likely to require a greater reliance on technology-supported home learning, as well as a renewed attention to "basic skills" for students who are perceived as "at risk" for falling further behind grade level standards. That is, the learning expectations and assumptions developed during the pandemic, with students and families taking on more of the content learning at home through a range of virtual learning platforms, is likely to continue as part of normal learning moving forward, with teachers needing to adapt accordingly. These new expectations have created both opportunities and challenges for teachers. This case study presents the complex intersection of several of these opportunities and challenges, including technology access, language support, and a greater reliance on basic skills curricula that leave elementary teachers with little time for supporting robust science learning experiences.

Sonia is a first-year bilingual elementary teacher working in a fourth-grade, dual-language classroom in Oregon's Willamette Valley. She is expected to make use of various technologies to promote online communication with her students and their parents and to support her students' home learning. Robust home learning opportunities, using a range of online platforms, are now seen as essential if students are

C. Buxton (✉)
College of Education, Oregon State University, Corvallis, OR, USA
e-mail: buxtonc@oregonstate.edu

K. Hale
College of Education, Western Oregon University, Monmouth, Oregon, USA
e-mail: halek@wou.edu

© The Author(s), under exclusive license to Springer Nature Switzerland AG 2023   21
S. Jeong et al. (eds.), *Navigating Elementary Science Teaching and Learning*,
Springer Texts in Education, https://doi.org/10.1007/978-3-031-33418-4_2

to "keep up with standards," a concern that has grown in public schools in the few years since the COVID-19 pandemic disrupted existing school structures. Sonia quickly recognized that her immigrant students from multilingual families had all been identified as at risk for falling behind by her school administration. Sonia's class of 28 students includes half Spanish-dominant speakers, who are mostly first- and second-generation immigrants from Mexico, Honduras, and Guatemala, and half English-dominant speakers, largely of European descent. Starting during preplanning, before Sonia had even met her students, she felt pressure from her administration to focus her instruction on basics skills in math and reading and to scale back or eliminate the more robust science curriculum she had been planning to make the centerpiece of her classroom. In addition, Sonia's early communication with some of her students' families raised questions about their access to the more dynamic, virtual content required for home learning, due to both technology access issues and a lack of language support provided with these tools.

As one of the Next Generation Science Standards (NGSS Lead States, 2013) early adopting states, Oregon science standards now call for a robust three-dimensional science learning experience for all elementary students. Sonia, like most of the elementary educators that our teacher preparation program graduates, was both excited and nervous about engaging her emergent bilingual students in rich science meaning-making. Sonia was committed to the goal that all her students would come to see science as a set of practices for better understanding the natural world and for solving and communicating about everyday problems in their lives.

One month into the school year, however, Sonia's focus had turned from how to ensure robust science learning for all students to expressing the concern that many of her students lacked the technological resources and support at home to participate in the expanded home learning that her administration expected Sonia to facilitate and monitor.

> One thing I learned was that parent connectedness and communication was harder than I thought, even as a Spanish speaker. I'd ask what do you have in terms of technology at home? And parents were telling me, yeah, we've got technology. And then I realized, well, no, they actually don't have a laptop. They have a smartphone, but that's really not sufficient to do the online work that my kids need to do. And then of course, not everybody has reliable Wi-Fi. So basically, I discovered that it's a much larger gap than I thought in terms of families and parents really even feeling comfortable getting on and using a laptop versus their phone. How is our school addressing this? Too slowly and without enough resources. Basically, it falls to us as teachers to help our students' families navigate these technology issues.

In addition to worries about access to technological resources, Sonia also worried that her Spanish-dominant students did not have the benefit at home of helpful bilingual language supports. Sonia worked hard to provide these resources in her classroom during the school day, and she knows that many of her students depend on them.

As we've asked families to take on more responsibility and support for their children's content learning, we're also asking them to support bilingual language learning, even if we don't say it explicitly. But as a dual language teacher, I've had lots of practice learning how to scaffold language bilingually and my classroom is full of bilingual language supports. What kinds of bilingual supports and resources do my students have at home? I can tell you that it's all over the map and completely unequal.

Mostly, however, Sonia was distraught because of the persistent pressure from district and school administrators to focus on "basic skills" in foundational reading and mathematics with little room for science investigations or for inquiry of any kind. As she related:

Admin is so worried about the kids falling further behind, but I'm more worried about helping them get ahead. All I hear from admin is fear about the reading and math gaps we've seen since we returned to in person classes after COVID. Of course, I'm worried about that too -- I'm not blind. But I mean, that's such a low bar. I want us to do weekly investigations of phenomena focused on the NGSS practices. I want us to use the talk moves and the lab roles with explicit language of science goals like we learned in teacher preparation. I want to do biographies of scientists of color and talk about why broadening participation in science matters. But I need to squeeze that all in between the mandatory math worksheets and basic reading comprehension exercises that I'm required to do, and that's just not working for some of my students, like Antonio.

Antonio is one of my Spanish-dominant kids who has been getting less and less engaged in class in recent weeks. At the start of the year, when I was talking about wanting to do a lot of science, Antonio seemed really excited about that. The other day at our morning meeting, he said he was sad because we weren't doing as much science as I had promised. He told me, "¡No podemos abandonar las ciencias porque es demasiado importante!" (We can't give up science because it's too important). I told him to hang in there, and I was trying to bring in more science, but honestly, I wasn't sure how to make time for it. Then I thought, could I send some of the science home? So, this week, in addition to sending home another packet of required worksheets and online assignments for home learning, I included a Ziploc bag with the materials to do a thaumatrope activity we learned in methods class.

Two weeks later, Sonia had new frustrations.

Antonio was so excited to have a science activity to do at home and was quick to complete the hands-on portion of the activity and bring it to school to show me. He was able to clearly explain the activity and what happened in class, but he failed to complete the write up about the activity that was posted and submitted on our class learning platform. I made a second activity to send home, and he did that too. But I still haven't been able to get Antonio to do the math and reading worksheets or the other written science assignments on the digital learning platform. So, I decided to do a home visit to talk with Antonio's family about how we could use Antonio's enthusiasm for the hands-on science activities to promote his engagement with the other learning materials.

When I arrived at Antonio's apartment, there were several extended family members gathered and interested in talking about how to support Antonio's learning. Antonio's two uncles, who worked in an auto shop and as a project manager in construction, were particularly excited to talk about how they had always supported his interest in science and had helped him with the two recent activities sent home. They felt confident in their ability to support him with these hands-on activities because they had the benefit of using their home language to explain what they were thinking, and they described to me how they apply science and engineering in their work in practical ways. Antonio had spent many nights and weekends helping his uncles at their businesses. It was clear that Antonio has ample support at home for his science learning, but I was still left with my concern about support for the digital schoolwork that Antonio is now required to complete.

As I continued to discuss school expectations with Antonio, his mom and his uncles, it became apparent that one of the challenges was the use of multiple learning platforms and technology applications. Antonio's mom expressed frustration because she was unable to understand the specialized apps that he was being asked to use. Just when they thought they were getting it figured it out, something new was added or changed. It was clear that they needed support to engage in the multiple technology resources that were being incorporated on the digital learning platforms.

Thinking back to the challenges that teachers and students faced during the COVID-19 pandemic, I reflected on how neighboring families had sometimes joined together in community learning pods to support the fully online learning that had been necessary then. I raised the possibility of forming a new community learning pod consisting of neighborhood parents, extended family members and students to support each other in learning the new platforms and keeping up with the changing technologies. We also talked about how the school might be able to provide language resources and technology support for this learning pod. Antonio's family was excited by this idea and quickly agreed to participate.

I left Antonio's apartment more excited than I had been for months, optimistic that a learning pod approach could help me address multiple challenges, from technology, to bilingual language support, to motivation and building on families' funds of science knowledge. That night, however, as I lay in bed, I began listing all the steps it would take to get one learning pod in place but had to stop when I started feeling overwhelmed. Then I realized that this learning pod might serve the needs of four or five of my students who lived in the same neighborhood as Antonio. Then I thought of other students in other neighborhoods. How many learning pods could I support? I'm just one first-year teacher. Why would I think I could do this? Why would I think this is part of my job? But if I don't try to provide the support that I know my students need, then who will?

For Reflection and Discussion

1. For a number of years, teachers and schools have been challenged by equity issues around technology access, usage, and support, as learning through technology plays an increasingly central role both in school and at home (Dolan, 2016). This digital divide, which ranges from Wi-Fi access, to available hardware and software, to support for troubleshooting, has been increasingly visible as greater access to technologies at home becomes necessary for access to learning. As a classroom teacher, what ideas do you have for how you can help your school address these technology challenges?

2. Sonia became increasingly excited about the idea of starting her own multilingual learning pods to better support her students outside of school but began to wonder about the range of challenges that starting such a venture would raise. She started listing questions and concerns until she had to stop, daunted by the list she generated. Make your own list of challenges that you would think about in Sonia's situation.

3. During the COVID-19 pandemic, families came together on their own to support each other in online learning. As a teacher, how can you continue to promote this kind of community action to support learning and incorporate the many assets of your students' families and communities?
4. While schools have increasingly recognized the challenge of resource inequalities and are trying to address these issues, too often, they continue to frame these issues as deficits. New teachers are not blind to these challenges but have also been trained to be more asset-oriented in their thinking (Fránquiz et al., 2011). How can you simultaneously be realistic about the needs of your students and families while maintaining an asset orientation?
5. Beyond the idea of learning pods, what other ways can families and communities help support authentic hands-on science learning, multilingualism and technology in home and community contexts?

References

Dolan, J. E. (2016). Splicing the divide: A review of research on the evolving digital divide among K–12 students. *Journal of Research on Technology in Education, 48*(1), 16–37.

Fránquiz, M. E., Salazar, M. D. C., & DeNicolo, C. P. (2011). Challenging majoritarian tales: Portraits of bilingual teachers deconstructing deficit views of bilingual learners. *Bilingual Research Journal, 34*(3), 279–300.

NGSS Lead States. (2013). *Next generation science standards: For states, by states.* The National Academies Press.

Cory Buxton is a Professor of Science Education at Oregon State University. His research focuses on how teachers can provide more equitable science learning opportunities for all students and especially for multilingual learners. His most recent work and new NSF grant are on developing and testing a model of teacher professional learning that connects students' language development, cultural connections to science, knowledge of science for solving social problems, and STEM career pathways.

Karla Hale is an Instructor in Science Education at Western Oregon University and was a previous K-12 teacher in elementary and secondary science. Her research focus is inspiring new teachers to provide equitable classroom environments that promote inquiry-based STEM instruction for all students. Karla recently completed her PhD in STEM Education at Oregon State University.

Commentary: *La Ciencia en el Arte de hacer Tamales*: Finding Other Ways to Support Families

3

Diana M. Crespo-Camacho

Abstract

This is a commentary to the case narrative, *"No podemos abandonar las ciencias: Supporting Science Home Learning with Multilingual Families"* written by Cory Buxton and Karla Hale.

Antonio is right, *no podemos abandonar la ciencia* (we cannot abandon science). With this idea in mind, my job as Community Liaison, before the COVID-19 pandemic, was focused on looking for ways to empower and involve parents in their children's science learning at a bilingual elementary school. Last year, I organized an event called *La Ciencia en el Arte de Hacer Tamales* (Science in the Art of Making Tamales) to connect families with the science already in their lives. I invited a group of Spanish-speaking parents, which I am part of, and their children to cook sweet tamales while they were learning science. It turned out to be a good way to support students' science learning by incorporating families' funds of knowledge.

However, if we think about the possible future that Sonia is experiencing, she needs not only to find ways to support students' science learning but to support families to navigate technology issues to help students get ahead and not just keep them from falling further behind. For example, Sonia visited Antonio's home to communicate individually with his parents to learn firsthand how she could support Antonio to complete other school assignments. She also considered the idea of starting community learning pods where parents can support each other in learning. These strategies have potential to engage Latino parents in addressing the issues Sonia is facing because both reflect core values cherished by us such as

D. M. Crespo-Camacho (✉)
College of Education, Oregon State University, Corvallis, OR, USA
e-mail: Diana.Crespo@oregonstate.edu

caring and collaboration. Yet these strategies also come with multiple challenges, as Sonia reflected.

But there is another strategy that may complement Sonia's efforts to engage parents in the process of schooling and encourage them to participate. This strategy would involve Sonia's school reaching out to parents and making arrangements to assist them in learning (Delgado-Gaitan, 2012). For this strategy to be successful, Fuentes (2005 as cited in Baquedano-López et al., 2013) noted that it is important that parents are partnering with each other and with other organizations before they partner with the school.

An example of this strategy is the group of parents that attended the tamales event mentioned earlier. This group is the Latino Parent Teacher Association (PTA) of the school. The group was created by a concerned parent to provide a connection to the existing PTA for Spanish-speaking parents. The group not only provides this connection but also serves as a social network that brings together parents with familiar social and cultural ties.

Before the COVID-19 pandemic, the group met every Friday in the school cafeteria, and it was regularly attended by six to ten moms and one dad. The meetings were in Spanish, lasted for an hour or so, and always included coffee and Mexican food that some of the moms brought to share with the group. During the meeting we were regularly informed about PTA and school activities, once a month we helped teachers to prepare materials for their classes, and twice a month, somebody from the school or from outside came to talk to the group about a topic of interest. When there was a guest speaker, more parents attended the meeting than regularly.

In one of these meetings, the school introduced the new math curriculum for Grades 3–5 that the school had recently adopted. We learned about the structure of the curriculum and where to find the lessons and homework. During the meeting, a couple of moms expressed their concern about their difficulty in helping their kids with math. After the meeting, the PTA and the school started conversations and they decided to hire a teacher to support parents' own mathematical learning so they could help their kids. A group of moms attended a couple of evening work-shops and they were very happy to be able to understand math. In the next meeting, one mom said satisfied, "*ya puedo ayudar a mi hija con las multiplicaciones*" (Now I can help my daughter with multiplication).

In addition to these connections with the school, the group meetings provided us with the opportunity to learn and enjoy ourselves together, while we shared our daily life, ideas, problems, and experiences with each other in a respectful way. In Spanish, we call this *convivencia*. In the school year, during weekly meetings and school events, there is always time to chat and share coffee and food. During the summer, we had the opportunity to participate in a series of field trips to visit the forest and natural areas in Oregon with the support of a local NGO. Through these *paseos*, as the mothers called the field trips, families were able to learn about those places and enjoy nature together.

Today, closer to the end of the COVID-19 pandemic, we can begin to imagine a new normal future. Recently, I asked Rosa, the leader of the Latino PTA group

¿Cómo imaginas el papel del grupo en apoyar el aprendizaje de nuestros hijos cuando regresemos a la "normalidad"? (How do you envision the role of the group in supporting our children's learning when we return to "normality"?).

> *Vamos a estar en comunicación continua con la escuela y es importante que nos sigamos capacitando para apoyar a nuestros hijos en aspectos básicos como es lectura y escritura, y las matemáticas. Además, vamos a mantener la invitación abierta al resto de las familias en la escuela para que se unan al grupo porque juntos podemos hacer más y mejores cosas por nuestros hijos.* (We will be in constant communication with the school and it is important that we continue learning how to support our children in basic aspects such as reading and writing, and mathematics. In addition, we will keep the invitation open to the rest of the families in the school to join the group because together we can do more and better things for our children).

Rosa's response shows a strong commitment to helping families and collaborating with the school. Commitment, collaboration, and communication are three of the five elements necessary to build trust and effective partnerships between parents and the school (Delgado-Gaitan, 2013). The other two elements, consistency and cooperation, have been present in the way the group has been functioning. These five elements and the space the group creates for *convivencia* motivate all of us to participate and be present.

If the future brings us the challenges that Sonia is enduring, our parents' group will be ready to partner with the school and help teachers to support learning. Teachers should not feel that these challenges are theirs to solve alone. But in Sonia's case, where such a group is not already in place, starting one brings another set of challenges. What challenges do you imagine facing if you wished to start such a parents' group?

References

Baquedano-López, P., Alexander, R. A., & Hernandez, S. J. (2013). Equity issues in parental and community involvement in schools: What teacher educators need to know. *Review of Research in Education, 37*(1), 149–182.

Delgado-Gaitan, C. (2012). Culture, literacy, and power in family–community–school–relationships. *Theory Into Practice, 51*(4), 305–311.

Delgado-Gaitan, C. (2013). *Creating a college culture for Latino students: Successful programs, practices, and strategies.* Corwin.

Diana M. Crespo-Camacho is a Ph.D. candidate in Language, Equity and Educational Policy at Oregon State University working as a Doctoral Research Assistant on the NSF project Supporting Students' Language, Knowledge, and Culture through Science (LaCuKnoS) in the College of Education. LaCuKnoS seeks to develop and refine a model of justice-centered science education, bringing teachers, students, families, and researchers together as co-learners. Diana's research interests include understanding how multilingual family engagement enhances their children's science learning.

Case: The Day the Dirt Flew

4

Catherine Citta

Abstract

Ms. Cat is a first-year teacher in an inclusive PreK classroom in rural northeast Georgia. The majority of her students have goals involving communication, social skills and self-regulation. Ms. Cat created a lesson for all her students to plant seeds in their school's courtyard as a part of their "plants and flowers" unit. The content of the lesson is in the forefront of her mind as she begins the lesson, but then individual students' needs take precedent. What happens when you have to take a step back from the content and focus on developmental skills, but still make the science learning activity meaningful for everyone?

My Pre-K students and I just watched a time-lapse video showing a seed growing into a flower and are reading a book about farming. We are walking outside to plant our seeds, but maybe watching that video wasn't the greatest idea since now all my students are wondering why there aren't any flowers growing from the soil. "Ms. Cat, where are the flowers?" "Why is the dirt empty?" I promise them we're going to explore but they first must go stand around the garden beds. "Let's spread out. I know you want to know where the flowers are, so let's get to gardening!"

My assistant, Mrs. Sarah, and I divide the students into groups so that everyone is circled around one of the four garden beds. We start off the activity by having everyone dig their hands into the dirt: "How does the dirt feel? Is it cold or hot? It is wet or dry?" Caitlin answers, "It's cold. I like it." Some of the other students nod along humming their "yeah, it's cold" and "mmmhmmm."

While the students are digging into the dirt with their hands, I glance over at Marcus, who is not a fan of soft textures. Anticipating that this experience might

C. Citta (✉)
Department of Communication Sciences and Special Education, University of Georgia, Athens, GA, USA
e-mail: ccitta@uga.edu

be challenging for him, Mrs. Sarah and I already planned for Mrs. Sarah to help him put his hands in the dirt: "Marcus, it's time to touch the dirt. We're going to feel it with our hands." I try to verbally prepare him for the experience, but my prompting does not help. Mrs. Sarah leans over to guide his hands into the dirt. "Ompf." Oh no! I knew Marcus might not want to give it a try, but he bumps into Mrs. Sarah as he goes off running…and there goes Mrs. Sarah after him. Turning quickly to make sure everyone else is on task, I look back at Mrs. Sarah and Marcus, and whew, she has his hand and they begin walking back over to the group—which is good because my other students are getting restless. "Ms. Cat, I'm bored!" Uh-oh. It is definitely time to move on: "Okay everyone, hands out of the dirt! Do you remember how in our book the farmer had to plant the seeds? That's what we're going to do. Some of us are going to dig holes for the seeds to go in."

Students chorus around me:

I want a turn!

My turn!

Let me dig!

Me! Me! Me!

"Okay, so we are going to use trowels to dig into the dirt," I tell them. "A what?" asks Blake. "A trowel," I respond, "It's a small shovel you can hold in your hand."

I step up to the first garden bed. "Blake, why don't you and Quniayh take the two trowels. You're going to hold it like this." I put my hand over Blake's to help him dig somewhat of a trench. "We're going to take our trowel and drag it across the dirt to make a spot for the seeds to go." Quniayh follows what I model and then I walk over to the next garden bed to give a trowel to another student.

Blake is raking the dirt back and forth. "No Blake, just one line toward you. Now you need to pass it to Caitlin." He walks over and hands it to her. Giving Blake positive praise for following the direction, I tell him, "Thank you, Blake, that was a great job giving it to Caitlin."

Not every student is going to get a turn, so I tell them, "Alrighty, some of you are going to pour out some of the seeds from the packets. Then if you haven't had a turn digging or sprinkling seeds, you will get to help push the dirt to cover the seeds. After that, we will take turns watering our garden so our plants can grow."

I demonstrate how to sprinkle the seeds into the line of dug out dirt and then hand off the seed packet to a student. I move to the next bed and show the students how to use the watering can.

I look over at Marcus's garden bed, and I see that Mrs. Sarah and Marcus are sitting on the ground with the dirt in Mrs. Sarah's hands. She begins to rub the dirt onto his hands. "Mrs. Sarah, that is a great idea." I am glad she thought of that. I do not want him to be upset, but it is important for him to engage in what we are

Fig 4.1 Fruits of the students' labor

doing. He can experience the feeling of the dirt but does not have to be elbows deep in it.

I walk around and see that a few of the students are starting to sprinkle the seeds into their garden beds. I turn back to the first bed and see Caitlin dump the entire packet out. Whoops, I forgot that she is going to need help with this part. Self-control is not her strong suit. And now she is trying to dig them all out with her hands.

"Woosh!" I hear the sound of water hitting the ground. Adrienne, a student at the furthest bed from me, is screaming and laughing in a high pitch giggle only a 4 year old can make. This cannot be good. I turn and look behind me. She is soaking wet, water dripping from her hair, and the watering can dangling in her hand. "Adrienne, did you just dump the water on yourself?" "Mmhmm." If I wasn't so stressed out right now, I'd laugh.

"Ahhhhhhhhhhhhhhhhhhhhhhhhhhhh!" I turn back around, and now Blake is running around his garden bed with Quniayh chasing after him with a handful of dirt. She raises her hand and cocks her arm like she's going to throw a baseball.

There it goes… The dirt is flying (Fig. 4.1).

For Reflection and Discussion

1. In what way(s) was Ms. Cat proactive in preparing for this science activity?
2. What do you think were Ms. Cat's science-specific learning goals for this activity?
3. What strategies could Ms. Cat have implemented to support the success of all students with this science activity?
4. Some students are not able to participate in every part of the science activity (digging, sprinkling seeds). What are ways in which the activity could be structured/differentiated to accommodate varying levels of participation?
5. Should all students be required to participate (e.g., Marcus and the dirt) in "hands-on" science learning? Why or why not?
6. What might have happened if Ms. Cat did not have the support of an assistant teacher?

Catherine Citta is a doctoral student in the Department of Communication Sciences and Special Education at the University of Georgia. Her focus areas include education and intervention for children aged from birth to 5 years and their families. She is a former early childhood special education teacher and early intervention supervisor and trainer.

Commentary: Tend the Soil Before Planting the Seeds

5

Rachel A. Larimore

Abstract

This is a commentary to the case narrative, *"The Day the Dirt Flew"* written by Catherine Citta.

Ms. Cat's science lesson was on the right track in many ways. She took the children outside so they could experience natural phenomena first-hand. She also recognized that individual children were in different places in their social development, so they engaged in different ways with the lesson. However, the lesson did not fully align with the development of the children. The result was an experience that did not achieve the goal of the teacher and was stressful for some of the children. Yet, with some minor shifts, this lesson could more fully support the individual children's needs and help them to achieve goals related to science concepts.

The essence of developmentally appropriate practice is recognizing the current development of children and building on their experiences, knowledge, and skills for play-based learning (National Association for the Education of Young Children, 2020). In other words, teaching is about acknowledging where children are and then identifying how can we support them while integrating the learning goals we have identified for them. In this case, the primary learning goal was related to plants and flowers. Small shifts in the structure of the activities, preparation of the physical environment, and the interactions with individual children can better support the individuals while also helping them make sense of plants and flowers.

The structure of an activity impacts the success of any teaching. In this case, Ms. Cat provided children an opportunity to experience plants first-hand by digging in the soil and planting seeds. However, this activity was done after reading

R. A. Larimore (✉)
Samara Early Learning, Midland, MI, USA
e-mail: rachel@samarael.com

a book and watching a time-lapsed video. This structure focuses more on learning about plants and flowers rather than making sense of their own experiences with the phenomena of seeds growing into plants. What if digging in the soil with trowels and examining seeds came first? Then, the more formal teacher-led activities (e.g., reading a story) can intentionally build on the observations and questions that emerge from their personal experiences. In other words, by focusing on experience with phenomena first, children are figuring out the science ideas rather than being told the information.

Structuring an activity is not just about when the experience happens, but also how it happens. When structuring activities, it is important to not only think of the phenomenon as the foundational step, but also consider the experiences children will want to explore related to that phenomenon. It is also important to give them plenty of time for this exploration. For example, Ms. Cat recognized exploring the soil was an important part of the process. However, there was limited time for children to really dig in (pun intended!) to exploring the soil—to explore the textures, to notice how it crumbles or clumps in their hand, and to see how the trowel can break through the soil. Instead, Ms. Cat was feeling rushed and directed the group to move onto planting the seeds, which meant hastening through a foundational experience in making sense of the phenomenon of seeds growing into plants.

While teacher-led activities can be structured to allow for time with phenomena, so can providing a physical environment that will encourage curiosity and questions to emerge during free play. Free play can be rich with opportunity for science sense-making if we are intentional about the materials that we provide both indoors and outside in the play area. For example, trowels for digging in the soil, inside and/or outside, could be available for days or weeks leading up to the teacher-led plant lesson. This intentional planning would allow every child to experience digging with the trowels, whereas in the lesson the digging was limited to a few children. Then, during Ms. Cat's lesson, the children would know what a trowel is, its purpose, and how to use it, which would make it easier for children to self-regulate when they watch others rather than dig themselves. Other materials such as house plants in the indoor classroom and outdoor play area, whether growing or fully grown, can spark observation of plants—within the context of children's everyday lives. Bowls with a variety of seeds to sort, count, and create art with can spark ongoing conversations about the sizes, shapes, structures, and even purposes of seeds. These play experiences exploring materials related to plants and seeds lay the foundation for actually planting the seeds in the ground outside. Further, since the children will have physically and cognitively explored the soil and seeds, they will be better able to self-regulate when it comes to putting seeds in the soil. Preparing the physical environment with materials recognizes that young children need multiple experiences, over time, to make sense of the world around them.

It is not enough, however, to simply provide materials and hope children use them appropriately. The teacher–child interactions during free play and teacher-led activities matter in children's sense-making. First, it is okay for children to make mistakes. Making mistakes is part of the learning process and opportunity for teaching. For example, if a child puts multiple seeds in a hole that they dug, the child could be afforded an opportunity to make the connection between seeds and plants. In the moment, the teacher could facilitate counting the number of seeds; document the number of seeds verbally, written or even through photos; and then return to the concept later through additional observation. The child and teacher could return two weeks later, for example, and count the number of plants growing from that hole, comparing the number of growing plants to the photos from planting. These investigations over time are now more personal and meaningful to that individual child than the time-lapse video, but they address the same phenomenon Ms. Cat was trying to demonstrate (i.e., change over time).

The most important factor in interactions with children is to remember that we are teaching children first, then content—that is, the whole child and their current development in all developmental domains. How we structure activities and prepare the environment will help set up children for success. However, there will be times when a child simply is not interested in our teacher-led lessons…and that's okay! If a child is not ready to explore soil and would rather toss it in the air (an action exploring soil, just not in the way that the teacher had anticipated), then find a way to support and use the child's action as a strength for learning. For example, Ms. Cat could establish a zone for soil throwing away from other children. Or she may put out a giant piece of paper on the ground and let the child drop the soil to see how it splatters. How would she have known this was a potential need? From days or weeks of free play and even small group activities leading up to the teacher-led, whole-group activity.

Additionally, interactions with children must honor their choices about their own bodies. It was already known Marcus did not like soft things. The teachers can acknowledge this by offering him an opportunity to touch the soil and then honoring his decision by *not* forcing him to touch the soil and finding other ways for him to engage in the activity. For example, maybe he would have preferred holding and counting the seeds. Just as plant growth is part of a system (e.g., seeds, soil, water, and sunlight), we must recognize children are made up of a system of physical, cognitive, and social-emotional needs. Growth cannot happen in one domain without the other two domains, and together they make up the whole child.

In summary, this case included positive elements such as going outside to experience natural phenomena first-hand. The case also highlights, both literally and metaphorically, the importance of tending the soil before planting the seeds. The conditions must be right for children to grow. We as educators can facilitate those conditions for growth by recognizing children's individual needs as well as our curricular goals. We then can work to align those two things in the way we structure activities, the materials we provide, and in how we interact with children.

Reference

National Association for the Education of Young Children (NAEYC). (2020). *Developmentally appropriate practice national association for the education of young children* (Issue April).

Rachel A. Larimore is an educator, speaker, consultant, author, and former nature-based preschool director. As the founder and Chief Visionary of Samara Early Learning her work focuses on helping early childhood educators start nature-based schools or add nature-based approaches into their existing program.

Case: Can We Start With Science Today?

6

Brianna Wallace and Michael Svec

Abstract

At the beginning of her first year of teaching, Meg was surprised that inquiry-based science education was not promoted in her new school. As a low performing, high-poverty school, the school's focus was on core subjects that were tested, such as math and reading. Facing challenges of minimal resources, little time dedicated to planning science instruction, and so much to teach, assess, and analyze, most teachers felt swamped. Under such pressures, teachers expressed assumptions that promoted a deficit perspective—for example, Meg, the new fourth-grade teacher, was told the narrative that students at this school were not capable of doing higher-level thinking and carrying out scientific experiments. In this case, Meg took the steps to challenge these deficit assumptions by engaging her students in authentic science to dispel a deficit model.

I remember, on my first day at my new school, how excited I was to be teaching in a school that served a familiar community. My new school was named after its founder—a local African American civil rights activist and civic leader. Many parents had a sense of pride in the school due to the school's dedication and commitment to serving the students and community members. Located in southern United States, the school served approximately 550 students, all of whom received free/reduced lunch, with racial/ethnic demographics of 85% African American, 7% Caucasian, 6% Hispanic/Latinx, and 2% two or more races.

B. Wallace (✉)
SC, USA
e-mail: b.burnette95@gmail.com

M. Svec
Department of Education, Furman University, Greenville, SC, USA
e-mail: michael.sve@furman.edu

© The Author(s), under exclusive license to Springer Nature Switzerland AG 2023
S. Jeong et al. (eds.), *Navigating Elementary Science Teaching and Learning*,
Springer Texts in Education, https://doi.org/10.1007/978-3-031-33418-4_6

My excitement as a first-year teacher also stemmed from my undergraduate education, which promoted inquiry- and project-based learning and higher-order thinking. However, this school had performed poorly on state mandated testing, resulting in a narrowing of the school curriculum and instructional strategies. I soon learned that most of the science instruction relied on interactive notebooks, PowerPoints, and traditional tests. There was an assumption that students could not handle higher-order thinking and inquiry-based experiments, promoting a deficit perspective among some teachers. I often heard, "The kids simply cannot understand or handle experiments like that." In private conversations, I was told that science was an abstract subject that teachers did not focus on because in the elementary classroom, math and reading held more importance.

I was not comfortable with these assumptions about our students—that they only needed to work harder, that they were not motivated, or that they were not capable of meeting higher expectation. I sought to motivate my students with more authentic learning opportunities. My first steps were two-fold: building trust and implementing strategies. First, I built a trusting rapport with the students that they would be able to succeed in my classroom. Second, I wove three strategies into the science curriculum to engage the students in more critical thinking and authentic science: offering rich, inquiry-based learning experiences; employing a guided workshop model; and promoting student voice and choice. I implemented those strategies one at a time into an inquiry-based fourth-grade plant and animal unit focused on the characteristics and growth of organisms.

"Miss M, I don't understand how birds can do this!" students in science centers exclaimed as they try to work through a bird beak modeling activity. River struggled to pick up water using a toothpick as Quinn struggled with picking up marshmallow materials with a straw. Asked to explain their challenges, River stated, "This is impossible! You can't pick up water using a toothpick, at this point my bird is going to be thirsty and never get water!" Quinn said, "My problem isn't as bad as his, but it takes me a longer time to pick up the marshmallows with a straw and sometimes I drop them and have to start over. If my bird has a family, then their babies are going to be hungry for a while! Can we just switch? A toothpick would be easier for my bird." Throughout this, the students were examining, making observations, and taking notes.

Both River and Quinn were investigating the importance of animal adaptations and their crucial need for survival. The toothpick is similar to the beaks of birds like heron and tern, who spear their prey, such as water creatures and insects. The straw is similar to the beak birds like hummingbirds use to feed on different liquids. Without realizing it, both River and Quinn recognized that this task was about more than just picking up the food and water that were necessary for survival. Through this investigation, the students understood the importance of structural adaptations for animals. The students were highly engaged in these hands-on investigations, which motivated them to continue learning. Inquiry labs helped students be successful as well as learn, as the standards suggest, how to develop and use models. They could take risks, try and retry, and explain their thinking without focusing on getting a "right answer."

Next, I began using the guided workshop model with a mini-lesson of guided practice to introduce new concept, procedure, and use of technology. This strategy helped students clearly outline their daily tasks, provided a good example for the students to emulate, and allowed them to track their progress. Building on the success of the modeling investigations, I added more intentional modeling of procedures, giving the students more opportunity to independently apply skills in new contexts for their research with their peers.

"Just like the birds in our experiment, giraffes have special things about them to help them survive," River explained about her animal for her on-line research project. I replied, "That's correct, and we learned about some special vocabulary words to describe these 'special things.' Look at your notes, what do we call this?" She exclaimed, "structural adaptations!" This activity was a prime example of the guided workshop model. A mini-lesson of guided practice introduced the new concept of structural adaptations. The students first investigated structural adaptations with bird beak studies, then were challenged to research structural adaptations of an animal that they selected. As the classroom teacher, I stated the objective of this lesson: how to find credible websites/texts. The students took notes and created visual illustrations of how their animal's structural adaptations help the animal survive. After the students found examples, I encouraged them to compare and contrast their structural adaptations with their peers' examples to explain similarities and differences. The use of the guided workshop model minimized behavior problems, increased student motivation, and allowed students to learn new information without the traditional slideshow note-taking. It also helped the students construct explanations based on modeling, observations, and informational texts, consistent with the state standard for this unit.

For the final topic in the unit, seed growth, I integrated voice and choice, first by having students select their own seeds and three conditions for comparisons. Students engaged in daily data collection and observations of their investigation using their previous success with inquiry learning experiences. I modeled how to use the rulers, make sketches, and take digital pictures for their qualitative and quantitative observations.

The second integration of voice and choice was for the format of the final unit culmination presentation. The students chose how they wanted to present the results of their work to the larger school community, including their parents and peers. Public presentation of learning added authenticity, shifting attention away from getting the right answer on a test to explaining to others what you have learned. Students had the opportunity to select how they wanted to present and communicate their information from their science labs and informational text research. Among the suggested options were videos or posters.

When it was time to present their projects to the school, families, and community members, the students clearly and effectively explained their learning and were proud of their progress. School administration congratulated me upon my students' successes. By trusting in my students and slowly scaffolding new strategies, the students found success beyond a test score. More importantly, the students learned how to think through inquiry and discovery in order to build on their scientific schema. Students reported with excitement, "My mom loved my project!" and began asking, "Can we start with science today?"

For Reflection and Discussion

1. How can educators, especially first-year teachers, balance the time to provide more opportunities for authentic learning within the science content?
2. Building on her successes in her first year, how could Miss M expose more student-centered science activities for students in her second year of teaching?
3. How could Miss M share her successes with her peers to encourage higher expectations for all the students?
4. How can schools provide more support for first year educators with integrating authentic science activities for all students?

Brianna Wallace is an assistant principal at an elementary school in South Carolina with years of experience in teaching elementary. Her scholarship and teaching includes teaching all core subjects and leading professional development on culturally responsive and culturally relevant teaching.

Michael Svec is a professor of science education at Furman University with over two decades of experience in preparing elementary and secondary science teachers. His scholarship and teaching includes history of astronomy, science fiction, and international comparisons.

Commentary: Partnering with University Outreach Experts

Cynthia Canan

Abstract

This is a commentary to the case narrative, *"Can We Start With Science Today?"* written by Brianna Wallace and Michael Svec.

Having never worked in a K–12 education system, I found Miss B's experience both frustrating and enlightening. I was disappointed on Miss B's behalf because of the lack of support she faced to promote science in her classroom. Furthermore, while I can understand the school's rationale to shift resources to core subjects that were tested, I felt saddened for her students, who were victims of the deficit perspective.

For the last 6 years, I have been working for a program that promotes safe medication practices for all ages and while my background is in infectious disease research, I have always been passionate about public science education. In fact, in my graduate school application, I wrote that my career goal was:

"...to become a public health program coordinator in order to educate the public about disease prevention and major health issues. In this position, I will work directly with researchers to provide novel ground-breaking scientific information in a manner that is easily understood by the public."

So, if, like Miss B, you crave more authentic learning experiences but lack support to develop content, what type of resources are available? If you want students to connect science in the classroom to the most up-to-date science in the world, who could work with you? While I do not know all resources that may be available in

C. Canan (✉)
College of Pharmacy, The Ohio State University, Columbus, Ohio, USA
e-mail: canan.20@osu.edu

your area, I encourage you to investigate available K–12 outreach programs from nearby colleges and universities.

At my university, my colleagues and I, from a range of diverse disciplines, provide free, fun, and age-appropriate learning activities and modules for K–12. Some activities are short (30–60 min), while others are long-term unit plans developed with teachers. From my experiences, those who run university outreach programs are incredibly enthusiastic—almost obsessive—about public outreach. Here is what my colleagues have to say about why they are in outreach:

> Courtney: I ended up pursuing a career in outreach education as a happy accident. Growing up I loved animals and knew from an early age that I wanted to work with animals in my future career. But I had no idea what a that career path looked like. I did well in science in middle and high school, but my most memorable experiences with my favorite subject, animals and nature, occurred outside of the classroom. I studied wildlife biology in college and after graduating found myself teaching preschool-aged students at a zoo about animals and conservation. I quickly realized that working as an outreach educator gives you the potential to create memorable out-of-school experiences that can inspire and grow a person's love of science. That's why 20 years later I'm still developing programs that invite my audience to see that science is fun and that there is a place for everyone in science.

> Katherine: I started out my career getting a Ph.D. because I thought I wanted to do science. However, through the experience of graduate school and all the education that comes along with it, I realized what I really liked was sharing science. And public outreach is just a whole career built on sharing how science intersects with people's lives. In one of my projects, our goal is to create watercooler moments and it is the idea that you'll learn something and be so excited about what you learned that you can't help but to share it at the watercooler the next day at work with your colleagues.

> Wayne: You know, I never thought I was passionate about outreach until I started my Ph.D. program and it literally carried me through eight years of graduate school. I would have to refuel by doing outreach which helped me to see the big picture perspective of why I was still getting my Ph.D. So, while I do outreach for personal and selfish reason, there's also a lot of relationship building with the teachers and students that I enjoy. Being able to unlock the things that students already know and letting them come to these natural discoveries or setting up a scaffold so students can walk the path that they need to go energizes me. I think that outreach is a great way to interact with other people, and I think it's a lot of fun and it drives me forward.

Being at a research-focused institution has allowed me to stay up to date on scientific discoveries in my field. Working with higher education students has brought me new perspectives on the types of careers and opportunities available and of interest to students in STEM. Because of the richness of my environment, I have shaped my outreach content to be diverse and engaging: (1) I continuously evolve and update my activities due to advantageously receiving new scientific understanding directly from the experts in the field; (2) my programs are often designed and facilitated by different types of university students, including undergraduate, graduate, and professional students, who provide a different perspective in K–12 education; (3) I incorporate career exploration into my outreach design through my connections to past students and former colleagues; and (4) if I don't know the answer to a scientific inquiry by the K–12 student or teacher, most often, I can find the most appropriate answer through my university.

Finally, and this is important to know, most of us working in university outreach would love to work with teachers like you. We know that for an activity or learning module to be effective, we need to work together to incorporate ideas from both the outreach expert and the teacher in the classroom. While there might be different levels of opportunity to engage, to me, any engagement with the teacher is mutually beneficial: I know I will be developing content that the teacher wants, and the teacher will be implementing activities that are up-to-date and relevant to the classroom. Here is what my colleagues have to say about engaging with teachers:

> Courtney: In my previous position, much of my time focused on developing resources for and working with K-12 teachers. While we had a suite of ready to teach lab modules that teachers could download from our Website, the model organism we worked with (Arabidopsis thaliana) and the subjects we covered (various topics in plant science including Mendelian genetics) were outside of many teachers' comfort zones. Whenever possible I would make personal connections with teachers via workshops, education conferences, science events, and through content on our Website. I always encouraged teachers to reach out with questions or to request support in developing learning experiences for their students. When someone reached out, we would brainstorm ways to meet their teaching goals. That could include something as simple as emailing with students to help prepare them for a science fair, to co-teaching labs in the classroom, to co-developing experiments with advanced biology students.

> Katherine: I teach science communication to undergraduate students and for their final project, they are required to review Ohio's Learning Standards for Science and create content that specifically align with what teachers are teaching and what students are learning. Outside of the classroom, I've helped built kiosk activities for public libraries and we always take time to align our content to what students are learning in the classroom so that they can take their learning to the next level.

> Wayne: I feel like if there is a back-and-forth discussion between me and the teacher about our expectations and goals, that's when I can create the best experiences. Sometimes, the goals are very amorphous—teachers don't have to be super specific and don't need to dictate the words or phrases I should use, but if they can relate the expected takeaway, I can better tailor my engagement to what they are doing in the classroom. It's actually harder to connect and engage with the students when I don't get a chance to partner with the teachers. Our program becomes stickily a field trip experience and unfortunately, not every student is inherently engaged with field trips. However, when I can discuss the classroom goals with the teachers, there's always a lot more engagement from the students, and the students will come up with their own ideas and think about concepts in their own ways, moving them further.

I hope Miss B knows that she is not alone in wanting students to have engaging and authentic science learning. Unfortunately, university outreach might be a little hard to find since universities do not often have consolidated lists of outreach programs or experts. But I promise, the effort you put into finding us will be worth it because we are passionate about K–12 and public education, and we would love to create long-term partnerships with you.

Cynthia Canan is a Lecturer in the College of Pharmacy at The Ohio State University. She is the Director of the Generation Rx Laboratory, an innovative drug education research hub that promotes science learning, health career exploration, medication safety, and prescription drug misuse prevention. She holds a Ph.D. in immunology. Currently, she is interested in effective public communication methods when discussing controversial science.

Case: I'm Not Stupid in *Your* Classroom!

8

Deborah Hanuscin

Abstract

This case raises questions regarding the inclusion of all students in the science classroom, and the way in which teaching helps shapes students' view of themselves as learners who are capable of doing science. Robin, a beginning elementary teacher, is having a conversation on the playground with a fourth-grade student, Kai, who has just returned to Robin's classroom after attending another school for a period of several months. Kai receives special education services and has an IEP for language arts, and this case highlights the dilemmas Robin faces in creating a science classroom that is equitable and inclusive. The case is followed by a commentary written by Professor of Special Education, Delinda van Garderen.

Robin stood at the edge of the playground surveying her students when she noticed Kai ambling up to her. Kai had started the year in Robin's classroom, then transferred to another school in the district several months ago when his family moved. Just this week he had reappeared in her classroom—his family having moved yet again. She had worried about Kai in the interim. He was among a group of students who received services from a Special Educator during Robin's language arts instruction. Kai had a keen sense of humor, bright smile, and a knack for showmanship, but also a tendency to get discouraged easily when he didn't understand something.

"It's so nice to have you return, Kai!" she greeted him.

"I'm glad to be back," Kai responded.

"Why's that?" Robin probed, keeping her eyes on the playground.

"Well, I'm not stupid in *your* class," he offered.

D. Hanuscin (✉)

Science, Math and Technology Education, Western Washington University, Bellingham, WA, USA

e-mail: hanuscd@wwu.edu

47

S. Jeong et al. (eds.), *Navigating Elementary Science Teaching and Learning*, Springer Texts in Education, https://doi.org/10.1007/978-3-031-33418-4_8

Kai's comment had *not* been the answer she was expecting, and for a moment she felt overwhelmed with this reminder of the influence of a teacher. "What do you mean?" she asked casually, one eye dutifully on the playground.

"Like…" Kai paused thoughtfully, "I am *good* at things in your class. Like SCIENCE—I can do science!"

She smiled to herself. "Well, you'll get to do science when recess is over, so you had better go take your turn at four-square now before you miss out!" she chided playfully. She watched Kai scurry off to the court.

She recalled how Kai had shone during their very first science investigations about energy. Their culminating project, an engineering challenge, had been to *apply scientific ideas to design and test a device that converts energy from one form to another* (NGSS Performance Expectation 4-PS3-4) (NGSS Lead States, 2013). Kai and his partner had watched numerous YouTube videos about Rube Goldberg devices, and were determined to create one of their own. In the end, they had decided that their device would pitch the kickball to players at recess, kind of like the pitching machines in batting cages. Almost the entire fourth grade had lined up to be testers on the chosen day! Kai and his partner, Liam, had asked testers to rate the device using a star system (5 stars being best) and offer suggestions for improvement. Liam had gotten the idea from his mom's online shopping, and the attention she paid to reviews from other customers. Robin had worried that Kai and Liam would get caught up in the product and lose sight of the concepts involved, but had been so impressed with the humor in their "shark tank" style pitch that explained how their device worked, and how it converted energy from one form to another.

As she thought ahead to the upcoming science unit, she felt wary. She had been so pleased with Kai's recognition that he was capable when it came to science, but now she worried that his confidence would be dashed— Kai's struggles related to reading and writing. In this particular science unit students would be *constructing an argument that plants and animals have internal and external structures that function to support survival, growth, behavior, and reproduction* (NGSS Performance Expectation 4-LS1-1). They would be reading many nonfiction books about plants and animals, in addition to their science textbook. They would also be doing lots of writing related to structuring arguments with claims, evidence, and reasoning. Her culminating project for the unit was for students to each publish a book about a group of organisms of their choice, comparing and contrasting their structures, behaviors, and ways of surviving. As she started listing off all the language arts skills involved, she grew increasingly nervous. Adding to her concern was the knowledge that her Special Education colleague, Mrs. Branch, would not be in the classroom during science time to support Kai.

Luckily, today's activity would be focused on gathering her students' preconceptions and eliciting their initial ideas and experiences related to plant and animal adaptations. She would have time to speak with Mrs. Branch after school, and to revisit her plans for the unit.

For Reflection and Discussion

1. Kai's comment about not being stupid in Robin's class reminded Robin of the influence of a teacher. What do you think she meant by that? In what ways can teachers influence their students? In what ways have teachers influenced you?
2. Kai's challenges relate specifically to reading and writing. Why do you think he didn't encounter barriers related to these challenges in the first science unit he completed with Robin's class? How are these challenges likely to affect him in the upcoming unit?
3. In what ways might Robin respond to the dilemma of ensuring that Kai feels confident as a learner while creating a lesson to meet the performance expectations of the standard? Consider how the way in which she responds can impact Kai's feelings of being a capable science learner.

Reference

NGSS Lead States. (2013). *Next generation science standards: For states, by states.* The National Academies Press.

Deborah Hanuscin is a former elementary teacher and informal science educator. She is currently a Professor in Elementary Education and Science, Math, and Technology Education (SMATE) at Western Washington University. Her research focuses on elementary science teacher learning and the design of professional development and curricula to support that. She is a past president of the Association for Science Teacher Education, and recipient of the ASTE Outstanding Science Teacher Educator award.

Commentary: Designing Equitable Science Instruction for All Students

9

Delinda van Garderen

Abstract

This is a commentary to the case narrative, *"I'm Not Stupid in Your Classroom!"* written by Deborah Hanuscin.

According to national demographic statistics, over 63% of students who receive special education services are receiving 80% of their instruction in the regular education classroom (U.S. Department of Education, 2020). This suggests that as a teacher, you are more than likely to have a student such as Kai in your classroom taking science with you. A close look at the NGSS suggests that science is to not only be for "all students," but there is an expectation that teachers are to create learning opportunities that enable all students, including students such as Kai, to meet "all standards" (NGSS Lead States, 2013, Appendix D). Based on the research, if given equitable learning opportunities, students with special needs are capable of engaging in science (e.g., Therrien et al., 2011, 2014). The question, then, becomes *how can instruction be designed in equitable ways to promote the success of special education students without "watering down" the curriculum or lowering expectations?* One approach that can help is Universal Design for Learning.

Universal Design for Learning (UDL) is a framework that provides a systematic and intentional way to design, or perhaps (re)frame learning to meet the individual needs of all learners including students with special needs (CAST, 2018). UDL proposes that we not view the *students* as limited, but rather that we recognize that *the way in which we design instruction* can limit the ways in which students

D. van Garderen (✉)
Departement of Special Education, University of Missouri, Columbia, MO, USA
e-mail: vangarderend@missouri.edu

Table 9.1 UDL principles and guidelines (available http://udlguidelines.cast.org/)

Provide multiple means of engagement	Provide multiple means of representation	Provide multiple means of action and expression
• Provide options for recruiting interest	• Provide options for perception	• Provide options for physical action
• Provide options for sustaining effort and persistence	• Provide options for language and symbols	• Provide options for expression and communication
• Provide options for self-regulation	• Provide options for comprehension	• Provide options for executive function

can succeed. Kai's own feelings of "not being stupid" in Robin's class, while feeling that way in another class are an illustration of that. Learning environments should be designed to support high expectations for all students, including student with special needs (Meyer et al., 2014). Too often, however, our instructional decisions can present barriers for our students, resulting in them feeling disengaged, unmotivated, and discouraged.

The UDL framework provides a way by which teachers can reshape and guide the design of learning environments in science that are flexible and responsive to learner needs (Meyer et al., 2014). The UDL framework comprises of three principles along with 9 guidelines for implementation (Table 9.1). Combined, the three principles are focused on the "why," "what," and "how" of learning taking into consideration that learning involves more than facts and skills but also emotion and affect (Meyer et al., 2014).

If we revisit Kai's success in Robin's previous science unit through the lens of the UDL framework, we can identify specific features of the learning environment and Robin's instructional choices that enabled this success by reducing potential barriers Kai might experience. For example, giving Kai a choice of what to design for the engineering project led him to something personally relevant—which helped generate his interest (Principle 1). Working with the support of a partner may have also helped sustain his effort and persistence and minimize any potential discouragement or frustration related to reading and writing (Principle 2). Having access to information in video format, rather than just in text format, helped him comprehend the task and design process (Principle 2). Finally, creating a shark-tank video pitch enabled Kai to express and communicate his understanding in a way that drew on his strengths, versus challenges (Principle 3).

Robin is rightly concerned about Kai's potential to succeed in the next science unit—not because *Kai is not* capable of doing so, but because she recognizes *the way the unit is designed presents barriers to Kai's learning.* Encountering these barriers will likely be discouraging to Kai and can influence how he views himself as a learner. For teachers, being able to identify and anticipate barriers their curriculum might pose for students is essential. The UDL framework can help teachers find solutions to reduce those barriers. Table 9.2 shows several solutions Robin might consider in light of the barriers Kai experiences related to reading and writing.

Table 9.2 UDL solutions that could be applied to Robin's upcoming science unit

Learning activity	Possible barriers for Kai	Possible solutions
Reading nonfiction books and science textbook	Difficulty reading (decoding) Understanding key concepts and vocabulary	*Representation* • Provide videos with the same type of information as found in the printed materials • Partner read or use a text-to-speech program • Define words using student-friendly definitions • Concept mapping
Writing structured arguments	Difficulty composing	*Action and Expression* • Sentence starter or scaffold (e.g., My claim is …. My evidence for my claim is ….) • Talk-to-text software • Work with a partner/small group to write sentences • Provide a graphic organizer to model text structure
Writing and publishing information	Difficulty synthesizing ideas Organizing information Sustaining interest and motivation	*Action and Expression* • Provide choice for alternative products (e.g., video report) • Work with a partner to summarize key concepts/ideas learned • Provide graphic organizers (e.g., compare & contrast table) or graphic organizer (e.g., Venn diagram) to record information • Provide book template or model • Checklist of information to be included in the book *Engagement* • Provide choice for animal/plant of focus • Provide choice to work with a partner

Even though Mrs. Branch doesn't push in during Robin's science instruction, Robin's plan to seek out support from her is important. A strength of UDL as an approach for learning environments is that it provides all teachers, special education, and general education, a common platform to work from. It can serve as a way to collaboratively plan for and design instruction that not only benefits students with special needs but makes learning accessible to all learners in the classroom. By contributing *both* of their expertise, they can ensure an equitable and inclusive science learning environment.

References

CAST. (2018). *Universal design for learning guidelines version 2.2.* http://udlguidelines.cast.org. Accessed September 29, 2020.

Meyer, A., Rose, D. H., & Gordon, D. (2014). *Universal design for learning.* CAST Professional Publishing.

NGSS Lead States. (2013). *Next generation science standards: For states, by states.* The National Academies Press.

Therrien, W. J., Taylor, J. C., Hosp, J. L., Kaldenberg, E. R., & Gorsh, J. (2011). Science instruction for students with learning disabilities: A meta-analysis. *Learning Disabilities Research & Practice, 26*(4), 188–203.

Therrien, W. J., Taylor, J. C., Watt, S., & Kaldenberg, E. R. (2014). Science instruction for students with emotional and behavioral disorders. *Remedial and Special Education, 35*(1), 15–27.

U.S. Department of Education, Office of Special Education and Rehabilitative Services. (2020). *41st annual report to congress on the implementation of the individuals with disabilities education act*, 2019, Washington, D.C. Office of Special Education Programs.

Delinda van Garderen is a former elementary teacher and current Professor in the Department of Special Education at the University of Missouri. Her research interests focus on students with learning disabilities, struggling learners, and teachers in the content areas of mathematics and science. In particular, she studies how students and teachers use representations to solve mathematics word problems, student development and intervention in number (e.g., conservation of quantity, numerical magnitude), and use of Universal Design for Learning to plan instruction to meet the needs of diverse learners in science.

Case: When the Sun Doesn't Shine

10

Wahyu Setioko and Afif Alhariri Pratama

Abstract

With support from a local science center Mr. Harry, a fourth-grade teacher at an elementary school in a rural school district, embraced a rare total solar eclipse phenomenon as an informal science learning opportunity for his students. Despite the positive engagement throughout the observation from those who attended, he was concerned about the absence of some of his students. After the event, Mr. Harry was surprised to learn that his students' diverse cultures, beliefs and previous experiences intersect with their potential science learning experiences. He wondered what he could do differently next time to elicit and navigate these diverse beliefs that play a role in student learning experiences.

Darkness slowly emerged as the sun became increasingly covered by the moon, depicting a dusk scenery without a reddish tinge. A silhouette of a ring appeared in the sky. The corona was visible on the edge of the solar disk. Suddenly, the atmosphere became quiet. The birds stopped chirping; the roosters no longer cackle; the skinny black cat that had been pacing around then stared blankly under the tree. In one minute and twelve seconds, the universe was mute. A handful of Mr. Harry's fourth-grade students, their parents, and Mr. Harry were admiring the whole mystique.

W. Setioko (✉)
Department of Teaching and Learning, The Ohio State University, Columbus, OH, USA
e-mail: setioko.1@osu.edu

A. A. Pratama
Karang Makmur 2 Elementary School, Musi Banyuasin District, South Sumatra, Indonesia

"Nice! This is so cool!" Brian exclaimed with joy. He removed his special glasses he had been using to observe the eclipse and kept staring at the perfectly round total solar eclipse.

"Right. It's beautiful, like a ring!" said Tiana, no less excited.

A few other students and their parents who came to observe the eclipse looked up at the sky with infinite amazement and were thoroughly impressed with the phenomenon they were seeing. Then the ring in the sky began to fade. They put the glasses back on. Slowly, the dark atmosphere brightened. Birds chirped again, and roosters cackled. The universe was cheerful at once as the light appeared to erase darkness.

"Well, this total solar eclipse is a rare phenomenon. It only occurs when the sun, the moon, and the Earth are in a straight line. That is what causes a solar eclipse." Mr. Harry, the science teacher, drew a model on the ground for his students.

"How can it be like a ring?" asked Brian, still looking up at the sun. He had been paying attention to the eclipse with the utmost interest. Mr. Harry instructed, "Brian, you should not be staring at the sun for more than 2 min in a row, even with the glasses." He then continued, "Why don't you think about the ring in relation to the distance from the moon to the Earth versus the distance from the moon to the sun, and the sizes of the Earth, moon and the sun." Brian gave Mr. Harry a puzzled look. Mr. Harry continued pointing with his finger, "Here, take a look at our signboard standing over there. Which one is bigger? The signboard or my thumb?" Mr. Harry held up his thumb as if he was trying to cover the board.

"The board!" students answered in unison.

"Believe it or not, our thumbs can cover the board." Mr. Harry illustrated this by placing his thumbs close to his eyes while looking at the board.

"You're right," said Tiana, another student, while mimicking Mr. Harry. The other students took their thumbs and followed Mr. Harry.

"Although the moon is smaller than the sun, its surface can seem to 'cover' the sun's surface because of its close distance to the Earth, like our thumbs seem to be covering the signboard that is bigger. The remaining sun rays on the circumference are what produces a ring-like appearance." Mr. Harry continued with his explanation, "There are two ways to observe an eclipse. We can use the special eclipse glasses as we did today. Or we can use the projection method using a pinhole box. There!" Mr. Harry pointed to the cardboard boxes standing on tripods that he had prepared prior to his lesson. Students walked to the boxes, and Mr. Harry invited them to look at the sun's projection in the boxes. They then discussed in-depth the eclipse phenomenon, including its process and why we were not allowed to look directly at the sun.

After his lesson, Mr. Harry had mixed feelings about the event. He was happy that the students who participated in the observation were engaged, excited, and asked questions. Mr. Harry was certain that students learned about the solar eclipse and how one occurred and was delighted they were able to share their experience. However, a number of students and their parents did not show up that day—which Mr. Harry did not anticipate and left him dismayed. He wondered what he could have done to encourage the students who chose not to participate in the observation and began to reflect on the whole experience, starting from his planning day.

* * *

Before the observation day of the solar eclipse.

A total solar eclipse is a rare sight visible from a very narrow path on Earth as it takes 375 years for a total solar eclipse to happen again at the same location. On August 21, 2017, scientists predicted that a total solar eclipse would occur. Mr. Harry, who teaches fourth grade at an elementary school in a rural school district, embraced this event as a learning opportunity for his students. His idea for this lesson was supported by a local science center whose mission was to provide science learning opportunities in informal settings to socio-economically and geographically disadvantaged students. With the science center's support, Mr. Harry received curricular materials, including information booklets about the solar eclipse, special eclipse glasses for all the students, and the pinhole boxes for observation.

Mr. Harry was excited and spent a few days informing all of his students' parents about this activity. Mr. Harry thought it would be a memorable learning moment for the students and their families to witness this extraordinary phenomenon nature has to offer. An informal science learning opportunity like this will promote children's interest in science. He was also excited that he received the support of the science center to enrich his lesson. With much anticipation, he dedicated a lot of his time planning for this lesson.

* * *

After the observation day of the solar eclipse.
As class began few days after the eclipse event, Mr. Harry still felt a bit disappointed, thinking about the low attendance of his students. To his curiosity, Mr. Harry asked some of his students who missed the eclipse event. He was surprised to learn some of their responses.

Andy, a Southeast Asian student, said, "My mom said it was a jinx. The Buto (a gigantic mythical creature) was swallowing the sun. So, it couldn't shine. We had to stay at home and prayed for him to spew out the sun back." Li, a Chinese American student, chimed in, "No! It was a giant dragon, not Buto! And I helped mom and dad beat all the pots and pans at our home to scare the dragons away with loud sounds. It was so much fun!"

Mr. Harry was intrigued and asked Reyansh, whose family was from India, why he wasn't able to come to the solar eclipse event.

Reyansh said, "Don't you know about Rahu's story, Mr. Harry? So, there was a demon trying to steal an elixir for immortality. But the sun and moon saw him. They reported his crime to the god, Vishnu. He was beheaded before the elixir passed his throat. His body died, but his head turned immortal. His head is called Rahu. To this day, Rahu keeps chasing the sun and the moon through the sky to get revenge. Sometimes he gets them and eats the sun. That's why we have solar eclipses for a moment. Since Rahu has no neck, the sun slips out of the head's bottom and shines again. We avoid going out or eating food. We are Hindus, so we fast during an eclipse."

"I am sorry, Mr. Harry. When the eclipse occurred, my family and I were praying with other people in the mosque. I am taught that eclipses are a sign and a reminder of Allah's (the god) power. As Muslims, we are encouraged to worship Him together during the eclipse," said Ahmad.

"Mr. Harry, I wanted to go, but my dad said I couldn't. The thing is, my dad experienced another solar eclipse when he was in Indonesia about twenty years ago. He remembered how scary it was. So, he would not let us go," told Ana.

Mr. Harry knew what Ana was talking about, as Ana's father called. On June 11, 1983, a total solar eclipse happened and was observable in Java, Indonesia's most populous island. A few weeks before the eclipse, Ana's father said that the government intensively broadcasted a campaign on TV, radio, and newspapers for people to stay at home when the solar eclipse happened and not to look at the sun as it would potentially cause blindness. Moreover, the Indonesian village officers sent instructions to its residents about protecting their livestock, applying a 2–4 h curfew policy during the eclipse, and stressing to watch the eclipse on television only. This type of messaging that the sun's rays would become dangerous during the solar eclipse took a strong hold in the beliefs of the people in the Javanese community like Ana's parents.

Mr. Harry overlooked these various beliefs of his students' diverse cultures and previous experiences. He wondered what he could do differently next time to elicit and navigate these diverse beliefs that intersect with and play a role in students' experiences.

For Reflection and Discussion

1. What can Mr. Harry do to elicit and navigate different beliefs that play a role in the way that students experience natural phenomena and intersect with their science learning experiences?
2. How can you build upon the assets of your students in this solar eclipse lesson?
3. What are ways in which science teachers can create a space that acknowledges diverse cultural beliefs and helps students navigate potentially difficult conversations at the intersection of current events and culture? How can you encourage your students to be respectful of each other's perspectives?
4. How might your students' antecedent experiences impact their experiences with school science?

Wahyu Setioko is a Ph.D. graduate in STEM Education from The Ohio State University. His interests center on informal science education, elementary science teacher education, informal-formal science education collaboration, and teaching and learning in virtual environments. Over the past decade, he has been a science museum educator, a rural elementary school teacher, a K-12 science curriculum developer, and a non-profit professional providing professional development for elementary school teachers in Indonesia.

Afif Alhariri Pratama is a community development specialist, passionate about improving education quality for underprivileged children. He began his career as an elementary teacher and a community development facilitator in a remote village in Musi Banyuasin district, South Sumatra, Indonesia. He then joined the Indonesia Teaching Movement as a community development manager and was responsible for managing volunteers, facilitators, and community partners to work together in providing professional development for teachers in rural areas.

Commentary: Cultivating Cultural Knowledge

11

Ashlyn E. Pierson

Abstract

This is a commentary to the case narrative, *"When the Sun Doesn't Shine"* written by Wahyu Setioko and Afif Alhariri Pratama.

Making Western scientific knowledge and practices accessible can help children ask and answer questions about natural phenomena. For example, learning about controlled experiments can help children design investigations or make sense of cause-and-effect relationships as they observe and explore their world. However, focusing exclusively on Western science can marginalize minoritized students and erase their cultural practices from science. To create inclusive learning environments, science educators must find ways to invite, value, and cultivate students' cultural knowledge and ways of knowing (Warren et al., 2020).

In "When the Sun Doesn't Shine," Mr. Harry shared a powerful moment with students and their families as they observed a solar eclipse together. Mr. Harry created an environment where his students could observe the eclipse safely, ask questions about their observations, and make sense of complex concepts (e.g., how the moon can "cover" the sun because of its close distance to the earth). This was a meaningful and memorable experience for students who attended; however, not all students in Mr. Harry's class participated in the event.

Later, Mr. Harry learned that some students avoided the event based on their cultural practices or their families' experiences. Students readily shared theses explanations when Mr. Harry asked about them, but they did not volunteer these explanations when Mr. Harry initially invited his students to view the eclipse. This is not surprising—when students feel that their language, practices, or beliefs do

A. E. Pierson (✉)
Department of Teaching and Learning, The Ohio State University, Columbus, OH, USA
e-mail: pierson.199@osu.edu

not fit with the dominant culture at their school, they can feel hesitant to share their resources (Cole et al., 2016). However, when teachers explicitly invite and value students' resources, students may be more willing to leverage and share them (Pierson et al., 2021). For example, framing science as a discipline with multiple "languages" (e.g., tables, graphs, simulations) and emphasizing the value of describing phenomena in multiple languages (e.g., English, Spanish, Korean) can legitimize the use of language other than English in science classrooms.

To this end, how could Mr. Harry have invited all his students to share their resources? Full participation in eclipse viewing would not necessarily have been the goal, because this would have required some students contradict their family's beliefs and traditions. Instead, Mr. Harry might have asked students to share their stories and experiences before planning his eclipse events. Hearing these stories, Mr. Harry might have chosen to host multiple events related to the eclipse (perhaps one during the eclipse and one before or after the eclipse) so that all students and their families could participate. Events before or after the eclipse could make the phenomenon accessible to the students; for example, Mr. Harry could have prompted similar questions and investigations using models of the eclipse. Rather than attempting to replace students' stories with Western scientific concepts, Mr. Harry could frame the phenomenon as having multiple explanations, helping students navigate multiple epistemologies by comparing and contrasting explanations and their implications.

Discussing these stories could benefit all students; for example, hearing cultural stories could help students notice or recall specific aspects of the eclipse phenomenon. However, it is possible that some students from dominant culture backgrounds could dismiss their peers' stories about the eclipse. In such contexts, it is especially important for the teacher to establish norms for respect, to model these norms for students, and to closely monitor students' discussions to ensure that all students' contributions are valued.

This approach to science teaching—cultural cultivation rather than cultural assimilation—is relatively new for many educators. Still, this perspective is critical to making science teaching and learning more equitable. There is still much to learn about how teachers can work with their students to navigate multiple ways of knowing in science classrooms.

References

Cole, M. W., David, S. S., & Jiménez, R. T. (2016). Collaborative translation: Negotiating student investment in culturally responsive pedagogy. *Language Arts, 93*(6), 430–443.

Pierson, A. E., Clark, D. B., & Brady, C. E. (2021). Scientific modeling and translanguaging: A multilingual and multimodal approach to support science learning and engagement. *Science Education, 105*(4), 776–813.

Warren, B., Vossoughi, S., Rosebery, A. S., Bang, M., & Taylor, E. V. (2020). Multiple ways of knowing*: Re-imagining disciplinary learning. In *Handbook of the cultural foundations of learning* (pp. 277–294). Routledge.

Ashlyn Pierson is an assistant professor of STEM education. She earned her Ph.D. at Vanderbilt University's Peabody College of Education, specializing in Mathematics & Science Education and Learning Sciences & Learning Environment Design. Her research merges theory and practice by using design studies to explore the interplay between disciplinary STEM practices and students' everyday practices and diverse linguistic resources.

Case: Clarissa Says God Doesn't Exist

Ryan S. Nixon

Abstract

Mr. Rabe had deeply held religious beliefs and felt that these beliefs were a defining part of who he was and how he lived his life. While he avoided explicitly sharing his religious beliefs, Mr. Rabe knew that many of his students were raised in religious homes. Therefore, he wanted to let his students live their religious beliefs, or lack thereof, in his classroom without advocating for or against any of them. The dilemma presented in this case occurred at recess right after a science lesson in which Mr. Rabe had been teaching his fifth graders that the apparent brightness of stars in the sky depends on their distance from Earth, with more distant stars looking dimmer and closer stars looking brighter. In response to a student's question, Mr. Rabe emphasized the vastness of space and the distance between the stars. This idea led to a conversation among students at recess in which one student, Clarissa, claimed that this vastness was evidence that God did not exist. Jasmine, upset by this declaration, left the conversation. In a brief, personal conversation Mr. Rabe informed Jasmine that some people see the vastness of space as supporting their belief in God and suggested she talk with her family about these questions. As they returned from recess, Mr. Rabe wondered how he had done and if he should address this issue with the whole class.

"Well, I guess it's good they're talking about what they learned in class," thought Mr. Rabe ruefully as Jasmine, one of his fifth-grade students, walked back out onto the playground. He really hadn't intended to open the "separation of church and state" can of worms today, but there it was. He felt like he had responded well to Jasmine. He had tried to let her believe what she wanted to believe and not

R. S. Nixon (✉)
Department of Teacher Education, Brigham Young University, Provo, UT, USA
e-mail: rynixon@byu.edu

advocate for any religious position, while also letting her know that it was okay to believe in God if she wanted to. As a person who believed in God himself, he felt this was particularly important.

Mr. Rabe had always had deeply held religious beliefs and felt that these beliefs were a defining part of who he was and how he lived his life. Most of the good things in his life were tied to his belief in and relationship with God. However, he knew in many circles, especially scientific circles, religious beliefs were looked down upon and were thought to contradict scientific knowledge and progress. Mr. Rabe didn't know all the answers, but he felt that he had reconciled his religious beliefs and scientific knowledge to the point that he could authentically believe and understand both.

While he avoided explicitly discussing it, Mr. Rabe picked up things about his fifth-graders' religious beliefs just by spending seven hours a day with them. Mr. Rabe knew that many of them were raised in religious homes. Therefore, he wanted to let his students live their religious beliefs, or lack thereof, in his classroom without advocating for or against any of them.

The conversation he had just had with Jasmine was tied to the lesson that occurred just prior to recess. Mr. Rabe had been teaching NGSS performance expectation 5-ESS1-1: "Support an argument that the apparent brightness of the sun and stars is due to their relative distances from the Earth." He wanted them to understand that some stars that were really bright looked dim because they were so far away from Earth, while other stars that were actually dimmer looked brighter because they were closer to Earth.

The lesson included an investigation using flashlights in the gym. The students explored how the brightness of flashlights differs when they're at different distances. The investigation question for the day was: "How does the brightness of a flashlight change with distance?" They had turned off most of the lights in the gym and had both used their eyes and Google's Science Journal app to measure the flashlights' brightness.

After finishing in the gym, Mr. Rabe led the students back to the classroom and gave them some time to construct their answers to the investigation question. As the students shared their responses, Mr. Rabe periodically interjected an idea or a question to challenge the students' thinking and help them solidify their understanding. He was, as usual, impressed with their willingness to engage with the ideas and think deeply.

During this discussion, Alejandro asked a question that Mr. Rabe always hoped someone would ask. The question showed that the students were thinking deeply, and it allowed Mr. Rabe to give them a little glimpse into the vastness of the universe. He never addressed the question in religious terms but thinking about the vastness of the universe filled him with wonder at all that God had created.

Alejandro raised his hand, his face thoughtful. When called on, he started slowly. "So, the Sun is super bright in the sky, right? Brighter than all the other stars. But that's not because it's *actually* brighter than all the other stars out there, it's just because it's really close…right?".

"That's right, Alejandro," Mr. Rabe responded, encouraging him to continue.

"But then how far away are all the other stars?" Alejandro continued. "They don't look very bright. Our flashlights didn't get that dim even when they were all the way across the gym."

"That is a great question, Alejandro," said Mr. Rabe. Then, using a talk move familiar to his class, he said, looking around, "Who can restate his question for us?"

Ariel volunteered, "He asked how far away the other stars were since they have to be really far away to be so bright but look so dim."

"Did she get that right, Alejandro?" Mr. Rabe asked turning to him. Alejandro nodded, and Mr. Rabe continued. "Let me see if I can help you understand how far away the stars are with a little demonstration." He walked to the light switch and flipped it off, making the room dark except for the light from the window. The students watched him walk to the front of the room and grab one of the flashlights.

Pointing the flashlight to the class, he explained. "In a moment, I'm going to turn this flashlight on. I want you to pay attention to how long it takes for you to see it. Raise your hand when you can see it. Does that make sense?"

After seeing heads nod, Mr. Rabe reached down and clicked the button. Hands immediately flew up across the room.

"Whoa!" exclaimed Daniel. "It's like you can't even tell how fast it went!"

"It was just there," Jasmine added with a snap of her fingers.

Mr. Rabe repeated the demonstration several more times, pointing the flashlight at different parts of the room so everyone had a chance to be in the direct beam. Students, in various ways, observed that the light gets to them, and everyone in the room, really fast. Faster than they could even tell.

"So, how fast is light?" Mr. Rabe asked.

"Really, really fast," said McKenzie.

"Super fast," said Alejandro.

"Yeah, light is really fast," Mr. Rabe continued. "Now, think about going that fast. If you were going as fast as that light, you would have to go that speed for over four years to get to the closest star. That's the closest one!"

He paused to let that sink in for a moment, then asked, "As you're flying through space for those four years, what are you passing? If you looked out the window, what would you see?"

"Planets?" offered Angela hesitantly.

"Good thinking, but that would only be for a little bit. Those are really close compared to the stars," Mr. Rabe responded.

He paused, allowing some wait time. It was Ariel that spoke up hesitantly. "You wouldn't see anything," she said.

"Why not, Ariel?" Mr. Rabe asked.

"Because there's nothing there," said Ariel.

"That's right," Mr. Rabe affirmed. "After you got past the planets right by us, you would pass *nothing* going as fast as light for four years. *Nothing* in that big huge space. Space is almost totally empty."

The looks on the students' faces mostly showed amazement, though some were concerned, and a couple were a little confused. Determined to return to the main thrust of the lesson, Mr. Rabe brought the conversation back to the apparent brightness of stars and wrapped up the lesson just in time for recess.

It soon became apparent that Mr. Rabe was not the only one reflecting on the science lesson during recess. He first noticed raised voices from a group of girls who were gathered in conversation. Jasmine, clearly upset, stormed over to the fence near Mr. Rabe. After giving her a moment to cool off, Mr. Rabe approached her, "Are you okay, Jasmine?"

After a moment of silence, the words tumbled out of Jasmine's mouth. "Clarissa says since space is so big and empty, God doesn't exist."

Mr. Rabe was surprised. While he'd had these types of conversations with students before, it hadn't yet been in this context. He took a moment to think about his response so he could respond carefully. He was grateful to have just Jasmine at this moment, rather than the whole class.

"Well, what do you think about that, Jasmine?" Mr. Rabe started. "Do you agree with her?"

"I don't know," Jasmine responded. "I don't like what she said. I think there is a God. That's what my mom teaches me."

Mr. Rabe waited, seeing if she had said all she wanted to before responding. "You know, some people think that the emptiness of space proves there's no God. Other people, though, see the vastness of space as supporting their belief in God. So, there are different ways of thinking about it."

"Oh," responded Jasmine, "okay." She sounded like she was relieved to see that she could still choose to believe in God.

"You might want to talk about this with your mom sometime. That can be a good place to figure out this kind of stuff," Mr. Rabe encouraged.

"Yeah," Jasmine said, nodding, as if making a note to do just that.

It was clear that Jasmine wasn't totally settled, but the brief conversation seemed to have calmed her some and given her an excuse to keep thinking about what she knew and believed. She gave Mr. Rabe a slight smile and then turned to walk back onto the playground. As he watched her return to Clarissa and the other girls, Mr. Rabe wondered if he had both kept within the bounds of the law and done what he felt he should do as a person of faith. He also wondered how the other students in the group, and in the rest of the class, were responding. As the class gathered in their spot at the end of recess, he wondered if he should intentionally address this issue with the whole class when they got back to the classroom or if he should let it lie until individual students approached him with questions.

For Reflection and Discussion

1. Should Mr. Rabe address this issue with the whole class? If so, how? When?
2. How did Mr. Rabe do at addressing Jasmine's question? In what ways could he have done better?

3. Clarissa's assertion that there was no God was a particularly sensitive issue for Mr. Rabe because of his religious beliefs. In what ways, if any, would this situation have been different if the teacher had not held strong religious beliefs?
4. How would this response need to be different if this had occurred in a whole class discussion? How would it have been different if it was with younger students rather than his fifth graders?
5. As teachers, our beliefs, religious or otherwise, strongly influence our interactions with students. What other dilemmas might a teacher face as a result of their deeply held beliefs?

Ryan S. Nixon is an associate professor of science education at Brigham Young University, a private university owned and operated by The Church of Jesus Christ of Latter-Day Saints. His research focuses on teachers' knowledge of science subject matter, specifically exploring how teachers develop this knowledge through teaching experience. He teaches future elementary teachers how to teach science.

Commentary: Clarissa Doesn't Speak for Science

13

Mark A. Bloom

Abstract

This is a commentary to the case narrative, *"Clarissa Says God Doesn't Exist"* written by Ryan S. Nixon.

Science content at any level can create tension for learners because of deeply held beliefs. Climate change acceptance (or denial) often falls along political lines: left leaning learners embrace the subject matter, while students from politically conservative homes exhibit doubt about global warming claims. Students exhibit wildly varied reactions to public health claims, such as the safety of vaccines or the benefits of wearing face masks to prevent spread of communicable diseases, as seen following the emergence of COVID-19. Students whose backgrounds include farming, hunting, or fishing might react differently to the prospect of dissecting an animal in a biology laboratory than would a student who believes animals should be cared for, not used for food or recreation. Science teachers must be attentive to students' beliefs and emotions when introducing content or activities that students might find controversial. Part of a teacher's job is to create an environment where all students can learn and grow. Failure to attend to students' concerns can lead to a breakdown in the learning process and could foster negative emotions toward science in general. This case concerns how a teacher tries to help when science content creates tension among students because of their differing religious beliefs.

Science has, for a long time, been in conflict with religious ideas. In the early 1600s, Galileo Galilei famously defied Catholic Church doctrine when he sided with the heliocentric model of the solar system put forth by Nicolaus Copernicus, asserting that the planets revolve around the sun. The facts were clear, but

M. A. Bloom (✉)
College of Natural Sciences and Mathematics, Dallas Baptist University, Dallas, TX, USA
e-mail: markb@dbu.edu

© The Author(s), under exclusive license to Springer Nature Switzerland AG 2023
S. Jeong et al. (eds.), *Navigating Elementary Science Teaching and Learning*,
Springer Texts in Education, https://doi.org/10.1007/978-3-031-33418-4_13

the results of the Roman Inquisition determined that the Copernican model defied scripture. As such, Galileo was labeled a heretic, forced to recant his opinion, and spent his remaining days under house arrest. Three centuries later, John Scopes, of Dayton, Tennessee, was found guilty of violating the Butler Act, which criminalized the teaching of human evolution in state-funded schools. He was fined $100 for the violation. Sadly, I have in my office a book published in 2004 that continues to promote a geocentric view of the solar system—a view supported by the author's biblical beliefs. Fights over evolution education in American schools continue today.

Despite the historical conflicts between science and religion, research has shown that our perception that scientists look down on religious beliefs (the concern of Mr. Rabe) is not factually based (Ecklund, 2010). What is true, however, is that, in addition to evolution and climate change, there are myriad socioscientific issues that continue to be sources of potential conflict to students' religious backgrounds—issues such as stem cell research, cloning, gene editing, and race and gender equity. So how does a science teacher successfully address these important issues without throwing fuel on an already uncomfortable fire?

Having taught biology for over 25 years, with nearly half that time at religious institutions, I have learned to be very attentive to the religious beliefs of my students so that I can successfully teach religiously sensitive science content. Despite consistently teaching perceived controversial science content such as evolution and climate change, I have encountered little to no pushback from even my most conservative Christian students. How I avoid conflict in my classes is through a careful explanation of the nature and limits of science, as well as by making a distinction between science and philosophy, typically at the outset of my classes. I believe that, had Mr. Rabe used this same strategy, he could have quickly resolved Jasmine's distress. Indeed, Jasmine's distress might have been avoided altogether.

A few salient characteristics distinguish scientific claims from non-scientific claims, foremost the fact that all scientific claims are rooted in empirical/ observable/measurable evidence. If no empirical evidence exists to support an idea, then it is not a scientific one. Scientists make observations of the empirical evidence and then make inferences to explain that evidence. Hypotheses are developed to test the fitness of the inferences using scientific methodology. Inferences that survive hypothesis testing persist as credible scientific explanations. So, for an explanation to be characterized as scientific, it must be based on physical (natural) evidence and be explained with testable (natural) causes. These are the limits of science.

Accommodating the concerns of religiously minded students when teaching religiously sensitive content requires one important additional step. When distinguishing science from non-science, a caring science teacher will also make clear that just because a claim is not scientific does not mean it is nonsense! There are many aspects of life that cannot be answered scientifically—what is the meaning of life? What is our purpose on the planet? What is beautiful? Should we save endangered species?—and in Clarissa's case, "Is there a God?" Each of these questions are important, but their answers are not derived using scientific methodology. Science advances by operating on the assumption that all questions can be answered with science, but this is just an assumption. Science limits its answers to empirically based natural causes because scientists *act as if* there must be a natural cause. This assumption keeps scientists looking further and deeper into nature in hopes of learning more about the natural world. I make clear to my students that *operating under the assumption* that natural causes can explain all phenomena is different from *believing* so. Belief that natural causes are behind all phenomena is called philosophical naturalism and, as the name suggests, is a philosophical position—not a scientific one. Once religiously minded students, like Jasmine, can make this distinction, they can recognize that claims of science have no authority over their religious beliefs and can, therefore, view science as non-threatening.

I applaud Mr. Rabe's commitment to respecting the religious views of his students and his wish to create a safe and inclusive learning environment. This case demonstrates that even content that seems completely non-threatening and uncontroversial can lead to tension when juxtaposed with the religious beliefs of students. Based on my many years teaching biology to religious students, I believe a little bit of careful instruction on the nature and limits of science can be critical to easing tension surrounding religiously sensitive science.

Reference

Ecklund, E. H. (2010). *Science versus religion: What scientists really think*. Oxford University Press.

Mark A. Bloom is a professor of biology and science education at Dallas Baptist University (DBU), co-executive director of the International Consortium for Research in Science & Mathematics Education (ICRSME), and co-editor of the Electronic Journal for Research in Science & Mathematics Education (EJRSME). His research focuses on the intersection of science and religion and the teaching and learning of religiously sensitive socioscientific issues.

Case: Andrea Draws a Scientist

14

Katie L. Brkich, Maria A. Rodriguez,
and Alejandro Gallard Martínez

Abstract

This case, told from the perspective of a student, discusses cultural experiential differences found in elementary science classes. Andrea, a third-grade student, recently immigrated from Matamoros, México, to live with her aunt. She began learning English last year in the second grade, so most of her communication has been done in Spanish both in school and with her friends and family. In contrast, Andrea's teacher, Ms. Johnson, is a monolingual English native from Michigan whose worldviews and experiences are very different from that of Andrea. A dilemma occurs when Ms. Johnson asks the class to "draw a scientist". Andrea's drawing is different from a traditional American laboratory scientist. The case raises questions regarding what happens when Andrea's worldview, grounded in her own experiences, is regulated by the teacher's worldviews.

This morning, Andrea looked out the bus window at the rust-colored steel posts by her school. The border wall—a constant reminder that she lived in two different worlds. During the weekend, she lived in México with her parents. She found it difficult to explain everything she was learning at school because some ideas

K. L. Brkich (✉)
Department of Elementary and Special Education, Georgia Southern University, Statesboro, GA, USA
e-mail: kbrkich@georgiasouthern.edu

M. A. Rodriguez
Department of Curriculum and Instruction, University of Texas at Rio Grande Valley, Brownsville, TX, USA

A. G. Martínez
Middle and Secondary Education, Georgia Southern University, Savannah, GA, USA
e-mail: agallard@georgiasouthern.edu

were difficult to understand. During the week, she lived in the Rio Grande Valley with Tía Linda so that she could have better opportunities with an "American education".

After lunch, all the students in her third-grade class at Pedro Cano Elementary walked back to class in a single file but not quietly. Most students in her class were Latinx with strong attachments to México, but Andrea was the most recent immigrant to join the class/school. Everyone looked forward to science class, and the excitement could be heard down the hall. Andrea liked science but not as much as the rest of her classmates. She found science class confusing sometimes as it required her to focus more than any other subject. Most of the time, it was the words that Ms. Johnson used that Andrea found particularly puzzling. Ms. Johnson used words like state, value, and energy in completely different ways that Andrea did not always understand. Andrea was sure Ms. Johnson thought she did not pay enough attention, but Andrea tried her hardest to decipher the information presented during class discussions by looking around to see what everyone else was doing and trying to blend in.

When the third-graders arrived back to their classroom, Andrea took her seat at the back of the class. It had been her seat since her first day of school. By the time her parents had made all the necessary arrangements with Tía Linda, school had already started in Texas. Therefore, Andrea had started later than the rest of the students, walked into class, and took the only available seat at the back of the room. She had thought about asking Ms. Johnson to move her to another seat closer to the front like the one she had last year. However, she was very hesitant to ask to be moved as she was a beginning English learner. When she sat in the front row last year, she was able to ask her teacher for clarification without the other students noticing. However, sitting in the back this year proved to be a nice spot after all. From her seat, Andrea was able to see the entire board, she could see Ms. Johnson no matter where she stood in class, and she was able to look around the classroom at the posters for hints when she had questions. Looking around at the work her classmates completed before she arrived at the beginning of school year was probably what made Ms. Johnson think she was distracted so often; in reality, she was just trying to catch up.

There was often confusion between Andrea and Ms. Johnson because Andrea was so quiet and spoke mostly Spanish. Ms. Johnson knew that Andrea started school at age four in Matamoros, México, in the state of Tamaulipas, but what she didn't know was that Andrea was an A student and known to participate very energetically and enthusiastically in class. Her teachers knew her as a person who, if she did not understand, would not hesitate to raise her hand and ask a question. In sum, before moving to Texas, her school experiences were very successful. All that changed last year when she started second grade at a school in Texas and experienced a whirlwind of confusion, self-doubt, and struggle as well as a lot of loneliness. Andrea finds that communicating in English is still very unfamiliar and difficult, as she only uses the new language the small amount she speaks at school. She has gotten much better at listening and reading in English, but it is still a struggle to speak or write without first thinking through her responses in Spanish and then translating them to English.

Most teachers at her new school look like Andrea and the teachers she knew back home. However, Ms. Johnson looks different. She is young, White, and her clothes look very expensive and stylish. Ms. Johnson left her home in Michigan for college where she majored in finance. She then came to the Rio Grande Valley as part of a special teaching program. Andrea's initial impressions of her new teacher were not negative but made her wonder why she was not placed in a classroom with a teacher who spoke and looked like her. Nonetheless, Andrea liked that Ms. Johnson tried making everything fun and engaging. Most of all, she liked that Ms. Johnson talked differently from how Andrea talked and because no one made fun of the way Ms. Johnson spoke, no one made fun of Andrea when she misspoke.

When Ms. Johnson taught a unit on biomes, she shared pictures of the place where she grew up. Andrea could not remember the name of the city, but she remembered it was up north where the leaves turned different colors in the fall, and it snowed during the winter. Stories about the changing of seasons and the pictures of snow were different and interesting to Andrea. She related this to back home, in Matamoros, which also has seasons and temperatures that can range from cold to hot.

During this unit on biomass, students learned about different types of scientists such as biologists, botanists, chemists, and environmental scientists. Ms. Johnson hoped this would provide students with a diverse view of what makes someone a scientist. Near the end of the unit, Ms. Johnson asked the students to "draw a scientist". She reminded everyone of the different topics they had discussed during the previous weeks, then wrote the directions on the board again: Draw a scientist. Andrea thought about the directions and everything she had learned and almost immediately knew exactly what she would draw. She did not have to look around at the other students for guidance like she did normally. She did not have to show her picture to her best friend in class for a reassuring smile, thumbs up, or nod. Finally, an assignment that was not confusing!

Ms. Johnson walked around the class asking questions and giving praise. She picked up and showed off some of the students' drawings. One drawing depicted a man with crazy hair and glasses. Another drawing included men in laboratory coats and goggles. Yet another student drew an astronaut. Ms. Johnson walked over to Andrea. Andrea held her breath waiting for Ms. Johnson to pick up her picture and show everyone her unique drawing. Instead, Ms. Johnson hunched over Andrea's desk, looked upset, and told her, "Andrea, you did not follow directions. Why did you draw whatever you wanted instead of what was assigned? What didn't you understand?" (Fig. 14.1).

Andrea could not really understand Ms. Johnson's English but did hear the disappointment in her voice. Andrea did not dare look at her teacher or glance around at her classmates, but instead looked at her drawing and kept her head down. How had she messed up? How had she gotten the assignment wrong? The directions had been simple, easy even! Yet, there was Ms. Johnson reprimanding her. Andrea could feel her eyes swelling up with tears. The tears stung but she did

Fig. 14.1 Student drawing of a scientist depicting a man with crazy hair and glasses

not blink so they would not roll down her cheeks. She had been repeatedly told by her parents to behave in school and not to talk back to adults, especially teachers because they were the leaders and role models within her community in México. Andrea remained silent and did not give Ms. Johnson a response. Andrea knew that if she gave the wrong answer or tried to explain her frustration, Ms. Johnson might request a parent–teacher conference. Andrea did not want to embarrass her family over a drawing.

On the bus on the way home, Andrea pulled from her backpack her scientist drawing. Ms. Johnson had given her the opportunity to take it home over the weekend to "fix it". Andrea examined her drawing and tried to find her errors. She had drawn an older woman standing in a living room, her workplace, surrounded by plants, herbs, candles, and crystals. She wore a dress and her hair up in a bun. This kept her comfortable while she worked all day seeing people. Her hands were small and stained because she used them to crush leaves, measure liquids, and make concoctions. She knew how to use the plants and herbs around her to make teas and potions. The crystals and quartz were rocks that could be found in the ground but were usually bought in the market. Andrea studied her drawing for mistakes, but she was convinced she had done the assignment correctly (Fig. 14.2).

When Andrea arrived home, Tía Linda took one look at her and knew something was wrong. "*Hay mija*, what happened?" she asked. Andrea told her what happened in school and handed her drawing over to her aunt to review. Tía Linda took one look at the picture and understood. She grabbed Andrea up in a tight hug and

Fig. 14.2   Andrea's drawing of a scientist as a curandera (healer)

stroked her long brown hair reassuring her that nothing was wrong. She looked at her niece's tear stained face and explained "Ms. Johnson has probably never been to a *curandera* and can only imagine scientists who work in western laboratories. You did draw a scientist, just not the ones that Ms. Johnson knows about".

For Reflection and Discussion

1. How could Ms. Johnson have improved how she went about the "draw a scientist" learning task?
2. How could Ms. Johnson build on Andrea's personal knowledge and experience (i.e., her "funds of knowledge") to enrich the learning experience for Andrea? For her classmates?
3. What advice should Tía Linda give Andrea about how to revise her assignment to turn in on Monday?
4. How does this experience and the teacher's reaction reinforce stereotypes about scientists? Consider multiple elements of identity including gender, socio-economic status, language, race/ethnicity, and level of education.

Katie L. Brkich is a professor of elementary science education at Georgia Southern University. Her teaching and scholarship center around bringing social justice and culturally sustaining pedagogies into the elementary science classroom. Her most recent work looks at the framework of "contextually mitigating factors" as a lens to examine situations and analyze data that might not usually be considered in academic research.

Maria A. Rodriguez is a doctoral student at the University of Texas Rio Grande Valley where she is studying curriculum and instruction with a specialization in science education. Her research interests are secondary science education, gender studies, and teacher education. Ms. Rodriguez has 14 years' experience being an educator in Texas public schools and currently works as a district administrator.

Alejandro Gallard Martinez is a professor and Goizueta Distinguished Chair at Georgia Southern University's College of Education in the Middle Grades and Secondary Education Department. The platform for his research is to explore equity and social justice in education in general and in particular science education. His sociocultural frameworks include global perspectives on differences, otherness, polyphony of voices, and meaning-making that reflects categories used to situate people in social life through contextual mitigating factors (CMFs).

Commentary: Andrea's Organic Scientist

15

Jennifer D. Adams

Abstract

This is a commentary to the case narrative, *"Andrea Draws a Scientist"* written by Katie L. Brkich, Maria A. Rodriguez, and Alejandro Gallard Martínez.

The Draw a Scientist Test (DAST) has been around for more than 50 years. Although children, female identified children in particular, are drawing more female representations of scientists, the drawings are still overwhelmingly male. Furthermore, stereotypes of what a scientist does and looks like persist. For example, most of the drawings in prior research depicted scientists wearing a laboratory coat and goggles or eyeglasses, working indoors in a laboratory, and Caucasian or white. An analysis of years of researchers and teachers implementing the DAST reveals that the notion of "a scientist" remains narrow. For the heck of it I Googled "scientist" and while the images that came up were surprisingly racially and gender diverse, certain stereotypes remained. All of the scientists were in laboratory settings with copious amounts of glassware and equipment as props, and all wore eyewear (yes, safety) and laboratory coats. There were no images of people working outside, in nature, in communities, in classrooms, in museums, or wearing everyday clothing; there was nothing offered to make one think that a scientist could exist outside of the sanitized place of the laboratory. With these images in mind, one could understand why Andrea's teacher was unable to accept her drawing as that of a scientist.

However, Andrea did draw a scientist. During the unit on biomes, she learned about what different kinds of scientists do; she learned about the work of biologists, botanists, chemists, and environmental scientists. When she was given the

J. D. Adams (✉)
Department of Chemistry, Canada Research Chair in Creativity and STEM, University of Calgary, Calgary, AB, Canada
e-mail: jennifer.adams1@ucalgary.ca

S. Jeong et al. (eds.), *Navigating Elementary Science Teaching and Learning*,
Springer Texts in Education, https://doi.org/10.1007/978-3-031-33418-4_15

assignment to draw a scientist, she drew an image of what she understood a scientist to be based on her lived experiences. She grew up in Mexico, so her experience with people who had specialized knowledge of plants, chemistry, and the environment were curanderas who are mostly women who play an integral role in health, healing, and well-being in their communities. Similar to Antonio Gramsci's idea of the organic intellectual, Andrea's drawing depicts an organic scientist whose knowledge and practices are developed based on the culture, needs, experiences, and emotions of their communities. The curandera uses her knowledge of plants and their unique properties to do physical, mental, emotional, and spiritual healing—the work of physicians, psychiatrists, counselors, and religious leaders. Western scientists who are trained within the disciplinary and practical constraints of the university have produced innovations that have transformed societies, yet they have not always been responsive to the needs of communities, especially the most marginalized. Furthermore, Western science, as practiced in schools, universities, and other institutions, often discounts the knowledge and wisdom of organic scientists, like Andrea's curandera. Curanderas and other organic scientists (shamans, traditional midwives, Pasifica navigators, etc.) learn through making observations, experimenting, documenting findings, and disseminating knowledge to other practitioners and to their community. They also share their knowledge through oral history, storytelling, visual art, song, and other ways that go beyond the written text. Similar to Western science, Traditional Knowledge is passed along and used to inform future practice. Interestingly, Western science is increasingly relying on Traditional Knowledge to understand animal behavior, resources management, and climate change due to the recognition of the value of generational, transdisciplinary, and place-based knowledge these organic scientists hold.

As a teacher of young children, it is critical that Ms. Johnson first be aware that Western science, as taught in schools, is not an objective, universal phenomena but is very culturally embedded. Images play a powerful role in shaping discourses and ideas in society, and from the Googled pictures we can see that there are many aspects of "a scientist" that are missing from those images. Engaging in dialogues with students, instead of immediately dismissing their ideas, could provide important learning opportunities for educators and offer them the chance to expand their own notions about science (and other ideas about our natural and built world). For example, Ms. Johnson could have asked Andrea the following questions: Please tell me a little more about your scientist, what does she do? What does she study? How is she similar to the scientists that you learned about in class? This approach would have opened up a dialogue with not only Andrea, but other students in the class so that their ideas of scientists could also be engaged. Perhaps this would have allowed Andrea and other students to begin to picture themselves as scientists within their own worldviews.

Young children are like Traditional Knowledge-keepers in the classroom. Their growing knowledge of community and the natural world is embedded in their day-to-day experiences that they then bring into the classroom to make connections to what they are learning there. Educators who do not immediately dismiss children's ideas are positioned to expand their own learning about diverse cultures and to value the different experiences and ideas that their students share. This will not only enrich learning experiences for all students but will also help us to collectively build practices and learning environments that allow all learners to flourish and cultivate minds that are able to think beyond disciplinary and conceptual boundaries.

References

Nicholas, G. (2018). *Western science is finally catching up to traditional knowledge.* The Conversation Canada. https://theconversation.com/its-taken-thousands-of-years-but-western-science-is-finally-catching-up-to-traditional-knowledge-90291

Terada, Y. (2019). *50 Years of children drawing scientists.* Edutopia. https://www.edutopia.org/article/50-years-children-drawing-scientists

Jennifer D. Adams is a Tier 2 Canada Research Chair of Creativity and Science and Associate Professor at The University of Calgary. She is an NSF Early CAREER award winner and the PI of the Creativity, Equity and STEM Lab where she leads her team in research on equity in STEM teaching and learning environments with an emphasis on anti-deficit, learner-centered, and justice-oriented approaches.

Case: Joseph Has No Money for Groceries!

16

Alexis D. Riley and Felicia Moore Mensah

Abstract

This case raises questions about teacher positionality in an elementary science classroom. Ms. Nelson, a second-year science teacher, is asked to consider how her privileged upbringing as a White woman from Long Island, New York City, predisposes her to ignore the structural issues her students experience, such as living in food deserts and having limited access to healthy foods. Questions of race and poverty, the issue of a colorblind curriculum, and proactive parent involvement come into focus for Ms. Nelson during a science unit on Food and Nutrition as she accidentally ostracizes her poorer Black and Brown students.

It's 4:30 pm on a Wednesday in March, and Aly Nelson sits in her classroom alone, contemplating where her Food and Nutrition unit went wrong. All weekend Aly had anticipated a joyful Wednesday event, where her 4th-grade students would engage in a hands-on activity about things we all love, food! Instead, her day was engulfed with unusually aloof students and concerned parents. Where did things go wrong?

In her second year of teaching in Flushing, Queens in New York City, Aly thought she was finally getting the hang of everything. From Long Island herself, she chose to work at Flushing Eagle Academy because of the culturally rich and racially diverse population of students she would teach. Almost a month ago, Aly and her 4th-grade students began the Food and Nutrition unit during their science block of the school day. Since most of the students in her class are Asian, Black,

---

A. D. Riley · F. M. Mensah
Science Education, New York University-Steinhardt, New York, NY, USA
e-mail: fm2140@tc.columbia.edu

A. D. Riley (✉)
Teachers College, Columbia University, Science Education, New York, United States
e-mail: alexis.riley@nyu.edu

and Brown, she was excited to have them discuss something they could all relate to: food. For three weeks, students learned about the food pyramid, the digestive system, and how to distinguish "healthy" and "unhealthy" foods.

"Broccoli is disgusting, Ms. Nelson! I think the food pyramid people have it all wrong. My body doesn't even want to digest that," exclaims Mia during the first week of the unit.

Aly's students are intrigued to learn about how their bodies digest food, and they love talking about why they think vegetables are gross and should be banned from the food pyramid all together! Aly feels really good about the progress of the science unit, especially since she did very little science teaching last year. She notices her students are engaging in a culturally relevant science unit that encourages curiosity and personal connections and is proud of the progress she is making in teaching the unit.

As a culminating activity for the unit, Aly thinks it is a great idea to have students bring in different foods. As a class, they will organize the foods into healthy/unhealthy categories, then build a real-life food pyramid! Ms. Nelson assigns different group members a food item on Monday and tells them that on Wednesday they will need to bring the food item to class for 20% of their end-of-unit project grade.

During the morning meeting on Wednesday, Aly notices a couple of sullen expressions where there are typically bright faces and smiles. Figuring that her students are a bit tired, she moves on with business as usual, hoping they will perk up after lunch and will be enthused to do the food activity. Right before lunch, she tells the students "Remember to take your food out of your bags and be prepared to share with your group for our science activity of the day!" Sullen faces turn to teary eyes for five students as they start to line up for lunch. Aly asks Mia "what's wrong?" after most of the students head to the lunchroom, "I couldn't get the food for the science activity, Ms. Nelson. My dad said we couldn't afford to drive all the way to Astoria for kale." Stunned, Ms. Nelson tells Mia, "It's alright. Go and meet your class for lunch."

Before Aly could consult her mentor teacher in the lunchroom, she receives a call from the main office telling her that three parents are "concerned" about today's food activity science lesson. Mr. Thompson, Luke's dad, catches Aly on the phone and expresses shame about not being able to help his child bring sunflower oil to class on such short notice. He explains, "I work evenings and double shifts this week, so I asked my brother to go to the local grocery store in the neighborhood, but they did not have that food item." Aly mostly listens to Mr. Thompson and says, "It's alright. I didn't mean to cause you stress." That joyful feeling Aly anticipated for the science activity is fading away.

Walking back into the classroom from lunch, the same five students still look distraught. Unprepared to alter her lesson, Aly starts the science activity as planned. Students begin to gather their food items to present to their small groups. She hears some chuckles around the classroom. Before finding the source of the problem, she hears, "Joseph has no money for groceries!" The classroom erupts in gales of laughter. Shame and embarrassment fill Joseph's eyes. Aly looks around and notices the same expression on the other four students' faces as well. Panicked, she studders, "That—That is enough of the food pyramid for today. Ok. Ok, let's just do math. Remember the problem set we had yesterday? Let's work on those problems. Get your math notebooks out and begin working on your math problems."

The students seem confused, and many are upset that they won't be able to share their food. Jacob complained, "But I practiced saying rutabaga for twenty minutes last night!" A flutter of relief seemed to cross Mia's face, but Joseph just buried his face in his math notebook until it was time for dismissal.

At the end of the day, Aly calls Joseph's mom to discuss the incident. Joseph's mom expresses her anger that Joseph was bullied in front of everyone. She shares that Joseph never even asked to go to the store to get food for his science project. "We bought groceries Saturday, but maybe Joseph thought it was too much of a burden to ask for money to buy more food on Monday." She elaborates, "we also have to take a couple of trains to get to the grocery store."

Around 5 p.m. while leaving school, Aly stops by the room of her mentor teacher, Ms. Wright, to ask for advice. With eight years of teaching experience at Flushing Eagle Academy, she was a great support to Aly during her first year of teaching. While Aly shares the events of the day, Ms. Wright listens intently and asks one question at the end, "Aly, do you notice any pattern, or can you make an observation about the students who didn't have anything to bring today for the science activity?" Aly takes some time to think but is overwhelmed by the question. She responds with a simple, "No." Ms. Wright reveals that all the students who had nothing to bring are Black or Brown students who live in poorer areas of Jackson Heights or Jamaica, Queens. Genuinely confused about the relevance of that information, Aly asks naively, "What does race have to do with this?" Ms. Wright explains, "some students in our school live in areas that do not have access to a wide range of healthy foods or large supermarkets. Most students eat from their neighborhood bodega, and the main meal of the day for many of the students is the school lunch we provide throughout the week."

"As a White woman from Manhattan, I also struggled at the beginning of my teaching career to learn about the different realities of my students," Ms. Wright began. "One of my most embarrassing experiences happened when one of my Asian students shared a home remedy her great-grandmother taught her for bruised knees. I exclaimed, 'That's not science!' in front of the whole classroom." Ms. Wright painfully confesses, "after that incident, many of the parents of color began to distrust me." She says, "I had to regain trust by first learning about the community I work in and asking students more questions about their everyday experiences." Ms. Wright finishes, "I invited family members and community members into my classroom to share their cultural knowledge."

Aly, feeling some shame and confusion, thanks Ms. Wright for the advice and heads home to plan her lessons for the next day. She continues to reflect on her students, their parents, and the conversation with Ms. Wright.

For Reflection and Discussion

1. For the Food and Nutrition unit, what planning shifts could Ms. Nelson make to ensure that none of her students feel ostracized?
2. How does Ms. Nelson's positionality as a White woman who grew up with more privileges than her students affect her science teaching?
3. After two years of teaching, should Ms. Nelson be expected to know about the neighborhoods and lives of her students? How could she learn about her students' lives outside of school?
4. What would a truly culturally relevant Food and Nutrition unit entail for Ms. Nelson's racially and economically diverse classroom?

Alexis Riley (@bigdocenergy23) is an Assistant Professor in the department of Teaching & Learning at New York University-Steinhardt in the Science Education program. She earned her PhD at Teachers College, Columbia University, New York City in Science Education. Her research positions the pedagogical practices and innovations of Black women teachers within science education by examining their personal narratives, science teacher preparation and professional development. Black feminist thought, critical race theory and culturally relevant pedagogy guide her research.

Felicia Moore Mensah (@docmensah) is the department chair of Mathematics, Science and Technology and a professor of science education at Teachers College, Columbia University, New York City. Her research addresses issues of diversity, equity, and identity in science teacher preparation and teacher professional development, with culturally relevant teaching, multiculturalism, and critical theories guiding her teaching and research. Her most recent work focuses on the preparation of teacher educators for racial literacy.

Commentary: Get to Know Your Students

17

Olayinka Mohorn

Abstract

This is a commentary to the case narrative, *"Joseph Has No Money for Groceries!"* Alexis Riley and Felicia Moore Mensah.

Ms. Nelson had good intentions when she created the food lesson for her 4th-grade class. She was excited and believed that her lesson would be educational, fun, and engaging to her students. In her zeal to try out this new lesson, Ms. Nelson forgot an essential part of her teaching. She did not know enough about her students. The *Danielson Framework for Teaching (FFT)* suggests that a proficient teacher purposefully acquires knowledge from several sources about students' varied approaches to learning, their knowledge, skills, special needs as well as interests and cultural heritages (Danielson, 2013). Therefore, Ms. Nelson had some serious work to learn more about her students and increase her cultural competency so that she could educate her students effectively.

Getting to know students is a crucial part of culturally relevant and culturally responsive teaching. It cannot be left out of planning for quality instruction. Teachers of culturally and linguistically diverse students should never assume anything about them and their home cultures. To plan lessons that appeal to her students' interests and backgrounds, Ms. Nelson needed to have information about her students, their families, and the communities in which they lived. Ms. Nelson's active inquiries of her students' likes and dislikes, needs, and home lives should be something that is ongoing. Strategies like morning meetings, circle time, or individual student lunch meetings could be done regularly to obtain this important information about students. Completing activities of this kind regularly would help to build

O. Mohorn (✉)
Curriculum and Instruction, College of Education, University of Illinois, Chicago, IL, USA
e-mail: omohor2@uic.edu

trust between Ms. Nelson and her students. It would also allow a space for student voices to be heard, giving them the chance to share their perspectives about relevant topics inside and outside of the classroom. Interest inventories, learning style assessments, and personality surveys are also good ways to obtain information on student preferences, which can all be used to inform instruction.

Communicating with parents is also an important part of getting to know students. Regular conversations with parents can provide information about their child's likes and dislikes and reveal how their child learns best. These conversations can also give teachers specifics on familial and community resources that are accessible to them. In our technology age, teacher–parent conversations can occur through emails, Zoom meetings, or via specific apps designed for this purpose, like Remind, Class Dojo, Google Classroom, ClassTag, and PhotoCircle to list a few. Several of Ms. Nelson's parents told her (unfortunately after she gave the assignment) that buying foods, especially certain mainstream "healthy" foods, was difficult for them because they lived in neighborhoods where access to those foods was scarce. What if she shared her idea about the assignment with them beforehand? Doing this might have saved her students from an embarrassing situation.

Cultural competence is another essential trait for teachers of diverse students. All teachers should be responsible for properly educating students from different cultural backgrounds, and at their core, they should have a genuine respect for all cultures represented in their classroom. They should possess a strong cultural curiosity such that they can move beyond celebrating "food and festivals" of diverse cultures to securing an authentic knowledge of diverse groups, which includes a deep understanding of institutional structures that create and perpetuate inequity and bias, specifically within education. Teachers need an unbiased lens when learning about the practices and beliefs associated with the cultures of students in their classrooms. Viewing culture simply as a collection of ways that a group of individuals make sense of the world and use available resources is a good place to start. Using information gleaned from getting to know their students, teachers can work to bring elements of students' cultures into their instructional practice. For example, cooperative group work can be an effective learning strategy to use with students from collectivist cultures of many Latin and African countries (Hammond, 2018). In another example, a chemistry teacher gave students the option to create a game to illustrate their conceptual knowledge about the periodic table instead of the standard paper and pencil assessment. Being willing to try new strategies and approaches in teaching, and maintaining a reflective stance, is key to becoming a culturally responsive teacher.

Ms. Nelson has a few things working for her. First, she is a reflective teacher and appears open to learning more about how she can improve her science pedagogy. Second, Ms. Nelson has a close relationship with an experienced mentor teacher, Ms. Wright, who can support her growth and development into a teacher skilled in culturally responsive teaching practices. Ms. Wright asked Ms. Nelson to think critically about who did not complete the assignment and what those students had in common. She challenged Ms. Nelson to be more reflective and to ask questions in her instructional practice when it does not meet the needs of all her students. Ms. Wright is someone whom Ms. Nelson can bounce ideas off, obtain constructive feedback, and learn from a more experienced teacher. Because of these qualities, Ms. Nelson is in a position to grow in her capacity as a culturally relevant and culturally responsive teacher.

References

Danielson, C. (2013). *The framework for teaching: Evaluation instrument.* Retrieved from www. danielsongroup.org

Hammond, Z. (2018). *Culturally responsive teaching and the brain: Authentic engagement and rigor for culturally and linguistically diverse Students.* Corwin Press.

Olayinka Mohorn earned her Ph.D. in Science Education from the University of Illinois at Chicago. She is currently a high school chemistry teacher in Chicago Public Schools. She studies science identity construction with a critical lens, with the ultimate goal of increasing diversity in STEM and healthcare career fields.

Case: Did I Inherit My Curly Hair from My Mom…, or from My Ma?

18

David Steele and Sophia Jeong

Abstract

This case points to tensions that arise when an elementary school teacher, Mrs. Biels, implements a life science lesson designed to introduce the inheritance of traits. The *Next Generation Science Standards* (NGSS) suggest that students as early as third grade should be able to recognize patterns of similarities and differences in traits shared between offspring and their parents, or among siblings. Too often, teachers are not aware of the hidden curriculum in their instructional practices, and when dealing with lessons on inheritance, do not reflect on the diverse landscape of the family structures of all their students. For instance, in this case, a dilemma emerges when Mrs. Biels begins a new genetics unit by engaging students with pictures of puppies and their parents and having students determine which parent the puppies inherited traits from. As the lesson progresses and students begin thinking about who they inherited their own traits from, Mrs. Biels is not prepared for how the lesson would impact one of her students.

"Hey, good morning, young scientists!" Mrs. Biels excitedly declares as her third-grade students make their way into her classroom for science. Mrs. Biels is always full of energy, as if she has just finished drinking an energy drink. She bounces around the classroom, greeting each of the students as they walk to their desks, while collecting their science journals from the storage bin. As students are returning to their seats, Mrs. Biels reminds them to answer the "Question of the Day"

D. Steele
Alder Graduate School of Education, STEM Education, Redwood City, CA, USA
e-mail: dsteele@aldergse.edu

S. Jeong (✉)
Department of Teaching and Learning, The Ohio State University, Columbus, OH, USA
e-mail: jeong.387@osu.edu

Fig. 18.1 Mom and two of her puppies

that is displayed on the SMART Board with a picture of an adorable puppy family (Fig. 18.1).

Students open their science journals and begin answering the question, "How are these puppies similar to each other and to their mom?" Mrs. Biels starts playing soothing classical music while students independently and quietly work on answering the prompt. A quick glance around the classroom allows Mrs. Biels to see that all students are eagerly writing a response. As soon as the music stops, students recognize this as a cue to meet with their group members to discuss their responses. Mrs. Biels enthusiastically monitors the classroom, walking around the classroom and listening in on groups as students share their thinking about how the puppies are similar to each other and their mom.

A huge smile covers Mrs. Biels' face as she listens in on the sense-making strategies her students use to answer the question. The smile is a mixture of happiness and relief. Mrs. Biels, a second-year teacher, has been reviewing curriculum resources for the past several months in order to learn new methods for implementing the Next Generation Science Standards (NGSS) in her classroom in a meaningful way that incorporates students' experiences and identities into her lessons. She believes that this is key to keeping students engaged in science learning. As a third-grade teacher, one of the topics she has to cover is the inheritance of traits and helping students recognize patterns of similarities and differences in traits shared between offspring and their parents, or among siblings. After consulting several resources, Mrs. Biels decided that using puppies as an example would be an excellent way to ensure students are engaged in the lesson. Who doesn't love puppies, right? The discussion so far seems to reinforce her idea that this will be a great lesson.

As students are finishing the discussions in their small groups, Mrs. Biels whispers, "If you can hear my voice clap once," and a handful of students closest to her respond by clapping their hands. She repeats, "If you can hear my voice clap twice," and a majority of the class responds by clapping their hands. As the remainder of the class quickly finish their last thoughts, Mrs. Biels moves along the desks to the front of the room and asks for a volunteer from each group to share one idea of how these puppies are similar to each other and to their mom. Several hands shoot up.

Mrs. Biels: Okay, let's start with Sarah. What is one similarity that your group decided on?

Sarah: Puppies are similar because they all have black noses.

Mrs. Biels: That's awesome Sarah. James, do you agree with Sarah that all of them have black noses?

James: Yeah, and they all have some white fur, too.

Timmothy: Yeah, but one puppy is brown and white, and one puppy is black and white, so they're not completely the same.

Mrs. Biels: That's true, Timmothy, that's a good observation. Even though they share some similarities, they are not identical. Can you share another similarity?

Timmothy: I mean, they're all dogs, and they have four legs.

Mrs. Biels: That's great, Timmothy. Shera, do you agree that these are all dogs?

Shera: Yeah, they're all dogs. There are two puppies with their mom.

Suzanna: Mrs. Biels, they all have white paws too!

Mrs. Biels: Excellent observations, young scientists! You've told me similarities with the puppies. What about how they're similar to the mom?

Luis: Well, one puppy is the same color as the mom. But the other puppy is black and white, so it looks different.

Aaman: Why are they different from one another if the puppies are brothers?

Mrs. Biels: Well, that's a good question, Aaman. Why don't you all take a few minutes to write down all the differences you observe and then create an explanation that answers Aaman's question.

Mrs. Biels writes Aaman's question on the board, "If the puppies are brothers, why are they so different from each other and their mom?" and starts her classical music, the cue for students to start their individual work time. Mrs. Biels continues her classroom ritual of checking in with students as they are working. As the music stops, students again transition to their small groups. After two minutes, Mrs. Biels uses an attention grabber to signal students to finish their conversation and to get ready for a whole-group discussion.

Mrs. Biels: Now class, when we left off, I asked you to write down all the differences you see between the puppies and to create an explanation for why the puppies look different from one another and their mom. I heard a couple of really good things while you were working, and I'll ask Luis to start by telling us one difference he observed between the puppies.

Luis: Well my group pointed out that one puppy is black and white, and the other puppy is brown and white.

Mrs. Biels: That's true, Luis. Is there another difference that your group discussed?

(20 seconds of wait time)

Mrs. Biels: If you only have one difference, you can call on another student who might be able to help you.

Luis: Okay, Lisa!

Lisa: We said the same about the puppies' colors, but also, I think their ears.

Mrs. Biels: Can you say a little more about their ears? What are you thinking is different?

Lisa: The brown and white puppy has ears that are pointy like the mom, and the black and white puppy has droopy ears.

Mrs. Biels: Oh, that's a great point Lisa. Does everyone see that? (As Mrs. Biels points to the differences in the ears, students' heads are nodding in affirmation, and a couple of them are mumbling "yes" in response). Lisa, who would you like to share next?

Lisa: Aaman, 'cause he asked the question last time.

Aaman: Well, I said that the one puppy has a white line of hair between the eyes, and the black puppy only has a small patch of white hair.

Mrs. Biels: Is there any other group that would like to share an observation they made about the differences between the puppies and their mom?

(20 seconds of wait time)

Mrs. Biels: Well, young scientists I think those are all great observations! What I'd like to know now, is if any group has an explanation as to why the brothers looked different from one another?

Aaman, without waiting on being called: Well, you told us that the big dog is the mom and so the brown and white puppy looks more like her, and the black and white puppy must look like the dad.

Mrs. Biels: Oh, so you're saying that the puppies look like their mom or dad? Joel, do you agree or disagree with Aaman and why?

Joel: I agree because you either look like your mom or your dad. I don't look exactly like my sister, but we do look alike because we are a family.

Mrs. Biels: Oh Joel, I think that is a great way to put it. We might not look exactly like our brothers and sisters, but we look similar because we are a family. Is there anyone in the class that disagrees with Joel and Aaman?

Timmothy quickly adds: The black and white puppy probably looks like his mom.

Mrs. Biels: Oh, remember Timmothy, we already said the brown and white dog in the photo is the mom.

Timmothy just shrugs his shoulders and puts his pencil down.

Mrs. Biels: Timmothy, do you see how the brown and white puppy looks like the mom that's in the photo?

Timmothy: Yea, I see that (in a flat tone).

Mrs. Biels: So, that means the black and white puppy would have to look like who?

Aaman blurts out, "The black and white puppy looks like the dad, I already said that," as the rest of the class starts murmuring in unison the same thing. Timmothy, slightly blushing, closes his notebook.

As Mrs. Biels gets ready to transition to the next phase of her lesson, she can't help but think about how engaged the students were in this opening activity and the number of students who wanted to share their thinking. It made her even more excited knowing that she planned the next activity for students to start thinking about where their own traits came from. Based on Timmothy's comment, she makes a mental note that she needs to be very clear that the adult dog in the picture is the mom.

As the class starts to settle down, Mrs. Biels shares a slide with a photograph of her standing in between her mom and dad and a Venn diagram. Students start pointing out that Mrs. Biels looks so much like her mom, when Luis says, "Yeah, but look guys, she got her nose from her dad. They're both really big." Mrs. Biels laughs a little, knowing they both do have big noses: "Okay, okay, okay. So we see I have a big nose and that I got it from my dad. Who knows what these two overlapping circles are called? If you remember, we've used them just a couple of times before." There's a distinct sound of paper rustling as students flip through their science journals to find the handout that has the two circles on it. Anna quickly finds the drawing and yells out, "It's a Venn diagram." Mrs. Biels lets all students find the Venn diagram handout they used last week when comparing and contrasting animals with and without bones.

Mrs. Biels: Now that we've all found the handout, Timmothy, can you remind the class how we used the Venn diagram?

Timmothy: We were comparing a grasshopper with a frog and looking at the way they are the same and how they are different. Oh, could we have used this to compare the puppies with the mom in the picture?

Mrs. Biels: What do you think Lisa?

Lisa: Yeah, you told us that we use this chart to compare and contrast two things to see what they have in common and what is different.

Mrs. Biels: Oh, I love that you used the terms compare and contrast! Those are excellent words to use.

Aaman: Yeah, when we made them last week, you told us to remember that the area where the two circles kind of make one circle is where you put things that are in common.

Mrs. Biels: Luis, do you agree with Aaman's statement?

Luis: Yes, you said it shows that traits can belong to both kinds of animals, And that's how they are similar.

Mrs. Biels: Does anyone disagree or want to add to what has been shared?

Timmothy: I agree with them, and you also said that things in the areas where the circles don't meet mean they are characteristics of just one of them.

Mrs. Biels: Excellent, Timmothy, thanks for sharing.

Mrs. Biels waits for about twenty seconds to see if anyone else wants to share before she changes her slide to reveal a Venn diagram that she completed prior to class. She knows students are going to want to investigate what is typed into the circles, so she gives them a minute to read and to discuss. She purposefully did not label each circle. After giving students a little time to review the diagram, Mrs. Biels was ready to engage in the next phase of her lesson.

Mrs. Biels: What do you notice about my Venn diagram?

Aaman quickly responds: You have stuff on both sides of the circle and in the middle where they overlap.

James: I see that you have the word nose on the right side, so I think that is where you put your dad.

Mrs. Biels: Lisa, can you tell me what James just said, but put it in your own words?

Lisa: You have some traits listed in each part of the circle, but James is saying that on the right side, you have traits that are similar to your dad.

Mrs. Biels: Raise your hand if you agree with Lisa.

All of the students shoot their hands up in the air. Mrs. Biels smiles because the students have been really engaged, and based on the classroom conversation, they are understanding that our traits are passed down from parents to offspring.

Sarah: So, I think that means that the traits on the left are what you got from your mom. But how did you decide what you put in the middle?

Mrs. Biels: Sarah, that's a great question. Who thinks they can answer that?

Timmothy: Maybe it's for things where you don't really know where you got it from.

Mrs. Biels: That's a great idea, can you give me an example?

Timmothy: Umm, well I could say I don't know where I got my curly hair from. I could have gotten it from my mom… or… my ma because they both have curly hair.

Mrs. Biels: If your grandma and mom both have curly hair, then your mom probably got it from her, and you got it from your mom.

Timmothy started to speak up and clarify that he wasn't talking about his grandma. He was trying to explain to her that both of his moms have curly hair.

Before he could explain, Aaman jumps in to give his own example. As Aaman begins talking, Timmothy pushes his paper to the side and pulls his hoodie over his head.

Aaman: Mrs. Biels, both my mom and dad have the same shaped big toe, and my big toe looks like theirs. So, I would put "big toe" in the overlapping area of the Venn diagram.

Mrs. Biels: I think that's an excellent example. What I want everyone to do now is take a few minutes to work individually on completing their Venn diagram. You should use mine as an example of how to complete the task.

Mrs. Biels once again turns on classical music while students are working. She notices that Timmothy still has his hoodie on and has not yet started working. She walks over and taps him on the shoulder and asks him to take his hoodie off and start working. Timmothy, still feeling unheard from his previous comment, begrudgingly lowers his hood and picks up his pencil. Mrs. Biels continues to walk around and monitor as students think about traits they have and who they look like. She stops the music, and students move themselves into their small working groups. Mrs. Biels directs them to share their Venn diagrams. As she is listening in on Luis' group, she hears Aaman laughing and turns to see what is so funny. She hears books thump to the floor and notices Timmothy getting up to walk toward the door, leaving his now-empty desk and the mess around it. Mrs. Biels starts to walk toward this group, and Timmothy, visibly upset, continues to walk out of the classroom.

Mrs. Biels: What happened to Timmothy?

Aaman: He kept doing his diagram wrong, and we told him that he needed to fix it.

Mrs. Biels: Why do you think it is wrong?

Aaman: Look what he did. He has mom on one side and ma on the other, and we told him it was supposed to be his mom and his dad.

The lunch bell rings, catching everyone off guard. Mrs. Biels, frazzled, quickly reminds students to place their Venn diagram in their science journals and to put those in their class bin for tomorrow. She wishes she could have spoken to Timmothy, but her administrator scheduled a meeting for her while her students were eating lunch. However, she makes a mental note to talk with Timmothy right after lunch. Lunchtime comes and goes, but Mrs. Biels is preoccupied with what happened with Timmothy and is nervously waiting for students to return to the classroom. She has never had a student walk out of class before and is not sure how she should handle this incident. The lunch bell rings again, and students quickly return their trays and head toward the hallway. Mrs. Biels hurriedly walks to the cafeteria and spots Timmothy. She walks up to him, and as they make eye contact, he drops his head, trying to walk very close to the student in front of him. Mrs. Biels, not trying to make a bad situation worse, does not want to call out his name in front of the other students. She finally catches up to Timmothy as he is about to walk into their classroom door. He is doing his best not to notice her and attempts to hurry through the door. Mrs. Biels gently addresses him, "Hey Timmothy, I was hoping that I could speak with you for a minute. Do you mind chatting with me in the hallway?" Timmothy does not verbally respond but nods in agreement. Mrs. Biels leans into her classroom and asks students to begin on a math worksheet. Together they walk down the hallway in complete silence, Timmothy two or three steps behind Mrs. Biels as they head to a bench. They sit in silence for a few minutes while Timmothy stares at his hands. Finally, Timmothy looks up at her with tears streaming down his face.

Timmothy: I'm sorry I walked out of your class Mrs. Biels, but Aaman kept making fun of my diagram.

Mrs. Biels: How so?

Timmothy (barely audible through the tears and gasping): I did my diagram like yours and put my mom and my ma on each circle, and Aaman kept telling me that I had to have a dad.

Mrs. Biels: Oh, I see. I'm sorry that made you upset. Did you mean to put your grandma on the diagram?

Timmothy, feeling exasperated and defeated, looks up with tears burning his cheeks and cries, "Mrs. Biels, I don't know why you don't understand me. My ma is not my grandma, she is my other mom. I have two moms."

For Reflection and Discussion

1. What were Mrs. Biels' assumptions about a family unit? In order for this lesson to be more inclusive, what should she have done?

2. What should Mrs. Biels do in the future to address the diverse landscape of family structures?
3. What role should teacher preparation programs and school leadership play in better preparing elementary teachers to address the wide variety of student identities they will encounter in their classrooms? What might this training involve?
4. How should teacher preparation programs handle preservice teachers who may not feel it is their job to include positive affirmation for these students' identities (representation, allyship, etc.)?

David Steele is a STEM Education faculty member at Alder Graduate School of Education. His work focuses on developing culturally responsive educators dedicated to diversity, equity, and inclusion. His research interests focus on preservice and in-service teacher pedagogy and LGBTQ+ issues in STEM classrooms and careers.

Sophia Jeong is an Assistant Professor of Science Education in the Department of Teaching and Learning at The Ohio State University. Her work draws on theories of new materialisms to explore ontological complexities of subjectivities by examining socio-material relations in the science classrooms. Her research interests focus on equity issues through the lens of rhizomatic analysis of K-16 science classrooms. She is passionate about fostering creativity, encouraging inquisitive minds, and developing socio-political consciousness through science education.

Commentary: It Never Occurred to Me to Ask

19

Stephanie Eldridge

Abstract

This is a commentary to the case narrative, *"Did I inherit my curly hair from my mom…, or from my ma?"* written by David Steele and Sophia Jeong.

While sitting at my desk one day before lunch, Shayde (pseudonym), one of my normally chipper students, approached me with sunken shoulders and an avoidant look. I raised my gaze from grading papers and asked them what they needed. "I'm so sorry, Mx. Eldridge. I lost my shoreline project. Can I get the stuff to start over?" The shoreline project was a weeklong mini-research project where students identified a coastline and described its features (like barrier islands, deltas and estuaries, and sea cliffs). They researched how waves and tides affected the formation of these features, using what they learned about wave motion and concepts like erosion and deposition. Now it was Friday—the day the project was due—and Shayde was asking to start over. I had students turn in assignments late all the time, but Shayde had never missed a due date. I couldn't hide my disappointment as I thought about how irresponsible Shayde had been. Nevertheless, I allowed Shayde to take all the supplies they needed, telling them they would lose ten points off their project grade per day late.

As Shayde worked through lunch, trying to recreate as much of their project as they could recall, their friend Chantel (pseudonym) joined them, and they began casually chatting.

Chantel: You know that's due today, right?

Shayde: Yeah, well I left my other one at the trailer.

S. Eldridge (✉)
Department of Mathematics, Science, and Social Studies Education, Mary Frances Early College of Education, University of Georgia, Athens, GA, USA
e-mail: Stephanie.Eldridge@uga.edu

© The Author(s), under exclusive license to Springer Nature Switzerland AG 2023
S. Jeong et al. (eds.), *Navigating Elementary Science Teaching and Learning*,
Springer Texts in Education, https://doi.org/10.1007/978-3-031-33418-4_19

Chantel: Oh. You staying with your grandparents now? What's going to happen to your mom?

Shayde: I don't know, but I'm not going back there. At least they're letting me stay with Oma.

Chantel: What about your dad?

Shayde: He's in Michigan. He might let me live with him. But I'm tired of moving and I like it here.

As I continued to eavesdrop on their conversation, alarm bells rang in my mind. Shayde had been bouncing between their mother, father, and grandparents, and their father moved out of state—and I had no idea. Now something had happened with their mom. I wondered, was Shayde safe now? Why was Shayde removed from their mother's home? How could I have missed that they were transient for so long? Shayde and I talked almost every day, and I still had no idea about their living situation. I knew what they shared with me—their hobbies, aspirations, favorite memes, and TV shows—but nothing about their home life. *And it had never occurred to me to ask.*

In this regard, I can relate to Mrs. Biels' situation. Mrs. Biels intended to create an inclusive classroom, and she conceptually understood the importance of incorporating students' experiences and identities into her class. She used an arsenal of great teaching strategies to center students' ideas and foster scientific discourse. Despite this, a dilemma arose when Timmothy contributed his ideas, grounded in his lived experience of having two moms. Mrs. Biels and Timmothy's peers invalidated those experiences through their responses to Timmothy, causing him to disengage.

Timmothy's story can be transposed upon a diverse array of family structures with similar impacts on the student's engagement and learning. Many students would not be able to participate in an activity that requires them to track their own inheritance. Consider students who are adopted, who don't know their biological parents, who live in single-parent households, who are being raised by grandparents or guardians, or who are otherwise unaware of who contributed egg and sperm to the zygote that eventually became them. If caretakers have not told their child that the child was adopted, the child discovering this through an inheritance lesson in their third-grade classroom might cause some issues.

Having been raised in a suburban household with a stay-at-home mother and a working father, I grew up mostly unaware of the possible family structures outside of the heteronormative nuclear family model. This model was all I ever knew, and it was always portrayed as the ideal, with any other structures being labeled "non-traditional" (and undesirable). The topic of non-heterosexual partnerships was so taboo that I did not know gay people existed until high school, when I realized I was one. It would have been life-changing for me, when I was Timmothy's age, to see even one representation of a queer[1] couple or have just one conversation about the existence of LGBTQ + people.

Although these conversations are happening today more often than they were, there is still reticence to have them with young children. I do not have enough fingers and toes to count how many times I have heard adults say that they support LGBTQ + people but think that elementary age kids are too young to understand the complexities of sexuality and gender. Meanwhile, developmental psychology shows that kids begin to solidify an inner sense of gender between ages 3 and 5. They also learn to police each other's gender expressions and may begin to tease each other for behavior outside of what they think is suitable for a "girl" or "boy."

Gender and sexuality, like dis/ability, socioeconomic status, and belief systems, can be invisible aspects of people's identities. Because of their invisibility, a teacher can easily assume that everyone in their classroom shares the dominant identities and beliefs that have been normalized as "default." For instance, Mrs. Biels may believe that all her students are cisgender (their internalized sense of gender matches the gender they were assigned by doctors at birth). She may also assume that all her students are heterosexual. Teachers may not realize they are making normative assumptions like these until their students challenge their assumptions; unfortunately, this is often after they have emotionally harmed the student.

If I had not come to recognize myself as queer, I may not have thought to question the way heterosexuality and binary gender are normalized in everything that we do, especially in schools, which are highly gendered spaces. As a child, my teachers scolded me for not ending my yesses with "ma'am" or "sir." In first grade, we were lined up according to gender—boys to the left, girls to the right. Sometimes we would be grouped as boys versus girls for activities. Throughout elementary school, girls and boys who played together were jokingly labeled girlfriend and boyfriend by parents and teachers. When teachers assigned roles in groups, the girls were often assigned writers, artists, and data recorders, while the boys were builders, data collectors, and leaders. The point is that our implicit assumptions reflect our social upbringings. If Mrs. Biels' upbringing was anything like my own, it is no wonder that it never occurred to her to ask what a family unit looks like for each of her students.

After my experience with Shayde, I made a habit of contacting the parents and guardians of every student, at the beginning and regularly throughout the year. Guardians understood that a call from me was not about behavior but relationship building. Critical reflection on my own assumptions about students' home lives and invisible identities has helped me build more authentic relationships with my students. Most importantly, my students learned that I care about them. I want to encourage other teachers to lean into conversations that cause us to challenge our assumptions about students.

---

[1] I use queer as an umbrella term for people with marginalized genders or sexualities.

Gender is one of many aspects of students' lives and identities. In 2019, the National Science Teaching Association (NSTA) released an updated position statement on gender equity in science education, which emphasizes the importance of science teachers' attention to gendered language and curricular and pedagogical choices. As we plan our lessons, we might ask ourselves, what assumptions are we basing this activity on? How might this activity interact with our students' invisible identities and lived experiences? How does the language that we are using reflect our assumptions about gender, sexuality, family structure, culture, and other factors important to our students' lives?

As teachers, we must challenge our default assumptions about students and understand that they enter our classroom with multifaceted identities and lived experiences that contribute to their worldviews. Early childhood teachers have a unique opportunity to play a part in interrupting children's gender-based assumptions. The questions for reflection in this case are a starting point for teachers to begin reflecting on all the implicit assumptions that they carry into their classrooms.

Stephanie Eldridge is a graduate student in science education at the University of Georgia. Steph has taught methods and curriculum courses for preservice science teachers and served as a supervisor during their student teaching internships and practicums. Steph's dissertation addresses ways to support preservice and in-service science teachers' lesson planning for gender inclusivity.

Part II
Designing Science Instruction

Lynn A. Bryan

Designing effective science instruction is one of the most consequential responsibilities of a science teacher. All students have the capacity to learn. Further, all students deserve the chance to learn through engaging in high-quality, worthwhile learning experiences. Designing those high-quality and worthwhile experiences is complex, as teachers need to take into account so many aspects of teaching and learning. In this part, I discuss a few aspects of designing instruction to consider—starting with a broad perspective and moving to specific considerations.

How teachers design science instruction reflects their orientation to teaching science, in other words, what they assume and believe—about how children learn, about students' characteristics, about what is essential to know in science, about the nature of science, etc. There are many ways to portray and differentiate orientations to teaching. The purpose here is not to provide an exhaustive review of all of them but to provide a few examples to illustrate what an orientation looks like to underscore how one's orientation influences the design of instruction. For example, a *traditional, didactic* orientation to science teaching involves a great deal of transmitting factual information to students, whether by the teacher, textbooks, or internet resources. Students may listen to lectures, take copious notes, fill out note guides, define scientific terms, and complete "cookbook" style lab activities that confirm what they hear in a lecture. This orientation is considered teacher-oriented because the teacher plays an active role in making sense of the content for the students as they passively receive information.

On the other hand, a *discovery* orientation to teaching assumes that students will learn by engaging with materials, and therefore, engagement in activities takes center stage. The assumption is that by doing activities that allow for manipulating experimental materials such as magnets, levers, thermometers, rocks, or

L. A. Bryan
Center for Advancing the Teaching and Learning of STEM, Purdue University, West Lafayette, IN, USA
e-mail: labryan@purdue.edu

microscopes, children will become interested in and develop confidence in learning science as they "discover" scientific principles on their own. In a discovery orientation to teaching, the teacher often observes, asks open-ended questions, and encourages exploration, but not in ways that purposefully scaffold concept development or challenge students' claims. Generally, almost any idea is acceptable so as not to squelch children's curiosity and observations.

Another common orientation is an *inquiry* orientation to teaching science. An inquiry orientation to teaching science reflects the assumption/belief that students actively construct and revise their understandings through learning experiences in which they engage with evidence that supports what they know and understand. An inquiry orientation suggests that the learning goals in science instruction include not only fundamental understandings of scientific inquiry and the practices involved in scientific inquiry (e.g., posing questions, using models, planning and collecting empirical data, analyzing and interpreting data, developing explanations), but also core science content knowledge generated and revised/refined through engaging in scientific inquiry. Teachers who take an inquiry approach to teaching science have an active role in supporting and guiding students' sense-making through scaffolding and discourse-rich interactions with students, as opposed to telling students information. While there is no one orientation or specific set of strategies for teaching science that research prescriptively and definitively dictates, decades of research in the cognitive and learning sciences as well as in science education provide evidence-based recommendations for designing science instruction; for example, see *How Students Learn: Science in the Classroom.* (National Research Council [NRC], 2005) and *A Framework for K-12 Science Education* (NRC, 2012). The spirit of an inquiry orientation is summarized in the National Research Council (2010) report, *Preparing Teachers: Building Evidence for Sound Policy:*

Instruction throughout K-12 education is likely to develop science proficiency if it provides students with opportunities for a range of scientific activities and scientific thinking, including, but not limited to: inquiry and investigation, collection and analysis of evidence, logical reasoning, and accumulation and application of information. (p. 137)

In this part, *Growing Wonder, Growing Crystals: Pedagogical Choices in Preschool Science Inquiry*, is an interesting case for prompting reflection on orientations to teaching science. The preschool teacher, Laura, engages her young science learners in a series of explorations about snow, ice, and crystals. However, the preservice teacher, Adelaide, becomes increasingly concerned as she observes not only more and more time being dedicated to "playing" but also hears children sharing scientifically inaccurate ideas. Some may say Laura's instruction suggests a discovery orientation, while others may argue for an inquiry orientation. Analysis of this case will undoubtedly stimulate discussion about the purpose of Laura's teaching and her decisions in designing the learning experiences for her preschoolers.

When designing science instruction, teachers just make an abundance of decisions: What content is essential to teach and why? What are the learning goals for my students? How will I know that my students have achieved the learning goals I set? What is developmentally appropriate for my students? What assets do students bring to the learning experience? What strategies will best facilitate students' learning? How do I organize my instruction to facilitate students' learning? What content, practices, and strategies do *I* need to learn in order to design effective instruction for my students? The cases in this section present dilemmas that address the myriad decisions teachers make in designing instruction. In *The Nature of Science is an Important Aspect of Science, but Can My Third Graders Understand It?*, a "science enthusiast" teacher named Lee wanted to design a lesson about the Nature of Science for her third graders—a topic that she felt was essential for students in developing science literacy. However, she initially questioned whether or not her third graders would be capable of understanding ideas that were "more philosophical than what she considered as science content." Leveraging her own learning from a professional development experience with her past experiences teaching science from an inquiry orientation, Lee was able to find out just how much her students could learn and that designing instruction is an iterative process.

The cases, *Broccoli, Bones, and Inquiry's Plight* and *Problem Amongst the Planets*, introduce a not-all-that-uncommon dilemma in designing science instruction: finding, organizing, and implementing fun and exciting activities is not sufficient for designing effective science instruction. The teachers in *Broccoli, Bones, and Inquiry's Plight* were excited to lead an after-school science program for second-through fourth-grade children. The teacher in *Problem Amongst the Planets* arranged and prepared for a field trip to the local planetarium that would align nicely with her unit on the solar system. Even though the teachers in both cases planned activities and had their resources and materials ready, the learning experiences went awry. One significant "take-home" message from these cases is the importance of teachers (1) clearly articulating their learning goals, (2) designing experiences in which children will have the opportunities to achieve the learning goals, and (3) determining ways in which students will be able to demonstrate that the students have met the learning goals.

The central dilemma in the final two cases of this section relates to teachers' designing instruction. Bailey, the teacher in *A Question I Couldn't Answer*, grappled with an important decision as she designed a unit on states of matter for her sixth-grade students—how do you translate your own content knowledge into instruction that is appropriate for elementary children? By all accounts, Bailey possessed a strong understanding of the science content from her high school and college science courses. In addition, she had a robust portfolio of pedagogical training as well as access to tremendous resources. However, all of her preparation in designing her lessons did not prepare her for Ava's question, "What color are oxygen atoms?" Is this one question essential for students to know? And if so, how should a teacher handle a seemingly innocuous question with a rather complex explanation? On the other hand, the three teachers in the case, *We are now a STEM school with a summer STEM program? How do we do THAT?*, found themselves grappling with a different teacher knowledge-related dilemma: How do you teach what you don't yet know? The team of three elementary teachers—Maddie,

Janet, and Daniella—were charged with hosting a yearly summer STEM program on short notice. However, like many elementary teachers, none had experience as learners or teachers with engineering design or makerspaces. This case highlights the complexity of designing integrated STEM instruction, especially when a confluence of factors only compounds that complexity! Maddie, Janet, and Daniella give it their all and, in the end, remind us that the beauty of designing science instruction is that the process is iterative, and teachers' learning and reflection can continually inform revisions of and enhancements to their design of science instruction.

References

National Research Council. (2005). How students learn: Science in the classroom. The National Academies Press. https://doi.org/10.17226/11102.
National Research Council. (2010). Preparing teachers: Building evidence for sound policy. The National Academies Press. https://doi.org/10.17226/12882.
National Research Council. (2012). A framework for K-12 science education: Practices, crosscutting concepts, and core ideas. The National Academies Press. https://doi.org/10.17226/13165.

Case: Growing Wonder, Growing Crystals: Pedagogical Choices in Preschool Science Inquiry

20

Julianne A. Wenner, Sara Raven, and Kelly Baldwin

Abstract

In a preschool classroom, Adelaide, a preservice teacher, is observing what seems to be an incredibly fun exploration about snow, ice, and crystals. Yet, she is struggling with the classroom teacher's pedagogical choices around this exploration. After a few days of observation, Adelaide finally asks the classroom teacher about her choices. Wouldn't it have been easier and more accurate to answer children's questions directly and provide scientifically correct information? Why would she decide to take so much class time for children to seemingly gain so little? Why doesn't she praise children when they are right and correct them when they are wrong?

Adelaide was so excited to *finally* be in a preschool classroom! She had been taking early childhood courses for almost two years now and had been eager to visit a preschool classroom so she could have an opportunity to see all the things they had been talking about in class. Armed with her clipboard, a pencil, and vague instructions from her professor to "Take notes on anything interesting," Adelaide was perched on a tiny chair in the corner of the room, ready to learn!

Laura, the lead teacher in the room, had been welcoming and encouraged Adelaide to ask questions and interact with the children during the week she was to be

J. A. Wenner
Boise State University, Boise, ID, USA
e-mail: juliannewenner@boisestate.edu

S. Raven (✉)
Texas A&M, College Station, TX, USA
e-mail: sraven@tamu.edu

K. Baldwin
Theiss Elementary School, Klein ISD, Klein, TX, USA
e-mail: kbaldwin1@kleinisd.net

observing in the classroom. One of the first activities the children engaged in was playing with snow at the sensory table. It was too cold and wet for the children to play outside, so Laura had shoveled great piles of snow onto the sensory table for students to play with. During Morning Meeting, the children wondered what would happen to the snow now that it was inside. Ethan asked, "Will it melt?" Several children assured him that it would, and Xander knowingly said, "Snow melts when it's warm." Touching the snow as it was passed around the circle on a plate, Ellie said, "Feel it! It's so cold!" and Kate said it felt like "burning ice." The children were given several different tools to use with the snow including little rakes and shovels, and eye-droppers with water. Using these tools and their hands, the children worked to move the snow, pack the snow, shovel and rake the snow, bury things in the snow, and Kate and Jonah even worked collaboratively to build a snowperson—complete with button eyes and a baby carrot nose (salvaged from the snack table).

As the children were interacting with the snow, Laura came by and started asking the children questions: "Has the snow changed since bringing it into the classroom?" and "How does it make you feel to touch the snow?" Xander volunteered that the snow was "fluffy" when Laura first brought it into the room, but now it was getting "chunky and a little bit hard." Ethan thought that the snow was "more sparkly" and it was fun to play with the snow, but, "My hands are really cold now. It makes my fingers hurt. I'm just going to use the shovel now." Then Ellie said that when she used the shovel, she felt like the "snow plow driver who clears my street." Laura picked up on this piece of information and asked the group questions about how snow plow drivers and other workers keep us safe. Kate said that sometimes her grandma spreads salt on the sidewalk to melt the snow "but not the kind of salt you eat." Laura, with a puzzled look on her face, crinkled her eyebrows and asked, "Why would you put salt on a sidewalk?! That sounds so silly!" Kate giggled and shared with everyone at the sensory table that salt melts ice so people don't slip and fall. Ethan then jumped up and down and shouted, "I want salt! I want to melt snow!" Laura smiled, and went to the cabinet to grab salt to add to the sensory table tools.

At the end of the first observation day, Adelaide had written pages and pages of notes and was thrilled to be coming back the next day. As she was about to leave the classroom, Adelaide enthusiastically said to Laura, "What a special treat to bring snow into the classroom this morning! The kids really loved it!"

"What do you mean, 'special treat,'?" Laura replied.

"Well, I just mean that it's not every day you have snow in your classroom. The kids really liked playing with it today and it looked like they had fun. I'll bet the kids just love when they get something special like that," Adelaide said.

Laura smiled a little, eyes twinkling. "Hmmm. I wonder if 'play' was the only thing that was going on with the snow?"

Adelaide smiled back and turned to leave, not really sure what Laura meant by that comment.

The next day, Adelaide returned to her tiny chair in the corner with fresh sheets of paper and sharpened pencil. To her surprise, there was more snow on the sensory table as well as ice cubes and larger containers of ice. *Why work with snow again today?* Adelaide wondered. *The kids played with it yesterday. It's everywhere outside. Why not fill up the sensory table with something else that's new and exciting?* At Morning Meeting, Laura shared with the students that she had left some small containers full of water outside last night.

"They turned to ice!" Jonah shouted.

"What makes you think the water would turn to ice, Jonah?" Laura asked.

"Because we've seen the water in the pond turn to ice when it gets cold. Water turns into ice when it gets cold," Jonah astutely observed.

Laura grinned at this contribution and went on to share that Jonah was indeed correct. She had put these ice-filled containers in the sensory table along with fresh snow so the children continue to play with the snow.

"And," Laura recapped for those who weren't present at the sensory table the day before, "I've added a new tool you can use at the sensory table: salt. Yesterday Kate and Ethan talked about how salt can melt snow and ice, so I thought you might want to check it out!"

There were murmurs of excitement throughout the circle. Laura quickly finished up Morning Meeting and the children had to be reminded to use "walking feet" to get to the sensory table so that no one would get hurt in the rush. Similar to the day before, Adelaide observed the children playing with the snow by moving it, packing it down, using the tools, and now, using the salt. Laura would stop by the sensory table every once in a while and talk with the children: "Tell me what you've been up to." "What have you noticed?" "What would you like to try?" Xander had figured out how to make "salt tunnels" through the piles of snow by strategically placing salt and letting it melt the snow. Kate predicted that colored ice would change colors as it melted: "Like it's blue when it's frozen, but melts clear." Laura then helped Kate set up blue water to freeze and try to melt later. Ellie noticed that the snow was "so sparkly and looked like crystals." Laura responded, noting, "that's an interesting observation, Ellie." Building on Ellie's observation, during the class read-aloud time, Laura read a book about snowflakes and how—if you look closely—they're really tiny little crystals. The children were delighted by this idea and decided they wanted to make their own crystals, "So we can kinda make snow, but different," Ethan said. Adelaide found all of the children's excitement and joy heartwarming, but couldn't help but wonder why Laura was dedicating so much time and space to snow.

The third day of Adelaide's observations, she noticed the sensory table was full of snow and ice. *Again*, she thought. *I don't get it. I must be missing something. It's fun, yes, but aren't kids supposed to be learning something here? This is school, after all.*

During Morning Meeting, Laura reminded the children of their discussion at the end of the day about crystals and their interest in making one of their own. Because they had been so interested in this idea, Laura had done two things for them. First, she had looked through the classroom closet of materials and found a small crystal collection someone had donated. These crystals were also accompanied by small bowls of table salt, rock salt, Epsom salt, and magnifying glasses so the children could take a closer look at everything. Second, Laura decided that as a class, they would make crystals. Kate suggested that to make crystals, they would have to "make a recipe." Laura then led a discussion of what that recipe might entail. The children discussed physical actions that might be taken to change the material nature of crystals—especially the use of heat, physical force (breaking things), and time. Kate proposed that we "start with crystals and water. Heat water in the microwave for 10 min. Then take it out. Then walk away."

After a brief discussion, Laura invited children over to the tables to follow a recipe for growing salt crystals that actually incorporated some of their ideas. Some of the children were surprised when the recipe asked them to add a bit of sand to their salt. When Laura asked why the recipe would suggest this, Xander said, "You add sand so the crystals can grow up. Then your own little crystals can stick on and grow all the way to the top 'til they stop! Like a beanstalk!" The children then added their salt/sand mixture to a mixture of warm water and liquid watercolor paint. They observed that stirring made the salt seem to disappear. Then, the children placed their mixtures in the refrigerator to see what would happen. Jonah checked on the work in the afternoon and said, "I see little crystals!… [They] need to grow more and more and more—all the way to the sky."

As they waited for the crystals to grow, Laura had one more activity up her sleeve: painting with an Epsom salt solution. Adelaide was excited to see that the children were going to get to do some artwork—this seemed much more in line with her ideas of preschool. As Laura mixed the Epsom salts into boiling water and then added paint, she asked the children questions about what they saw. Then she asked what their paintings might look like. Ethan thought his might be "crusty" because "salt feels like crumbs." After an afternoon of painting and letting the paintings dry while reading more books about snow, ice, and crystals, the children were amazed when they saw their salt crystal paintings. "It looks just like *Frozen!*" Kate shouted with glee.

By the fourth day of Adelaide's observations, she was starting to wonder about Laura's choices as a teacher. She had no doubt that Laura loved the children and wanted them to wonder about the world and have fun, but Adelaide wasn't sure if the children were actually *learning* anything. Laura never told them any facts or content! Adelaide decided that if she didn't see the children actually learning science today, she would ask Laura about it at the end of the day.

As children entered the classroom, Laura invited them to observe the crystals they had started growing the day before, telling them that they were going to talk about this at Morning Meeting. Once the children were seated and settled, Laura asked,

"So. What did you notice about your crystals?"

Right away, Xander blurted out, "Some of the crystals are growing and taller—like Ethan's! But mine are flat. They must have died."

Laura nodded solemnly and asked, "Xander, why do you think your crystals died?" *Great opportunity* thought Adelaide. *Now Laura will be able to inform them that crystals aren't really living.*

Xander thought about the question for a moment. Then he replied, "Well. We're *growing* the crystals just like people *grow*. And people can *die*. So if my crystals aren't growing, they must have died."

"Hmmm," Laura said, "That's a really interesting idea. What do the rest of you think?"

I give up, thought Adelaide. *I guess I need to ask her what's going on this afternoon.*

After the children went home that afternoon Adelaide worked up her courage to talk to Laura. She didn't want to be disrespectful, but she also didn't understand what was going on. Why didn't Laura correct students when they said something way off-base, like crystals being alive or dead? Why didn't she praise students when they got something right, rather than questioning "Why"? And why spend so much time on snow, ice, and crystals?

"Laura," Adelaide began timidly, "I was wondering if I could ask you some questions."

"Absolutely! Go for it!"

"Well, I thought the first day you brought in snow was really cool. And the second day with snow and ice was fun. And then you moved to crystals. And the whole time, the kids looked like they were having so much fun, but..." Adelaide took a deep breath, "I'm...not...quite...sure what the kids learned."

Laura smiled. "Hmmmm...Can you tell me why you think that?"

Uh-oh. I've done it now, Adelaide thought. *I've upset my mentor teacher and she's treating me like one of her students.* "It just seems like you never gave the kids much information. I mean, you read some books, and that was useful, but sometimes they would ask questions and you wouldn't give them the answers! Or say that they were right! Like Jonah knew that water left outside in the cold would turn to ice and rather than tell him how smart he is, you asked him why he thought that. Doesn't that make him question himself? And sometimes you just let kids give flat-out crazy answers and you don't correct them! Like today, Xander said that his crystals had died! Why didn't you just tell him that crystals aren't living? I just don't understand. It seems like the kids are having a lot of fun, but..." Adelaide trailed off and looked at her shoes.

"Adelaide, I'm so glad you asked me this. Truly. It's not often I get to explain the method behind my madness! I would argue that the children have learned *so much* through this exploration, and that their explorations will carry on for weeks to come. I have been following the children's interests, because that's what they want to know. And I have let them 'play' but 'play with a purpose'—that's why I ask all those questions, so they can really think about what it is they're doing and what connections they're making. The children are learning how to ask questions, try things out, pursue answers, and look closely at the world around them. If I give them the answers, or tell them they're right or wrong, they won't take the time to wrestle with ideas. Right now, it's about the journey and the investigation—not the destination or answer."

For Reflection and Discussion

1. Adelaide came into Laura's classroom with preconceived notions of what learning and science should look like at the preschool level. What do you believe about preschool learning and preschool science learning? Why do you think you hold those beliefs?
2. Laura noted that "If I give them the answers, or tell them they're right or wrong, they won't take the time to wrestle with ideas." At the same time, Xander thinks that crystals are alive and can die, and Kate thinks salt is put on sidewalks to melt the ice. Is Laura making a sound pedagogical choice by not correcting Xander and Kate?
3. What is a more accurate description of the scientific phenomena that are occurring when salt interacts with ice and as crystals "grow"?
4. Laura called what the children in her class were doing "play with a purpose." How might a preschool teacher be purposeful so that incidental/discovery learning can become more meaningful and rigorous? What pedagogical moves did Laura make to be purposeful about the snow/ice/crystal investigation?

Julianne A. Wenner is an associate professor in the Department of Teaching and Learning at Clemson University, where she conducts research in elementary science education, teacher leadership, and science teacher education. At the heart of Wenner's scholarship are taking an asset-based approach to creating strategies and systems that will support teachers and learners in their connections with science.

Sara Raven is a Lead UX Researcher at JP Morgan Chase where she focuses on products for children and families. Prior to joining Chase, she conducted research on (1) science teaching and learning in preschool and early elementary classrooms, and (2) issues of equity and diversity in science education. Dr. Raven has published manuscripts in a variety of highly respected peer-reviewed journals and books, and presented her research at international, national, regional, and state conferences.

Kelly Baldwin with nearly 20 years of experience in education, has a background in classroom education (public and private, in the United States and abroad), non-profit arts education, museum education, and adult education. She is deeply interested in early childhood and family advocacy/support, social justice advocacy, and the integration of quality children's literature into the teaching of all content areas. Ms. Baldwin is an early elementary teacher.

Commentary: Unstructured Play or Effective Science Inquiry?

21

Kathryn A. Baldwin and Allison Wilson

Abstract

This is a commentary to the case narrative, *"Growing Wonder, Growing Crystals: Pedagogical Choices in Preschool Science Inquiry"* written by Julianne A. Wenner, Sara Raven and Kelly Baldwin.

Preservice teacher Adeline thinks to herself, *"It's fun, yes, but aren't kids supposed to be learning something here?"* To new observers in a science classroom, be it preservice teachers like Adeline or parents, a science classroom might seem a bit chaotic and unstructured. However, the pedagogical choices of Laura, the mentor teacher in the case, are all well aligned with effective science instruction. Laura's class was practicing what the National Research Council calls three-dimensional (3D) learning, which focuses on students using Disciplinary Core Ideas, Crosscutting Concepts, and scientific practices to investigate and explain a phenomena.

K. A. Baldwin (✉)
School of Education, Eastern Washington University, Cheney, WA 99004, USA
e-mail: kbaldwin1@ewu.edu

A. Wilson
Early Childhood Education, Phyllis J. Washington, College of Education, University of Montana, Missoula, MT 59812, USA
e-mail: allison.wilson@mso.umt.edu

21.1   The Role of the Teacher in Effective Science Inquiry

The National Science Teachers Association (NSTA) position statement on Early Childhood Science Education (2014) underscores the important role that adults play in helping young children learn science through exploratory play-based opportunities. Laura's strategies demonstrate declarations put forth by NSTA on Early Childhood Science Education that recommend teachers:

1. Understand science experiences which are already a part of what young children encounter every day through play and interactions with others, but that teachers and other education providers need to provide a learning environment that encourages children to ask questions, plan investigations, and record and discuss findings.
2. Tap into, guide, and focus children's natural interests and abilities through carefully planned open-ended, inquiry-based explorations.
3. Provide numerous opportunities every day for young children to engage in science inquiry and learning by intentionally designing a rich, positive, and safe environment for exploration and discovery

Laura embeds many intentional strategies to direct children's attention and structure play experiences based on the interests, curiosity, and understandings of her students. First, Laura very clearly utilizes open-ended questions paired with sufficient wait time for her students to respond. Many teachers are aware of the importance of open-ended questions but sometimes follow-up too quickly with a closed-ended question before a child can respond. For example, when a teacher stops by the sensory table filled with snow they might ask, "What do you notice?" And then, before the child responds, "Is it cold?" The follow-up question reduces the opportunity for a more thoughtful response. Laura pauses after, "What do you notice?" "What are you up to?" "What would you like to try?" She does not follow up with a closed-ended question or an assumption of how she presumes her students may answer. Laura's intentional use of open-ended questions paired with appropriate wait time yields rich responses that provide insight into her students' prior knowledge and personal experiences. Her preschoolers therefore respond by discussing salt tunnels, their understanding of snowplow drivers, how ice changes color as it melts, and the various uses of salt as both a food and a safety tool.

These conversations drive Laura's planning for future learning opportunities that build on students' prior knowledge, interests and are integrated across activities. For children to learn, concepts should be integrated into experiences throughout the day. A strength of Laura's is her ability to elevate concept development around snow by drawing connections across classroom activities. Laura connects her students' initial experiences at the sensory table to daily morning meetings and follow-up activities that are responsive to her students' interests. These include read-alouds (book on snow and crystals) and small group learning opportunities (growing crystals, painting with an Epsom salt solution).

New teachers like Adeline may feel less comfortable with this approach and may want to explain concepts to children rather than promote inquiry. Adeline wonders, "Wouldn't it have been easier and more accurate to answer children's questions directly and provide scientifically correct information?" Teachers can shift their focus from explaining concepts to facilitating science inquiry. Rather than responding to children's questions with the correct answer, try asking the child to answer their own question. If a child responds with the common, "I don't know," the teacher can pose questions such as, "where could we find more information?" "who could we ask?" "how can we find out?".

Another way to build confidence with asking questions is to prepare potential questions in advance that help to move a conversation or interaction from one that is social to one that promotes students' higher-order thinking skills. Such skills focus more on a student's understanding rather than rote instruction. For example, when Laura asks "Xander, why do you think your crystals died?", she encourages Xander to think about his own thinking. Laura's strategy is even evident in her response to Adeline's statement where she is unsure of the students' learning, "Hmmmm…Can you tell me why you think that?" While pre-planning for questions may seem counterintuitive to following the lead and interests of your students, simple open-ended question starters, such as "What do you notice?" "What do you wonder?" "Tell me how you know this," are a place to begin.

In this case, Adeline exclaims, "Laura never told them any facts or content!" But Laura effectively engaged her science students in something called three-dimensional (3D) learning. In 3D learning, teachers integrate all three dimensions of the Next Generation Science Standards (NGSS): Science and Engineering Practices, Disciplinary Core Ideas, and Cross-cutting Concepts. Rather than focusing solely on content, students should apply all three dimensions in every science inquiry.

According to the National Science Teachers Association (NSTA, 2018) "Science and Engineering Practices should be used to actively engage students in science learning" and that teachers should emphasize the learning of the Science and Engineering Practices in the early childhood classroom (NSTA, 2014).

NSTA (2018) suggests the following to effectively engage students in science learning:

1. Engage in the Science and Engineering Practices, providing a "range of ways" scientists work.
2. All three dimensions (i.e., The Science and Engineering Practices, Disciplinary Core Ideas, and the Cross-cutting Concepts) should be integrated in all science learning.
3. Phenomena should be used to engage students in 3D learning.

21.2 The Role of the Preschool Student in Effective Science Inquiry

Laura's students apply many of the Science and Engineering Practices. For example, students are engaged in asking questions and defining problems. Ethan asked, "Will it melt?" an example of making predictions based on observations. In addition, Laura asked probing questions, such as "Has the snow changed since bringing it into the classroom?". Such probing questions, which may seem spontaneous to an outside observer, are often meant to shift the focus from teacher to students asking questions and eventually lead students to the driving question for the investigation. Laura also provides her students with many tools which students use when constructing explanations and designing solutions.

Laura's students also integrate the Cross-cutting Concepts of science, the themes that run through all science disciplines such as Patterns, Systems and Systems' Models, and Cause and Effect. In this case, Laura builds off the students' prior experiences and observations and provides salt when Kate shares that she knows salt is used to melt ice, so people don't slip and fall. By providing the salt, Laura is encouraging students to see what effect salt has on the snow, building a deeper understanding of the behavior of snow.

Phenomena, such as melting snow, can actively engage students in observing and trying to explain the world around them. A focus on phenomena can not only act as a "hook" for the lesson, but can also help the teacher to assess students' prior knowledge and lead to a driving question, all criteria to effectively engage students in any science investigation.

Laura's pedagogical choices are sound and align well with recommendations put forth by leading professional organizations in the field for teaching science at the early childhood level. In addition, Laura is helping to create a seamless transition into elementary school by using three-dimensional learning and the Science and Engineering Practices of the Next Generation Science Standards (NGSS).

References

Adair, A., & Hoisington, C. (20 July 2018). Promoting children's science learning one step at a time. Retrieved March 29, 2021, from https://www.naeyc.org/resources/blog/promoting-childrens-science-learning.

National Research Council (NRC). (2012). *A framework for K–12 science education: Practices, crosscutting concepts, and core ideas*. National Academies Press.

National Science Teachers Association (NSTA). (2014). NSTA position statement: Early childhood science education.

National Science Teachers Association (NSTA). (2018). NSTA position statement: Transitioning from scientific inquiry to three-dimensional teaching and learning.

NGSS Lead States. (2013). *Next generation science standards: For states, by states*. The National Academies Press.

Kathryn Baldwin is an associate professor of science education at Eastern Washington University. Dr. Baldwin teaches coursework in science methods and environmental and sustainability education. Her research interests focus on science education, earth and environmental education, outdoor learning, science teaching self-efficacy, problem and project-based learning.

Allison Wilson is an assistant professor of early childhood education at the University of Montana. Dr. Wilson teaches coursework focusing on family engagement, curriculum methods, early assessment and child guidance. Her research centers on promoting high-quality, intentional, language-rich interactions throughout classroom, community, and home environments.

Case: The Nature of Science is an Important Aspect of Science, but Can My Third Graders Understand It?

22

Valarie L. Akerson and Ingrid S. Carter

Abstract

This case describes how Ms. Lee Strong, a third-grade teacher, considers how to embed Nature of Science into her instruction after being inspired by a rich professional development experience. She questions the extent to which her students can conceptualize philosophical aspects of science, but then realizes that with explicit scaffolds and supports, third graders can engage in critical thinking about what makes science, science!

Lee Strong thought to herself "I have always loved teaching science. I remember being so surprised that not everyone loved it when I started teaching elementary school twenty years ago." Lee was one of the only teachers in the school that prioritized science instruction in addition to the other content areas in her elementary classroom. Most of her colleagues focused on literacy and mathematics instruction and she was unique in her passion for science. Over the years, Lee had led some efforts at her school to adopt a hands-on, inquiry-based science program, and developed an after-school science club for interested elementary students. "I remember meeting weekly with the other third grade teachers in my school to plan science lessons together," she reminisced, "and they would see my love for teaching science. I hoped they would enjoy teaching science as much as me. I can't believe I was even elected to the Board of Directors of the state science teaching association!".

V. L. Akerson (✉)
Department of Curriculum and Instruction, Indiana University, Bloomington, IN, USA
e-mail: vakerson@indiana.edu

I. S. Carter
Department of Elementary Education and Literacy, Metropolitan State University of Denver, Denver, CO, USA
e-mail: iweiland@msudenver.edu

© The Author(s), under exclusive license to Springer Nature Switzerland AG 2023
S. Jeong et al. (eds.), *Navigating Elementary Science Teaching and Learning*,
Springer Texts in Education, https://doi.org/10.1007/978-3-031-33418-4_22

One thing Lee always kept up with was looking for science teaching professional development opportunities. She thought,

> I love getting new ideas for teaching science and working with other teachers who are enthusiastic about elementary science teaching! But I wasn't ready for this most recent professional development program that focused on Nature of Science! When I signed up, I thought the topic would be something about nature, like trees and animals. In reality, it was about the ideas that actually *make science, science!* I had never thought about science in that way, and it was eye-opening. I was especially surprised to learn why there really was no single scientific method; rather, scientists engage in common practices. Also, science that we see in the textbooks doesn't really help students of any age to really conceptualize these ideas about Nature of Science. Some aspects of Nature of Science that I grasped pretty easily were ideas about empirical evidence and the distinction between observations and inferences. I had heard of the term "inferences" used in a literacy context with regard to reading comprehension, but it was new for me in science. I could see how I might easily add these ideas to my lessons as I generally have students make observations of evidence.

Other ideas were very new to Lee but made sense. One such notion was the idea that scientists strive to be objective, but because they are human, there is always an element of subjectivity in their work. Also new to her, but something that made a lot of sense, was the idea that scientific knowledge is developed within the social and cultural context that scientists work in. She thought, "These ideas are fascinating, and this might have been one of the best professional development programs I have been part of because of how it broadened my view of science" (Akerson et al., 2000).

Even though Lee felt excited about what she had learned, she questioned her previous teaching of science and wondered whether she had done her former students a disservice in leaving out teaching them about Nature of Science. She also felt a huge dilemma—despite the professional development staff reassuring teachers that elementary students could learn these ideas about science, she had huge doubts about her students' capabilities to understand ideas that were more philosophical than what she considered as science content. This was all just so new to her. She wondered, "Could my third graders actually realize that scientific explanations develop in a social and cultural context, and that scientists' background knowledge and experience influence their interpretation of data?".

Lee remembered the first time she taught science through inquiry to her third graders. She was having them light a bulb with a battery and a wire—the simplest of circuits—but this time she was having them figure out how to do it, rather than telling them. Lee thought, "It was very stressful for me to see whether they could do it. I knew I could guide students' thinking through rich questioning, such as 'What happens if…?' without telling them the answer." It turned out that not only could the students figure it out, but Ebaadah even stated that she knew why Ms. Strong had them figure it out for themselves—it was "because then we really learn how to do it, instead of just listening to someone tell us." Lee decided to muster courage and confidence in her students that were formed all those years ago when she was learning to teach science by inquiry. She decided to try some lessons that embedded Nature of Science by debriefing science content lessons using the poster

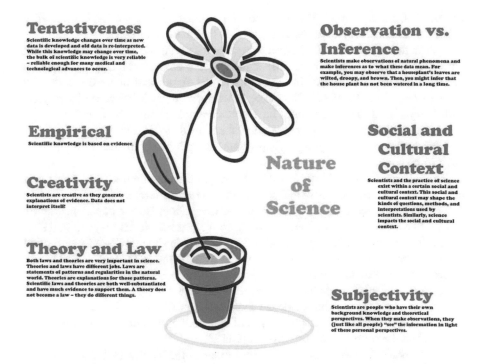

Fig. 22.1   Poster for Nature of Science debrief

that the professional development program provided for teacher use (see Fig. 22.1). She learned in the professional development program that explicit and reflective instruction helped students to make sense of Nature of Science. Lee decided she would again use electricity as the science content because it had served her well in the past when confronting doubts about whether a lesson would "work" for her students. She also intended to expand her use of science notebooks to include having her students reflect on Nature of Science within them, in addition to the content she already included. Her school had adopted the Full Option Science (FOSS) Energy kit (Regents of the University of California, 2018), and she decided to use the energy and circuits section of the unit with her students and to embed Nature of Science within the unit. Lee thought to herself, "Wish me luck!".

Interestingly, the FOSS Energy unit that includes energy and circuits begins with the simple circuit activity that Lee had used when she first started teaching through inquiry. She thought, "Because this is an activity I have used before, I just need to figure out how to embed some Nature of Science aspects. I think I should emphasize observation and inference in my debrief with students by using the Nature of Science poster after students light their bulbs." As the students began working on the problem, Lee knew they would be able to eventually "get" it. And in fact, after about ten minutes of exploring, Victor got the lightbulb to glow!

Karen exclaimed, "Wow! That's so cool! I wonder how it works?" Soon after, many others had completed the task and had built simple circuits.

Lee started this lesson as she had done before, asking students to light the bulb with only a battery and a wire. The students were enthusiastic about completing the task, and as usual were able to complete the circuit and make the bulb glow on their own! Lee directed, "Okay class, I want you to each list your observations and then make an inference about how you were able to light the bulb."

Megan said, "I saw metal on the wire, there needs to be metal."

Christopher added, "If you just leave them in a pile the bulb doesn't light—they have to be connected."

Melvin shared, "And I saw that if you connected them in a line, they don't light up the bulb. The battery, bulb and wires have to be in a circle."

Lee then directed them to think about the ideas about Nature of Science on the poster. She asked, "Are you certain you have figured out all the ways to make the lightbulb glow with a battery and wire?" The students shouted "No!" Lee replied, "You are very smart! there are other ways to light the bulb, and certainly that was an example of the tentative nature of science—scientists are able to change their ideas with new evidence, or even rethink current evidence!" Lee continued, "speaking of evidence, what evidence did we observe?".

Alec stated, "We saw that some ways of connecting the battery, bulb, and wires, would light the bulb, but other ways wouldn't."

Lee responded, "Great! So, the configurations of the battery, bulb and wire were our evidence that we observed, and then we made an inference about what kinds of configurations were successful in lighting the bulb, and what kinds were not successful in lighting a bulb." She paused a minute and saw the students nodding their heads. Hmmm.... maybe they are actually understanding some of this, she thought to herself. She decided to make a further connection to Nature of Science and stated,

> You know how artists are creative, class? Well, scientists are creative too! But instead of creating art, they create ideas of how to investigate ideas from the observations and inferences of the evidence they collect! Today we created ideas of how to light a bulb with a battery and a wire from the evidence we collected by observing which configurations worked, and which didn't, leading us to infer the circular structure that we can call a circuit! We are like scientists!

The students seemed very excited to be called scientists! Lee smiled and asked the students to record the drawings of circuits in their science notebooks. She also asked them to write a paragraph regarding how their circuits worked, using evidence, observations, and inferences to describe them. Additionally, Lee wanted them to write about how their circuit solutions were creative and how their work was similar to the work of scientists. She distributed tools for students to use in their writing, such as graphic organizers that asked for a beginning, middle, and end to their notebook entries. She said they could use sentence stems, such as "I observed_____and then I inferred_____ would make the lightbulb glow. My evidence was_____". Lee suggested, "if you want to draw pictures or diagrams, that would be a great way to help share your ideas. You can also write in your home language to help you describe your thinking."

At the end of the day, when Lee reviewed the notebook entries, overall, she was impressed. She saw that many students were able to distinguish between observation and inference and were generally clear about evidence and data. Several students had even written about how they were creating ideas, like scientists! She knew that this was only their first explicit exposure to Nature of Science in her class but thought that her students would likely be able to more fully understand these ideas about Nature of Science with continued emphasis.

Lee pondered how she would continue to embed Nature of Science in her lessons, and her colleagues still seemed to doubt Lee's approach. She was unsure whether her students would be able to grasp the idea that their own background knowledge and subjectivity might influence interpretations of data and inferences drawn from observations. She also wondered if students would grasp the idea that they may need to reinterpret evidence (tentative Nature of Science). She had some thinking to do about how she could help her third-grade students understand that all science operates within a social and cultural context. She breathed a sigh of relief when she remembered that the distinction between theory and law was not in her third-grade curriculum! But honestly, would the students be able to conceptualize the subjective Nature of Science, and its social and cultural context? Lee still wasn't sure but was determined to try.

22.1 For Reflection and Discussion

1. How might Lee continue emphasizing Nature of Science ideas with her third graders?
2. What strategies and supports did Lee use to facilitate students' thinking about Nature of Science?
3. How might Lee's confidence in explicitly teaching Nature of Science grow? What might she do to improve her confidence?

References

Akerson, V. L., Abd-El-Khalick, F. S., & Lederman, N. G. (2000). The influence of a reflective activity-based approach on elementary teachers' conceptions of the nature of science. *Journal of Research in Science Teaching, 37*, 295–317.

Akerson, V. L., Carter, I., Pongsanon, K., & Nargund-Joshi, V. (2019). Teaching and Learning nature of science in elementary classrooms: Research based strategies for practical implementation. *Science and Education, 28*, 391–411.

Harlen, W. (2001). *Primary science: Taking the plunge.* Heinemann.

Regents of the University of California Berkeley. (2018). Full option science system (FOSS) energy module. Delta Education.

Valarie L. Akerson is a professor of science education at Indiana University and a former elementary teacher. Her research focuses on preservice and in-service elementary teachers' ideas about Nature of Science, as well as their teaching practices. She is a past-president of the Association for Science Teacher Education and a past-president for NARST: A worldwide organization for improving science teaching and learning through research, and the 2021 Recipient of the Distinguished Contributions to Science Education Research Award.

Ingrid Carter is a professor in the Department of Elementary Education and Literacy at the Metropolitan State University of Denver. She earned her Ph.D. in curriculum and instruction with a focus in science education from Indiana University, Bloomington. Most of Dr. Carter's scholarly work focuses on preservice teacher education, in particular in the area of elementary science education.

Commentary: Professional Learning and Taking Risks: How Understanding the Nature of Science Influences Instruction

23

Melanie Kinskey

Abstract

This is a commentary to the case narrative, *"The Nature of Science is an Important Aspect of Science, but Can My Third Graders Understand It?"* written by Valarie L. Akerson and Ingrid S. Carter.

As a former elementary school teacher and current elementary teacher educator, the case of Ms. Long resonated deeply. One component of this case that stood out to me was Ms. Long's experience with her colleagues who did not share her passion for science. Ms. Long's passion is, unfortunately, not always common in elementary schools, which sometimes results in minimal science, if any, being taught. When I began my teaching career, I was much like Ms. Long's fellow teachers. I had an aversion to science and would avoid teaching it whenever possible. When I did teach science, my approach was often to take out the textbook and have my students read and answer questions for the duration of our allocated time. In hindsight I realizes what I was experiencing is what many new elementary teachers experience: a fear of teaching content I did not comprehend and a lack of understanding what my students were capable of. What was pivotal in my teaching career, however, was similar to Ms. Long's newfound understanding of Nature of Science: professional growth opportunities.

Ms. Long's attendance in professional development advanced her understanding of the Nature of Science. In my personal experience, my confidence and passion for science developed in my third year of teaching, while I was working a summer job at an informal environmental science camp. The instruction I was expected to

M. Kinskey (✉)
School of Teaching and Learning, Sam Houston State University, Huntsville, TX, USA
e-mail: mxk069@shsu.edu

S. Jeong et al. (eds.), *Navigating Elementary Science Teaching and Learning*,
Springer Texts in Education, https://doi.org/10.1007/978-3-031-33418-4_23

facilitate exposed me to the Nature of Science through authentic, outdoor, inquiry-based investigations. Much like Ms. Long's students, I gained an understanding of the aspects of Nature of Science while being trained in how to facilitate inquiry-based instruction for the environmental camp. Through the instructional materials I was provided, and once I started to engage my students in the inquiry-based activities, I realized what science ACTUALLY looked and felt like, and I began to develop the passion Ms. Long started her career with.

Returning to my classroom for the academic year following that summer teaching experience was transformational for my students and myself. Now that I knew what science was, I was willing to take risks in my classroom and challenge my students through inquiry-based experiences, which brings me to another important point made in the case about Ms. Long: dispelling common myths about science. Many elementary teachers, like my former self, hold uniformed, or naïve, views of Nature of Science. Like our students, we can hold onto myths that science is objective, always striving for proof and confirmation of a right way or answer. By holding onto these false ideas, we foster them in our students, continuing a viscous cycle. During her professional development, Ms. Strong, even with a passion for science, expressed how these ideas about science being subjective, creative, and always changing were new for her. She may have understood these ideas subliminally, but the awareness brought on by explicit Nature of Science training is what made the difference. Just like Ms. Long, once we develop this understanding of Nature of Science, we can begin transferring that into our instruction and help our students begin learning about the uniqueness of science.

Research shows us that once students are provided opportunities to understand that science is always changing with new evidence, they begin to feel more confident in their abilities to try new things, collect new evidence, and make new discoveries. Children are naturally curious and creative, and as teachers we have a wonderful opportunity to help our students unpack that creativity through scientific investigations—once we become informed ourselves. But taking the time to engage in professional learning experiences where we can interact with the Nature of Science is critical in making this transformation within our own understanding.

A third and possibly the most important point that arises in this case is Ms. Long's concern for her students' abilities to understand the philosophical aspects of the Nature of Science. This concern is common with elementary teachers, many of whom hold the misbelief that elementary children are not capable of engaging in challenging science content, despite much research taking place in early grades highlighting young children's abilities to engage in highly cognitive-demanding tasks. This misunderstanding often results in direct-instruction teaching that lacks opportunities for our students to act independently, which ends up inhibiting our students' abilities to think critically. As we see with Ms. Long, though, understanding our students' needs, such as when more scaffolding is needed or when they are able to engage with content independently, is developed with experience as we try new instructional approaches and learn what our students are capable of accomplishing. If we never take the opportunity to take risks with our own instruction and challenge our students, we will never learn what they can do.

While this case takes place in a third-grade classroom, this exact experience could occur in a kindergarten room as well. Once we, as teachers, understand the aspects of Nature of Science (that science is creative, tentative, subjective, evidence-based, and differentiate between observation and inference), it becomes easier to see the Nature of Science in our inquiry-based science lessons. If we are able to identify them, we should explicitly tell our students, in real time, how they are engaging with them—just as Ms. Strong did. An easy way to help our students understand the Nature of Science in earlier grades is to teach the aspects out of the context of science content.

For instance, taking time to teach kindergarten students how to use their five senses to make observations, before teaching them how to make sense of their observations for inferencing, will help them apply this understanding successfully when learning science content, such as how to classify objects based on the physical properties they observe. By only focusing on key aspects of Nature of Science, you remove the pressure of also learning science content, which can be valuable in improving confidence with teaching and learning science. Just like Ms. Strong, once you begin to explicitly tell your students how they are engaging with the Nature of Science, you will be surprised at how well they are able reflect on how they used Nature of Science to act as a scientist in your content-focused lessons.

I challenge you to be more like Ms. Long and less like me when I began my teaching career. Think about the science standard that you are expected to teach and consider how the Nature of Science is present within that standard. As you plan a lesson, think about how you could help your students to be creative as they work to master that standard; collect and/or analyze data to draw conclusions from; and/or practice the tentativeness of science by changing their conclusions based on new data-driven discoveries. And of course, don't forget to be explicit with your students about how they are engaging with the Nature of Science as they participate in inquiry, and give them opportunities to reflect on those components of science.

Melanie Kinskey is an assistant professor of science education in the School of Teaching and Learning at Sam Houston State University. She teaches undergraduate science methods courses to elementary preservice teachers and graduate courses in the master's and doctoral programs. Her research focuses on developing elementary preservice and in-service teachers' confidence and abilities to enact interdisciplinary science instruction that includes an emphasis on nature of scientific thinking and socioscientific issues.

Case: Broccoli, Bones, and Inquiry's Plight

24

Heather F. Lavender

Abstract

What do broccoli and bones have in common? To find out, the dynamic duo of a scientist-turned-science-education-doctoral-student Heather and neophyte-science-teacher Amy push their creative limits and the patience of their budding scientists during an after-school science program at a small private elementary school. The children come to find that their bones are in danger, their parents might get sick, and their least favorite food is their only hope. As Heather and Amy attempt to teach several intertwined science concepts, their ambitions and personal experiences derail the school's most popular extracurricular activity, "Fun Science"—proving Fun Science to be anything but fun for the teachers on this day.

Amy and Heather have been collaborating on elementary science both inside and outside of Amy's elementary classroom for over a year. Heather, an African American, is finishing her final year as a science education doctoral student. Prior to pursuing her doctorate, she had 12 years of experience in a microbiology laboratory. Two years ago, Amy received her master's degree in education with a concentration in science teaching. The two met in a science education course focused on project-based learning pedagogy and forged a collaboration leveraging their passion for science to instill in children an excitement for science. Heather often had ambitious lists of creative science activities, "Let's create magnetized slime, design an experiment to extract strawberry DNA, and do an activity using spherification," she would say. Amy often responded (or at times interrupted) with pragmatic planning, "Let's pair experiments according to my classroom's

H. F. Lavender (✉)
School of Education, Louisiana State University, Baton Rouge, LA, USA
e-mail: heatherl@lsu.edu

schedule and topics that align within our curriculum and save the other topics for after-school Fun Science."

Amy's first-grade class conducted inquiry-based investigations once a week. Heather often assisted by using her connections from her former microbiology lab to obtain additional supplies and invite guest speakers who were local scientists. When Amy's principal asked her and Heather to lead a weekly after-school science program, the two enthusiastically accepted. This after-school program would enrich the science learning experiences for children in second through fourth grades.

Heather and Amy organized their after-school program based on "student voice/ student choice" in project-based learning. Students could choose to repeat a previous experiment or to have a new experiment designed for them by Heather and Amy based on their interests. At the end of each Fun Science session, Heather and Amy met alone with the student whose experiment was to lead the following week. Nia, an African American girl in the third grade, chose to have Heather and Amy design a new experiment. While sitting with Heather and Amy, Nia voiced her choice and said, "I want to talk about bones. What do our bones do? How many do we have? What are they made of? My Momma says that she has big strong bones like all black women and that I need to have strong bones too." "Oh, okay," Amy responded, "I think we can work with that" and gave an approving nod to Heather. Amy asked Nia, "Is there anything else, like a backup idea that you can give us?" Nia said, "What about what's in outer space?" Amy stood to motion Nia toward the door and walked with her while responding, "I think we will stick with the bones idea, and I think you will be excited about what Ms. Heather and I come up with." Nia exited the room and as Amy walked back toward Heather, Heather passionately told Amy, "There are a lot of ailments plaguing the Black community that can be attributed to food availability and food preparation. An experiment on bones can segue into health, dietary habits, and the immune system!" In agreement, Heather and Amy prepared and planned to bring Nia's interest to light for the next meeting of the Fun Science participants.

The following week and before the school day ended, Heather prepared the room while Amy finished teaching. At each kids' table were a spaded spatula, a 5×7 unlined notecard titled "Coloring with Bones," a cartooned diagram of a sectioned bone displaying the bone layers, and a zip-lock bag holding four bones. The back of the room contained a long countertop stretching the entire width of the room and holding the remaining materials for the investigation: brown paper bags, 50 ml plastic conical tubes, a food pyramid sheet, a chart titled My Bone Record, eye goggles, and a blood cells' coloring sheet. A few minutes after the final bell rang, Amy entered the room and asked Heather, "Are we ready?" And Heather enthusiastically replied, "They're going to love this!".

The children began entering the room, and after taking their seats, Amy announced, "Alright my scientists! Today's experiment is brought to you through the interest of Nia." Amy began by asking the children, "Have any of you ever wondered what is in our bones or how they are made? When we spoke with Nia last week these were a few questions that she had about our bones. Do any of you have any questions about our bones?" One student asked about the bones on the table, "Where did these bones come from?" Amy responded, "These are all chicken bones." Another child asked, "Why are they all so different?" Amy responded, "That's a great observation! Let's all hold up a hand. Do you think the bones in our fingers are the same size as the bones in our arms or legs?" The students shook their heads from side-to-side, indicating a resounding "no." Amy echoed their motions with, "No, our fingers have different size bones that are also different from the size of the bones in our arms, legs, neck and all other bones of our body. So, those chicken bones come from different areas of a chicken's body and they too are different sizes." Amy pointed to a raised hand and a student asked, "We have different bones like chickens have different bones, is what's inside our bones like what's inside chicken bones?" Amy said, "I think we will be able to answer that question before we leave today," and a student rapidly shouted, "You have some people bones here too?" Amy tempered the class's excitement by patting her hands gently in the air and calmly said, "No, we don't have any human bones on display, but now Ms. Heather is going to show us a little more about our bones and how what's inside of them helps us."

Heather began a slide show demonstrating nutrients along with food groups and sources that help to make healthy bone tissue. Heather interjected, "What are some other foods that might have the nutrients known for helping bones to be healthy?" As students chimed in answers of string cheese, yogurt, and spinach, Amy unveiled a tray with fresh broccoli, milk, kale, and wedges of cheese. One student frantically asked, "Are we about to eat broccoli?!" Another student shouted, "I'm allergic to milk!" Heather continued the slide show corresponding with Amy's tray and the nutritional value of each item. The students began to poke at the items on their table, even though the rule of no touching until instructions always applied. Some students asked, "What are we doing with these and why are there so many?" Heather, still focused on finishing the slide show, answered, "We'll get to those. Let's talk more about what the bones do towards helping us to not get sick," showing how blood cells are made in the bones and their role in helping us to stay healthy through the immune system. Before Heather could finish, one student said, "So if my Mom don't eat right, her bones won't do that, and she could not have enough blood? How will she live?" Before Heather could respond, another student answered, "You have to have blood; she would die." More students began to murmur their concerns of having blood and eating certain foods that they did not like. Amy walked over to a few tables to address the children's concerns, while Heather spent a little more time assuring them that their parents were okay and that we could talk to our families about food choices. Heather ended the slide show with only a few minutes of Fun Science remaining and motioned for Amy to move on to the next step.

Amy quickly instructed the children to break some of the bones at the center of the table using their hands and to use the spaded spatula to scrape the bone marrow out onto the 5×7 notecard. Heather walked through the room and encouraged the children to rub and smear the bone marrow onto the notecard and between their fingers. One student asked, "We eat chicken. Will eating this help my bones?" only to hear the "ewe" from his classmates. As the children inquisitively investigated the bone marrow with several bones still unbroken, Amy called each table to go to the back countertop and take one of all the remaining items as well as put their goggles on. Though the children continued to break the bones, smearing even more marrow onto the notecard and squishing it between their fingers, Heather instructed the students to write their names on the plastic tubes and their brown paper bags. Heather and Amy moved frantically from table to table and across the room helping the students write names on bags, put bones in tubes, and color on coloring sheets. Heather and Amy crossed in the middle of the room, acknowledging that there were fewer than five minutes remaining in Fun Science. Amy instructed Heather to grab a stapler, extra brown bags, the parent note informing "contains vinegar" and meet her at the door. Amy clapped twice to get everyone's attention and instructed the children to line up next to Heather with any bone inside of one of the plastic tubes. Amy met the children at the front of the line, filled and sealed the plastic tube with vinegar before instructing each child to go to Heather to staple the parent note onto their bag. Children removed their goggles as they exited the room. Several children asked if they could add extra bones to their bag to take home and break. Amy was flustered by having gone over the allotted time for Fun Science and was aware that parents were downstairs waiting: "Of course, take as many bones as you like."

After the last young scientist left, Heather in despair said, "We did not send them with any of the coloring sheets, nor the food pyramid, nor the bone record sheet. How will they do what is on the parent note and check the bone each day? Plus, they don't have a bone in water to see the impact of what happens to the bone in vinegar. This was a disaster." Amy responded, "I think this particular experiment meant more to you. They broke a lot of bones, so when they see the rubbery-like consistency of the bone in vinegar, they will recall that it does not feel like all the bones they broke here. They will see that regardless of whether or not the bone in their tube was broken or unbroken. Ultimately, don't we want them to see a cause and effect? We can discuss other aspects with them next week, especially since Emily's choice is a repeat of slime." Amy and Heather laughed at the thought of how much the children loved to make slime and proceeded to clean up the room.

For Reflection and Discussion

1. Both Heather and Amy seemed to be well-prepared for Fun Science—they had solid background knowledge and experience, found engaging activities to do, and prepared and organized all of the materials in advance. Yet, they were not able to do everything they planned. What do you see as the essential dilemma? What could they have done differently?
2. In what ways did Heather and Amy foster students' interests and experiences in the Fun Science session?
3. At the end of the Fun Science session, the students took home a bag of materials to conduct the investigation with bones and vinegar. What are ways in which

Heather and Amy might follow up with students to see what the outcomes were of their investigation and help students make sense of their findings?

Heather Lavender received her Ph.D. in Curriculum and Instruction—Science Education at Louisiana State University in Baton Rouge, Louisiana. Her research focus is identity and science identity development pertaining to elementary age children and girls.

Commentary: Expecting the Unexpected of the Inquiries

25

Mutiara Syifa

Abstract

This is a commentary to the case narrative, *"Broccoli, Bones, and Inquiry's Plight"* written by Heather F. Lavender.

As a teacher, we might understand various fascinating concepts in science and believe that these concepts are essential to teach to the students. However, teaching science is not only delivering science concepts. There are more complex interactions in the classroom than just "teaching." These interactions might happen between students, students and teacher, and students and their own minds.

Besides mastering the subject-matter knowledge, Shulman (1987) proposes other vital categories of teacher knowledge including general pedagogical knowledge, curriculum knowledge, pedagogical content knowledge, knowledge of the learner and their characteristics, knowledge of educational contexts, and knowledge of educational ends. Pedagogical content knowledge helps teachers understand how to use the proper pedagogy to teach the subject matter to students. Therefore, when teaching some topics within the content area, the teacher knows how to bridge the students' diverse interests and background knowledge to the new concept.

Amy and Heather chose the Fun Science topic based on student interest, like Nia's ideas about learning bones. Nia and other students showed that they carried prior knowledge when asked about their interests. Nia connected two things to the topic she wanted to learn: her background identity as African American and what her mother taught about their strong bones. The students also showed their prior

M. Syifa (✉)
Inclusive Science, Technology, Engineering, Art, and Mathematics Education, Department Teaching and Learning, College of Education and Human Ecology, The Ohio State University, Columbus, Ohio, USA
e-mail: syifa.1@osu.edu

knowledge during the activity by asking questions and responding to Amy and Heather's lesson. However, some of the questions and responses seem unexpected by the teachers. Consequently, Amy and Heather missed some opportunities to respond to the students' questions in the classroom.

We as teachers generally believe that learning science should be fun so that the students want to learn more about science, and we hope that students can think critically as scientists, which resonates with Amy and Heather calling them scientists. However, what if we fail to notice the students' strengths behind their questions or responses? We think that we know how to teach science, and we think that we were well prepared, but sometimes we fail to center the students in our classrooms and consider the greater purpose of the lesson. The lesson might look fun and seem student-centered, but is it fun for the students when they are not showing engagement? Is it centering Nia's or other students' interests? Like Amy said at the end of the Fun Science: "I think this particular experiment meant more to you" not Nia, nor other students.

Undeniably, interaction in classroom activities tends to be spontaneous and develops rapidly, making it difficult for teachers to make a script of their lessons, even if they prepare it well. How do teachers then handle these multifaceted and complex interactions? First, teachers need to view students' ideas as assets and strengths. Providing a safe space for the students to share and express their ideas is incredibly challenging. When the students show that they are comfortable sharing their ideas, teachers must listen and respond to them on the fly. The teacher might not recognize these fine-grain moments when students show their strengths in science activities, but that does not mean that the teacher is not delivering high-quality teaching—it just means that they could improve their teaching.

Second, instead of making the students do experiments, we might need to experiment with and reflect on our teaching moves. Teaching experiments can help refine our teaching so that we can understand better how the students approach and learn science. Amy and Heather, or other collaborator teachers, can learn about each other's strengths and areas of growth. When one cannot attend to students' verbal or non-verbal responses, the collaborator can help interpret and respond to students' ideas.

Third, we can further prepare by visualizing potential unexpected events in the classroom activity. We can imagine ourselves as a student and begin to expect unexpected inquiries in our classroom activities. Then, when students bring ideas and questions about their experiences, the teachers are better prepared to interpret and respond to their students. Therefore, the learning process will be meaningful for the students. No matter how developed the teacher is in science content, pedagogical content, and other types of knowledge bases, if teachers are not aware of the teaching dynamic for an extended period, they and their students could be harmed.

Fourth, teachers might need to reflect on their biases. Critical consciousness about our biases is vital. We can attend to these biases by reflecting honestly about what we know, what we do not know, what we can do, and what we cannot do. Besides that, we can ask for feedback from our peers, like in Amy and Heather's collaborative work. We might hesitate to practice this because asking for peer advice can be uncomfortable. However, we must remember the greater good: to help us and our students. Is that not more valuable?

After improving our teaching as described above, if a student shouts that they have an allergy to milk or frantically asks questions about eating broccoli, we can notice these as opportunities to build on students' ideas. We can interpret that moment and decide between continuing with the planned material or responding to the students' valuable ideas. As we strive to be better teachers, we must keep the student–teacher relationship in mind. When our students are diverse, compassionately listening and supporting their ideas is the least we can do.

Reference

Shulman, L. (1987). Knowledge and teaching: Foundations of the new reform. *Harvard Educational Review, 57*(1), 1–23

Mutiara Syifa is graduated from the Indonesia University of Education, Indonesia, and Kangwon National University, South Korea. She is currently enrolled in The Ohio State University, pursuing her doctoral degree in STEM Education program. She currently works as a graduate assistant at OSU, where she is teaching and conducting research on K-12 science teacher education.

Case: Problem Amongst the Planets

26

Thomas Gaudin

Abstract

In this open case, Meg takes her third-grade students on a trip to the local science center to connect a new planetarium show with the solar system unit that she is teaching. The field trip is going great up until the planetarium show. During the show, Meg is surprised when her students start to misbehave and pay no attention to the show. As this important piece of the unit fails to connect with her students, Meg is left to wonder how she could have better planned her unit and prepared her class for their planetarium visit.

"Everyone please get back in your seats and quiet down!" Meg whispered exasperatedly. Meg thought that a planetarium show would be the perfect way to engage her third-grade students in their astronomy unit. She had been looking forward to this field trip for months, but what had begun as a fun, exciting field trip to the local science center had slowly descended into chaos. In the darkened theater, Meg strained to see the various ways that the students in her third-grade class were misbehaving. Some students were trying to climb out of their seats, while many more were talking over the narrator of the show. Not one student seemed to be paying attention to what was supposed to be the centerpiece of the solar system unit. Meg started to panic. This was not the reaction that she had expected. Her students were supposed to love the show and connect it back to what they had learned so far in class. They had a project coming up involving material from the show. How was she going to salvage her unit? Where did she go wrong?

This was Meg's first year teaching third-grade instead of fourth-grade; the science standards for her new grade excited her. For the first time, she would be

T. Gaudin (✉)
Sudekum Planetarium, Department of Physics, New Mexico Institute of Mining and Technology, Socorro, New Mexico, USA
e-mail: thomas.gaudin@student.nmt.edu

teaching a unit on the solar system! As she planned for the school year, Meg decided that it would be the perfect opportunity to incorporate the class's annual trip to the local science center. She planned to use the trip as the centerpiece of her unit. The science center had just come out with a new planetarium show called "Planetary Explorer" that seemed like it would fit perfectly into her lessons. The science center had even uploaded pre- and post-visit resources on their web-site. Meg would use these materials and the show itself to design a final project that would wrap up the unit.

As a ten-year veteran of teaching fourth-grade, Meg was no stranger to class field trips. She planned and led many trips over the years and accompanied classes on several others. However, she had no idea what to expect this time as her classes had never been to the science center before. As the day drew near, Meg told her students about their planetarium show and explained the plan for the big trip. "You are all going to love this trip," Meg explained. "The science center is a really fun place, and we are going to learn so much!" Several of her students were annual members at the museum, but some students had never been before. Stories were traded around the classroom of students' favorite parts of the museum and planetarium. The students seemed to engage well with Meg's solar system unit materials as well as the resources provided by the science center. Meg expected "Planetary Explorer" to give her students a chance to interact with the solar system in a much different way than she could do in the classroom. "What better way to learn more about each planet in the solar system than to be immersed in an interactive flight through space?" Meg thought to herself. This field trip would be the perfect way to give students a more personal connection to the material and prepare them for their big project.

When the big day arrived, Meg separated her students into groups and assigned parent chaperones. Upon arrival, the class was greeted by museum staff who gave a quick rundown of science center rules and set the students free to explore. The students seemed energetic and enthusiastic as they made their way between activities, and no major incidents occurred. Students seemed to handle the excitement of their museum trip well, but Meg noticed a rise in energy as the day progressed. Their planetarium show was the last item on their day's agenda.

Finally, the time came to enter the theater. Meg gathered everyone together to lay out expectations: "Now everyone, this show is going to help us learn more about the solar system. I want you to enjoy it, but remember, pay close attention, because we have a project on the planets coming up. We can learn a lot about them during this show." Her students were practically running to enter the dome they were so excited. The volume level slowly rose as more people entered the sold-out theater.

Trouble started immediately once the show began. It was not the show that Meg had expected. Instead of an engaging and interactive flight through space, the narrator talked slowly about each planet while the images moved slowly across the giant-curved screen. Meg worried that the material being presented was too complicated for her students to understand.

Around fifteen minutes in, Meg looked over and saw one student fast asleep and another shifting restlessly while staring at the screen. Just then one of Meg's students, Tim, leaned over to his friend Amanda and loudly said, "Look, my shoes are glowing in the dark!" She whispered to both students, "Amanda, Tim. Quiet down and pay attention!

Forty minutes of sitting in silence focus was proving to be an impossible task for the class. At twenty-five minutes, students were now having loud enough conversations that their few classmates still attentive were being distracted from the show. Some students were even trying to climb over their chairs and explore the dark room. "Everyone, please stay in your seats and quiet down! This is important." Meg whispered exasperatedly. Looking from student to student, Meg was starting to panic as the narrator continued to drone on about Saturn's rings.

This field trip was supposed to be the most important part of Meg's solar system unit. The rest of their activities including their big project on the unit were going to be built around the field trip. It was clear that the show that she had chosen to take her class to was not going to accomplish what had previously been hoped. Meg wondered to herself how she could have better prepared for this trip. Was there a different show in the planetarium's catalogue that may have been better for her intended purpose? Was the planetarium even the right choice for this unit?

For Reflection and Discussion

1. What were Meg's learning goals for the planetarium experience? In other words, what do you think she intended students to know and be able to do as a result of this part of the unit?
2. Describe what Meg could have done differently during her preparation for the unit. How could she have better prepared to avoid this mistake?
3. While a planetarium show is a unique way to connect to classroom material, how could a field experience be integrated as the focal point of a unit?
4. How could Meg have made sure that the show was an appropriate one for her unit?
5. How can Meg rework her original plans for using the planetarium show?

Thomas Gaudin is a graduate of Furman University where he received his B.S. in Physics with a minor in Informal Science Education. He has worked for the Creative Discovery Museum in Chattanooga, Tennessee and Roper Mountain Science Center in Greenville, South Carolina, and spent two years as a planetarium educator in the Sudekum Planetarium in Nashville, Tennessee. Currently, Thomas is pursuing his Ph.D. in Astronomy at Pennsylvania State University, where he also serves as a Graduate Teaching Assistant.

Commentary: Connecting Fieldtrips to Classroom Learning

27

Julia D. Plummer

Abstract

This is a commentary to the case narrative, *"Problem Amongst the Planets"* written by Thomas Gaudin.

Informal science settings, such as planetariums, science centers, museums, zoos, and gardens, provide opportunities for students to experience science in ways not afforded by the classroom. Many of us had positive experiences associated with school field trips and still remember details about our visit to a local museum or zoo from childhood. Including a field trip in the science curriculum has the potential to promote students' interest and motivation as well as conceptual learning about science topics.

There are ways that field trip experiences can be successful integrated into a classroom unit to best support learning across space and time (DeWitt & Storksdieck, 2008). First, student learning on a field trip is influenced by their prior knowledge about the topic. As the educators at the informal setting will be challenged to plan for each new group of students' prior knowledge, it is important for teachers to plan to mediate the experiences to meet their students' needs. Second, fieldtrip experiences are most successful when the students' experience is integrated into the curriculum. This appears to have been the goal of Meg's planetarium field trip as she was planning on leveraging the experience to begin the unit's big project. Students gain the most from fieldtrips when provided both pre-visit activities as an orientation and to clarify learning objectives as well as post-visit activities that reinforce the field trip experiences. Students need to be prepared for what they will experience in the learning environment. Meg prepared

J. D. Plummer (✉)
Department of Curriculum and Instruction, The Pennsylvania State University, University Park, PA, USA
e-mail: jdp17@psu.edu

the students by having a discussion where her students shared their planetarium experiences so that all students would know what the visit would be like. However, there seems to have been some mismatches between what Meg anticipated in the planetarium environment and what she and her students experienced during their visit.

Meg made some productive choices to help her students make the most of their visit from the planetarium by discussing the visit with the students ahead of time and selecting a program that she believed would support her objectives for unit's big project. But what other choices could she have made to make better use of her field trip opportunity?

First, one of Meg's goals was for students to "learn a lot" about the planets during the show, so that they would have content to utilize in the unit's big project. But was this approach a good way for students to learn detailed content about the planets? Field trips are not ideal for teaching extended factual content or complex concepts; they are not better classrooms but rather offer opportunities beyond what might be available within the classroom walls (DeWitt and Storcksdieck 2008). Thus, we might reflect on what is the best medium for learning about the planets, in general, versus what the planetarium might be able to offer that Meg could not provide through other resources in her classroom. Students can learn facts and details about planets by reading websites and books in the classroom and discussing these ideas with peers. But the planetarium offers several potential alternative opportunities for learners not easily experienced in the classroom. For example, the immersive, full-dome experience might allow students to view comparisons of planets in ways that facilitate a better understanding of size and scale or to simulate flying through the Solar System to appreciate the vast space between planets. Spatial concepts might be conveyed, such as the nature of the flat plane of the planets' orbits or how most planets rotate on their axes on that same plane. In addition, a teacher might take advantage of the planetarium's ability to inspire students' further interest that could be pursued through questions to the planetarium educator, their teacher, or their own research in the classroom. These opportunities might have required Meg to shift her own expectations for the planetarium and to better clarify this for her students.

Second, Meg reflects that the show she chose did not accomplish what she hoped. What more could she have done to match the experience to her goals? Reaching out to the planetarium educator ahead of time might have allowed Meg to determine whether the available programs would have met her students' needs or if there are other options. Teachers often believe they are integrating a visit to a planetarium field trip into their curriculum, but the level of this integration is often limited to simply covering some of the same material as their unit; they rarely utilize the planetarium educator as a resource in planning the integration (Schwarz et al., 2019). Meg and the planetarium educator could have selected a different show to fit her goals or worked together to prepare a program tailored to the goals of Meg's own unit in ways that would extend what Meg could otherwise do herself in her own classroom.

Third, the students appeared restless, distracted, and some fell asleep during the program. Could Meg have made other choices to produce a different outcome? Placing the planetarium visit after what may have been a long visit to the museum could have played a role in the students' attention span. Did Meg have a choice in when the students were able to visit the planetarium? Second, the students may have been more attentive if they had an initial understanding of the Solar System and were provided a more specific goal beyond "pay close attention." If students had already developed a framework for the Solar System and set goals for what they wanted to learn about during the planetarium to use for their unit project, they would have a greater purpose to attend to the program.

Finally, if these options were not possible for Meg, and she needed to make the best of the outcome of the current planetarium visit, what can she do? I would recommend that as soon as she is back in the classroom with the students, she take the time to debrief the students about the visit. Meg could lead the class in making a whole-class concept map by prompting students to share observations and ideas they remember from the planetarium or questions they now have about the Solar System. This post-visit activity could be used to prompt further research for their final unit project or to help Meg assess what additional areas she will need to help students learn about in the classroom.

As Meg is navigating her first year teaching third-grade and her astronomy unit, I hope that she will learn from this experience—like all of us who try new ideas, pedagogies, or field trips for the first time. She has many new strategies she can try next year to make her field trip a success for her and her students!

References

DeWitt, J., & Storksdieck, M. (2008). A short review of school field trips: Key findings from the past and implications for the future. *Visitor Studies, 11*(2), 181–197. https://doi.org/10.1080/10645570802355562

Schwarz, K., Ghent, C., & Plummer, J. (2019). Why do they come? The motivation behind field trips to the planetarium. *Planetarian, 48*(1), 20–22.

Julia D. Plummer is a professor of science education at The Pennsylvania State University, with a combined Ph.D. in Astronomy & Education from the University of Michigan. She spent more than a decade teaching children and adults in planetariums and other informal settings and continues to teach college-level introductory astronomy and science methods for preservice elementary teachers. Her research interests focus on the design of learning environments that support children's spatial thinking and science practices, primarily in the domain of astronomy.

Case: Should Student Exploration Always Come Before Teacher Explanation?

28

Elsun Seung

Abstract

Hannah and Emily, preservice teachers enrolled in an elementary science methods course, were requested to develop and teach science lessons in an elementary classroom for their early field experience. Their third-grade host teacher asked Hannah and Emily to teach a unit on animal adaption as a team. Hannah and Emily soon realized that they preferred different approaches in preparing their lessons. Hannah and Emily also realized that they shared a lack of understanding about the 5E model, which is a required instructional model for their lesson planning. They were wondering if the student exploration phase in the 5E model should always be before the teacher explanation phase in an inquiry-based class.

Hannah, an elementary education senior, is currently enrolled in a science teaching methods course. While taking the current methods course, Hannah has engaged in various discussions and activities regarding inquiry-based science teaching and learning. Hannah likes science, so she is excited to teach science lessons to elementary kids. More specifically, Hannah likes the idea of inquiry and scientific practices. Hannah agrees with the idea that, in an inquiry-based class, the learner should play an active role in constructing knowledge rather than passively reacting to external inputs (Llewellyn, 2013). Hannah also believes that for students to engage in inquiry-based learning, teachers need to treat their students like little scientists. Thus, through science classes, elementary students can get involved in various scientific practices such as asking questions, planning and carrying out investigations, analyzing and interpreting data (NGSS, 2013).

E. Seung (✉)
Center for Science Education, Indiana State University, Terre Haute, IN, USA
e-mail: elsun.seung@indstate.edu

© The Author(s), under exclusive license to Springer Nature Switzerland AG 2023   151
S. Jeong et al. (eds.), *Navigating Elementary Science Teaching and Learning*,
Springer Texts in Education, https://doi.org/10.1007/978-3-031-33418-4_28

Planning and teaching a science unit consisting of five sequential lessons is a requirement in the science methods course. Dr. Brana, the instructor of the course, requires that Hannah, Emily, and the other preservice teachers use the 5E model while developing their science lessons. During the methods course, Dr. Brana introduces the 5E model as a learning cycle consisting of five sequential stages (i.e., Engagement, Exploration, Explanation, Elaboration, and Evaluation) (Bybee et al., 2006). The preservice teachers in the class read an article about the 5E model and then discuss what teachers and students should do in each 5E phase based on their understanding of the reading.

According to the 5E instructional model, a teacher creates students' interest and curiosity and uncovers what students already know about the topic in the Engagement phase. Then in the Exploration phase, students engage in an investigation to collect evidence to answer their own questions. This Exploration phase is followed by the Explanation phase, where students explain their findings using evidence and the teacher formally provides scientific concepts. The last two phases of the 5E model are Elaboration, where students can apply their acquired knowledge to a new situation, and Evaluation, where the teacher looks for evidence that students have met the outcomes for the learning cycle.

Three weeks before their field experience, Hannah and Emily visit their host teacher Ms. Ross, a third-grade teacher, to discuss which topic she has in mind for their mini-unit. Ms. Ross asks Hannah and Emily to teach a set of lessons about animal adaption. She wants Hannah and Emily to develop and teach five lessons together as a team. Ms. Ross allows Hannah and Emily to use any teaching model or strategy they feel confident with.

After the meeting with Ms. Ross, Hannah and Emily share ideas about how to teach animal adaption using the 5E model. Emily believes, compared to physical science topics, it's difficult to develop appropriate hands-on activities for life science topics. Emily explains, "for animal adaptation lessons, we should prepare good visual aids such as pictures, animations, and video clips." After probing students' prior topic knowledge during Engagement, Emily wants to progress to the Explanation phase by using well-developed PowerPoint slides, including various visual aids. Emily insists that this strategy is more advantageous in drawing students' attention as well as targeting specific science concepts efficiently. After this Explanation phase, she also wants to provide some activities where students can elaborate and apply their acquired knowledge, which is Elaboration. From Emily's perspective, a well-prepared teacher explanation, using high-quality visual aids, can make a science class more interesting and efficient when it comes to teaching key concepts.

However, Hannah thinks Emily's plan does not follow the 5E model very closely because it does not include the Exploration phase. She wants to strictly adhere to the 5E model by including all 5E phases in each lesson. She also insists that, according to their reading and discussion, the 5E phases should be sequential. After the Engagement phase, Hannah wants students to be involved in the Exploration phase before she explains the target science concepts of the lesson. Hannah says students should collect data and develop their own explanations based on the evidence they collect. Hannah thinks using hands-on activities is the best strategy for student exploration. Hannah shares several science lessons with Emily that utilize the 5E model. She found the lessons in a practical journal for elementary teachers called *Science and Children* published by the National Science Teachers Association (NSTA). Most of the science lessons she found use hands-on activities for the Exploration phase. However, Hannah agrees with Emily's concern that it may be hard to develop student hands-on activities with the topic matter being animal adaption.

After realizing that they prefer different approaches in preparing lessons, Hannah and Emily decide to bring this issue to their methods course for discussion. In the following week's class, preservice teachers in the teaching methods course are supposed to share their lesson topics and teaching strategies they plan to use. Hannah and Emily will discuss the omission or inclusion of the student exploration phase before the teacher explanation phase with their fellow classmates. They also want to seek advice from the course instructor and classmates regarding types of exploration activities that could be useful for teaching about life science topics such as animal adaptations. Hannah and Emily also agree that they need clarity in their understanding of the 5E model itself. They decide to bring it up with Dr. Brana and the other preservice teachers in their methods class.

For Reflection and Discussion

1. How could Dr. Brana improve preservice teachers' understanding of the 5E model?
2. Should student exploration always come before teacher explanation in the 5E learning cycle? Why or why not?
3. What are the benefits of the student exploration when it is implemented before teacher explanation?
4. What resources would be useful when teachers have difficulty developing appropriate hands-on activities?

References

Bybee, R., Taylor, J. A., Gardner, A., Van Scotter, P., Carlson, J., Westbrook, A., & Landes, N. (2006). The BSCS 5E instructional model: Origins and effectiveness. BSCS.

Llewellyn, D. (2013). *Inquire within: Implementing inquiry-based science standards in grades 3–8.* Corwin Press.

NGSS Lead States. (2013). *Next Generation Science Standards: For States, By States.* The National Academies Press.

Elsun Seung is a professor of science education in the Center for Science Education at Indiana State University. She has taught science methods courses for K-12 preservice teachers for 16 years. In addition, she has been a middle and high school teacher for 15 years with emphasis on teaching general science and chemistry. She has conducted research projects to examine preservice teachers' understanding of inquiry-based science teaching, self-efficacy in teaching science as inquiry, and the effects of a cooperative mentoring program on beginning science teachers' inquiry-based teaching practice.

Commentary: Yes! Explore Before Explain

29

Joseph A. Taylor and Rodger W. Bybee

Abstract

This is a commentary to the case narrative, *"Should Student Exploration Always Come Before Teacher Explanation?"* written by Elsun Seung.

We begin this commentary with a historical perspective on the Biological Sciences Curriculum Study (BSCS) 5E Instructional Model, henceforth referred to as the 5Es. More than 30 years ago, a team of BSCS colleagues created the 5Es. BSCS had funding to design a new program for elementary science, technology, and health and needed an instructional model. With an awareness of the long history of instructional models, the BSCS team adapted a learning cycle described by Atkin and Karplus (1962), whose model was used in the elementary school program Science Curriculum Improvement Study (SCIS) developed at the Lawrence Hall of Science in Berkeley, California.

Designing the instructional model had several priorities. First, we wanted to begin with an instructional model that was research-based. Hence, we began with the SCIS learning cycle because it had substantial evidence supporting inclusion of each of the named phases and the sequence in which they were organized (Lawson et al., 1989). The BSCS additions and modifications to the SCIS learning cycle also had a research base. For example, we integrated cooperative learning (Johnson & Johnson 1987) as a complement to the original model for the SCIS program.

J. A. Taylor (✉)
Department of Leadership, Research, and Foundations, University of Colorado Colorado Springs, Colorado Springs, CO, USA
e-mail: jtaylo18@uccs.edu

R. W. Bybee
Biological Sciences Curriculum Study (BSCS), Colorado Springs, CO, USA

Second, we realized that a constructivist view of learning required experiences to challenge students' current conceptions and ample time and activities to facilitate the reconstruction of their ideas and abilities at more advanced, i.e., scientific, levels. Third, we wanted to provide perspectives for teachers that were grounded in research and had an orientation for individual lessons. We asked—what perspective should teachers have for a particular lesson or activity? Common terms such as engage, explore, explain, elaborate, and evaluate signaled the perspectives. In addition, we wanted to encourage coherence among lessons within an instructional sequence. That is, how does one lesson contribute to the next, and what were the learning outcomes for the sequence of lessons? Finally, we tried to describe the model in a manner that would be understandable, usable, and memorable for teachers. This was the origin of 5Es for the different phases of the model.

Before addressing the question of whether student exploration should come before explanation, it is important to clarify the types of activities that could constitute exploration. In the exploration phase, students participate in activities that provide time and opportunity to resolve the disequilibrium of the engagement experience. The exploration lesson or lessons provide concrete experiences where students express their current conceptions and demonstrate their abilities as they try to clarify puzzling elements of the engage phase. In addition, student exploration should include opportunities to formulate explanations, investigate phenomena, observe patterns, and develop their cognitive and physical abilities.

Consider the following vignette:

Kellan (age 22 months) announced to his mother, "Bye-bye. Going to Nana's and Grandpa's to see lizards." When Kellan arrived, we took him to the backyard. When a couple of lizards appeared, he stood silently and observed their behavior.

We offer this as an example of children's natural sequence of learning more about the natural world. Kellan's current understanding and interest in lizards was clearly engaged and he conducted an age-appropriate exploration, via his focused observation of their behavior. We suspect that Nana's or Grandpa's understanding of lizards, had those been shared with Kellan on the way to the back yard, would have been less impactful.

We note here that none of the student exploration opportunities described above require "hands on" manipulation of equipment or other physical materials, nor do they *require* primary (first-person) data collection. For example, exploration activities could also include simulations, analysis of secondary data and/or graphical representations of those data, or even guided readings/videos that do not circumvent the goals of exploration opportunities listed above. Perhaps with this wider view of student exploration, Emily would find it less difficult to use the 5E model in developing an instructional sequence.

The teacher's role in the exploration phase is to initiate the activity, describe appropriate background, provide adequate materials or equipment, and to question current non-scientific explanations. The teacher may also challenge students to think of the other possible explanations for phenomena under study. After this, the teacher steps back and becomes a coach with the tasks of listening, observing, and guiding students as they clarify their understanding and begin reconstructing scientific concepts. Progressing from the exploration to the explanation phase requires teachers to help students make connections between their current conceptions and those which were originally engaged and subsequently explored. The teacher directs students' attention to key aspects of the prior experiences and asks students for their explanations. Using students' explanations and experiences as the basis, the teacher introduces scientific or engineering concepts, briefly and explicitly.

At long last, we offer our short answer to the question—yes, student exploration should always come before teacher explanation, noting that student exploration can take many forms. Further, we follow the question to which we were charged with yet another: Where else might the explore be placed in the instructional sequence? Usually, uncertainty about where the exploration phase might otherwise reside in the sequence implies omitting the exploration phase. Either omitting or resequencing the exploration phase is problematic. Evidence for the effectiveness of the 5Es is provided by a significant body of research across three decades that suggests the integrity and necessity of each 5E phase, as well as the sum of the phases (e.g., Wilson et al., 2010), including a meta-analysis of 38 such studies of 5E impact (Cakir, 2017). This contemporary research builds on several prior studies of the SCIS learning cycle that observed decreased effectiveness when phases were omitted or *their position shifted* (e.g., Marek & Cavallo 1997).

Having now taken a stand on the question at hand, we turn to "the exception that proves the rule." What if the student needs an explanation to proceed? Some ideas are prerequisites to students understanding the primary concepts of a unit. Teachers will have to make a judgment about the priority and prerequisite nature of the concepts. One should maintain an emphasis on the primary or major concepts and practices of the unit and not digress with less-than essential explanations because they are efficient.

Finally, we close with recommendations for Dr. Brana and for additional readings and perspectives on the "explore before explain" question. With regard to further reading on this topic, we note here a series of popular NSTA publications by Patrick Brown with the title *Instructional Sequence Matters* and a subtitle *Explore Before Explain* (Brown, 2018, 2020, 2021). One substantial feature of Dr. Brown's series of books is the 5E Instructional Model. Also, the Predict-Observe-Explain [POE] model, a contemporary variation of the original SCIS learning cycle, is discussed in the series. A key highlight of Dr. Brown's series is his emphasis on how Explore Before Explain can change the mind-set of educators who might otherwise place teacher explanation before other activities, such as student exploration.

For Dr. Brana, we suggest that preservice teachers be exposed to the 5E model, applied in a variety of science disciplines and topics, and consider the model from both the teacher and learner perspective. A great exploration for preservice teachers is to inspect and analyze comprehensive, full-year curriculum programs that are organized and sequenced using the 5E or similar instructional model. Such programs will have multiple iterations of the 5Es, often with the model organizing an entire chapter and/or unit of instruction. This activity can help preservice teachers: develop a broader vision of how student exploration can happen, see examples of appropriate teacher scaffolding of student explanations, and observe how the model can contribute to conceptual coherence within and between chapters/units.

References

Atkin, J. M., & Karplus, R. (1962). Discovery or invention? *The Science Teacher, 29*(5), 45–51.

Brown, P. (2018). *Instructional sequence matters—Grades 6–8: Structuring lessons with NGSS in mind*. NSTA Press.

Brown, P. (2020). *Instructional sequence matters—Grades 3–5: Explore before explain*. NSTA Press.

Brown, P. (2021). *Instructional sequence matters—Grades 9–12: Explore before explain in physical science*. NSTA Press.

Cakir, N. K. (2017). Effect of 5E learning model on academic achievement, attitude and science process skills: meta-analysis study. *Journal of Education and Training Studies, 5*(11), 157–170.

Johnson, D. W., & Johnson, R. T. (1987). *Learning together and alone*. Prentice Hall.

Lawson, A., Abraham, M., & Renner, J. (1989). *A theory of instruction: Using the learning cycle to teach science concepts and thinking skills*. National Association for Research in Science Teaching.

Marek, E., & Cavallo, A. (1997). *The learning cycle: Elementary school science and beyond*. Heinemann.

Wilson, C. D., Taylor, J. A., Kowalski, S. M., & Carlson, J. (2010). The relative effects and equity of inquiry-based and commonplace science teaching on students' knowledge, reasoning, and argumentation. *Journal of Research in Science Teaching, 47*(3), 276–301.

Joseph A. Taylor is in the Leadership, Research, and Foundations department at the University of Colorado, Colorado Springs. Prior to this position, Joseph was Principal Scientist at BSCS Science Learning. Dr. Taylor's research interests include studying the effectiveness of science education interventions, as well as the research methods used for determining intervention effectiveness.

Rodger W. Bybee was the executive director of Biological Curriculum Study (BSCS) until 2007 when he retired. In the late 1980s Dr. Bybee led the BSCS team that created the 5E instructional model. Since retiring, he has continued consulting and contributing to science education.

Case: A Question I Couldn't Answer

30

Bailey Ondricek and Ryan S. Nixon

Abstract

Bailey is in her second year as a sixth-grade teacher. She has been preparing to teach a new science standard focused on matter. Bailey attended a district workshop to learn more about this standard and further studied on her own time. During the workshop, strategies she had seen modeled in her teacher preparation program were reiterated, and she felt confident in her ability to facilitate students' understanding with the support of supplementary materials provided by her school district. However, when the time came to teach her sixth-grade students about the particulate nature of matter, Bailey was at a loss when asked questions that she didn't know how to answer.

The students were gone for lunch as I frantically typed into the search engine: "What color are oxygen atoms?" The question was not one I had thought of myself, and I never wondered about or paid attention to the color of the atoms in textbook diagrams and illustrations. I hadn't worried about it. But Ava had. And because Ava had, a whole class of sixth graders were now wondering about it too.

The question arose during the unit on matter. On this particular Tuesday, we were working toward a discussion on the behavior of molecules in different states of matter. The previous lessons helped students understand that the fundamental unit of any given material was one molecule of that material, and that molecules could be broken up into individual atoms with names on the periodic table. Students began to conjecture that different combinations of the same atoms would

B. Ondricek (✉)
Davis School District, Layton, UT, USA

R. S. Nixon
Department of Teacher Education, Brigham Young University, Provo, UT, USA
e-mail: rynixon@byu.edu

be different molecules and therefore "different stuff." I knew these were challenging concepts, but my students were having "aha!" moments, like when McKenna expressed that she finally understood why water is called H_2O and carbon dioxide is called CO_2. They were intrigued by both the nomenclature and the visual representations of these tiny pieces of matter making up the world around us.

I knew these concepts were challenging because I had just spent hours preparing for this unit, making sure I understood the concepts and the ideas I could expect to see from my students. This was a new topic for me to teach and, frankly, a new topic for the whole sixth grade team. It was my second year of teaching, and I had just switched from teaching fourth grade to sixth grade. On the one hand, I was feeling nostalgic because I was at the same K-6 elementary school that I had attended over a decade earlier and teaching in the same classroom where I had been a student. On the other hand, I had heard other teachers comment that sixth graders could be intimidating, and I knew that the standards, especially in science and math, were extremely deep!

To prepare myself for teaching sixth grade, I had attended every applicable training I could find. I was taking endorsement classes, regularly meeting with teammates at school, and going to district workshops for the new state science standards. From the very first workshop, I could see that the inquiry-based teaching style being encouraged in my curriculum was almost exactly what I had seen modeled in my college courses. I played the student as our district trainer posed a scientific phenomenon, asked how we thought it worked, and had us revise our models as we were presented with additional information and vocabulary. After experiencing the lessons as a learner, I was provided with a district-created kit of materials I would need for teaching the standards on my own. There was a website mega-hyperlinked with an analysis of each standard, slide presentations with presenter notes of guiding questions for the students, assessments to administer, and even note pages for the kids.

The standards related to the properties of matter were familiar to me. I had learned about them in depth in high school and college; however, transferring that background knowledge into content for elementary teaching was its own challenge. In addition to digging in at the workshop and picking apart the standards to identify what a sixth grader should know about the subject, I spent time on my own watching videos on YouTube and reading up on the internet. After all of my preparations, I felt ready and excited to teach this concept to my students.

As the school year began, reality set in. I realized that the hour I was hoping to have to teach science had to be circumscribed to a 40-minute time slot which often shrank to something more like 30 minutes between the death march back from physical education and our hurried scamper off to lunch. Even within those 30 minutes, I was often interrupted by office announcements, or more frequently, the boisterous commentary of my animated class who simply couldn't wait to share their insights.

Despite these challenges, the students were doing very well. I was pleased with how they were able to engage with the science concepts and practices. They were so capable of constructing explanations to answer challenging questions about real-world phenomena. The practices I was taught were working!

On this particular day, I introduced the phenomenon by saying, "When I sit in front of a fan, I feel something." I then asked my students to break into paired discussions with their neighbor about why this phenomenon happens. The discussion was intended to slide naturally into a connection between the theoretical idea of elements making up molecules and the students' real-world experience with different types of matter in their daily lives. I proudly weaved between my students as they spoke in pairs, hearing snapshots of their conversations. I loved hearing them grapple with the concepts and experiment with vocabulary from previous lessons as they shared their explanations.

"Well, you can't really see the… *molecules* necessarily, but the air is made up of molecules."

"The little *atoms* of air are hitting you in the face!"

"Here comes the sun!" I sang when I wanted to regain their attention, and my students hummed back the appropriate, "doo doo doo doo" as they recognized our quiet signal and brought their eyes back to the front.

"I noticed you using the terms we talked about yesterday. Well done! Would anyone be willing to use the word 'molecule' to restate what their partner told them about feeling something when you sit in front of a fan?"

My precocious student Jay raised his hand, but couldn't wait to be called on and impulsively yelled, "Mrs. O! The molecules are what you're feeling! The molecules of the air! And they're made up of atoms! Uh… That's what my partner Emily said, I mean."

I passed out a blank paper and colored pencils and asked my students to draw a model of what they thought was happening when someone sits in front of a fan. I saw students drawing fans with lines sweeping out in front of the fan showing a whoosh of air (see Fig. 30.1).

Then I saw many of the students pause, seemingly unsure of where to go next. Sensing the hesitation, I leaned down by Julius, "It looks like you're stuck on something. How can I help?"

"I know the air is made up of different types of molecules, but I don't know what kinds. How can I draw them if I don't know what they look like?"

I could tell from the faces around the room that this was a question that was on the minds of many of my students, and potentially a great learning opportunity. Calling the class together, I asked my students what kinds of molecules make up the air. Confident responses of "Oxygen!" and whispered questions of, "Carbon dioxide?" rustled through different corners of the classroom. In my study beforehand, I had learned about the composition of the atmosphere. I explained that the air was made up of 78% Nitrogen, 21% Oxygen, < 1% Argon, and < 1% other trace gasses. I told my students to use the color green to connect two bubbles to represent the two nitrogen atoms that join to make a nitrogen molecule, and then two blue bubbles for the two oxygen atoms forming an oxygen molecule. We finished off with a single bubble with the red colored pencil to show the one atom of argon (Fig. 30.1).

Fig. 30.1 Student model showing what they felt when they sat in front of a fan

As my students filled their drawings with little bubble molecules, I silently celebrated a small victory. Because of that question, I was able to provide some information to help their models be even more scientifically accurate. Not only that, but the conversation had set up the perfect segue for the final point I wanted to make before wrapping things up for lunch. Pencils flew across the pages to the sound of the low chatter of learning.

However, we would not be reaching that final point today, because the class took an unexpected shift with Ava's question. I could first sense the change by the increased volume at Ava's table, while I was across the room challenging a different group of students to think about where the molecules blowing out of the fan were coming from. As I was wrapping up that conversation, Ava, clearly unable to hold her question in any longer called out to me (in a voice level I tried to discourage in my classroom), "Mrs. O, what color are oxygen atoms?"

My mind was blank for a moment, pausing on the unexpected question. I had never considered the color of atoms *at all*. As my brain slowly churned, I tried a classic stalling technique. "Tell me more about your question. What are you thinking?" Ava and her table explained that their table only had three blue colored pencils, even though there were four students. They were trying to decide how to share the blue pencils when Ava wondered if the color even mattered. Do you have to color oxygen atoms blue?

As Ava explained, my brain tracked back through everything I knew about atoms. I thought back to the district workshops and my studies to prepare for this unit and pushed back further in my memory through high school and college chemistry lessons...I could not remember any mention of the color of atoms. I knew that atoms were too small to see, even with a powerful microscope, but that didn't mean they couldn't have color. I had used the color blue to represent them in our model during the workshop, but was there a "right" or standard color for representing certain atoms? I really wasn't sure.

After a pause that felt like forever, I realized I'd better return my attention to the students or I would lose the moment. I started slowly: "This is a question I have never thought about before. I like that you're thinking deeply about what matter is made out of and what it looks like." I explained that I would tell them what I did know and explain my thinking, but that I didn't know the answer. The pieces I did know were that oxygen is clear when it's in the air. I see through oxygen in the air all the time. I also know that scientists sometimes use different colors to model that things are different—whether or not that's the color they actually are in reality.

"How would you feel about just using blue for oxygen for now, and letting me get back to you on that question later?" The students agreed that was an acceptable plan and, after finding another blue pencil, the class resumed drawing their models. We wrapped up our discussion about the phenomenon, but I couldn't get to the final point I had prepared before it was time for lunch.

Back in my classroom, using one of my precious moments alone during the school day, I found that the first results for the search "What color are oxygen atoms?" showed the official colors for representations of atoms in chemistry models and diagrams. It turned out that the official color for models of the oxygen atom is red. As I scrolled further, I found sites answering my question more directly. Several of these were from websites I was not sure I could trust. After reading a few brief articles from science organizations, I felt like I had a pretty confident answer, and I was ready to tell the students what I learned when they got back. As I finished the last few bites of my lunch, I reflected on Ava's question. It was amazing how, even after all my preparation, a child could ask a question that caught me so off-guard! As stressful as it felt in the moment, I knew that it was important for me to create a classroom environment where students were asking these deep scientific questions—even if I didn't always know the answer.

For Reflection and Discussion

1. How can a teacher determine the level of depth that is needed in responding to student questions?

2. Bailey did several things to prepare for this science unit. Was she sufficiently prepared? What did she do well to prepare? What could she do better?
3. What did Bailey do to respond to this unexpected student question? What went well? What could she do better next time?
4. Bailey referred to her challenges in transferring her knowledge of science as a *learner* to her knowledge of science needed as a *teacher*. What are some of the differences between these perspectives? How have you noticed this difference and process in your experience?
5. What should Bailey do if she can't find the answer during the break right after this lesson?
6. Bailey could avoid this question in the future by not specifying which color students should use for each type of atom. Should she do this? Why or why not?

Bailey Ondricek is a sixth-grade teacher in Davis School District's gifted and talented program. She is as passionate about learning as she is about teaching. Having graduated with a Bachelor's of Education from Brigham Young University, and a Master's of Education from Southern Utah University, her plan is to continue her education and pursue a PhD in the coming years. She sees her role as an educator as an opportunity to inspire children with the same love of learning that motivates her.

Ryan S. Nixon is an associate professor of science education at Brigham Young University, a private university owned and operated by The Church of Jesus Christ of Latter-day Saints. His research focuses on teachers' knowledge of science subject matter, specifically exploring how teachers develop this knowledge through teaching experience. He teaches future elementary teachers how to teach science.

Commentary: Some Questions Have No Correct Answer, and That's OK

31

Mu-Yin Lin and Anthony B. Thompson

Abstract

This is a commentary to the case narrative, *"A Question I Couldn't Answer"* written by Bailey Ondricek and Ryan S. Nixon.

Bailey Ondricek was excited about teaching the concepts of atoms and molecules to her 6th-grade students. She did all she could to prepare herself for introducing these concepts to her elementary-aged learners. She let students experience a phenomenon and develop a model to help them understand that matter is made of particles too small to be seen, such as air blown on the face. Even though she thought she was fully prepared, she still faced questions from students that she didn't know how to answer.

One of the challenges of being a teacher is that students will ask a wide variety of questions. In fact, in this era of information overload, students can easily learn a fact that the teacher does not know, or raise a question that the teacher cannot answer. It is impossible for teachers to guarantee that they know everything and can answer all questions. However, educators often have expectations with respect to the identity of "teacher" as an individual that knows everything.

Often, educators are uncomfortable admitting they don't have the answers to questions in front of students. Every subject of learning is profound, but it is difficult for teachers to know everything in depth and develop a robust understanding of science. In elementary schools, where reading and mathematics often receive

M.-Y. Lin (✉)
Dual Language Program, Guy B. Phillips Middle School, Chapel Hill, NC, USA
e-mail: muyinlin@chccs.k12.nc.us

A. B. Thompson
Technology Advancement and Commercialization, RTI International, Research Triangle Park, NC, USA
e-mail: abthompson@rti.org

a higher priority, teachers may not focus on developing that robust content base. Nevertheless, the perception can still exist that every question should have a "correct answer." A "teacher" is often seen as the provider of correct answers, not the guide of academic inquiry. However, should the nature of education really be this way?

Today's education tends to teach students how to answer questions "correctly," but rarely encourages students to "discover" problems. In fact, a good question does not necessarily have a good answer and often comes with a series of questions. The question posed to Bailey from her 6th-grade student was "what color are oxygen atoms?" This question can be interpreted in many ways. One possible answer is that chemists often use the color red to denote oxygen atoms in drawings and crystal structures; however, this does not address the actual color of the atom. Another possible answer is that a single oxygen atom has no color as it cannot be seen with the naked eye. In technical terms, color is a result of photon absorption, which depends upon the energy levels of the molecular orbitals of the molecule being observed. In other words, molecules can have color, whereas single atoms within molecules do not. Oxygen in its pure molecular form, O_2, has a pale blue color when liquefied, but is colorless as a gas. Other molecules containing oxygen can have a wide variety of colors. For example, CO_2 is colorless, NO_2 is reddish-brown, and permanganate (MnO_4^-) is deep purple. Other oxygen compounds, such as Cu_2O, can have different colors depending on crystal size. In a more complex interpretation of the question, if it were possible to isolate and measure a UV–Visible absorption spectrum of a sample containing pure single oxygen atoms, there may be photon absorption bands that could be assigned to a theoretical "color." However, this concept brings up questions that are far too complex for 6th-grade students, such as atomic orbitals and the instability of radical species. In summary, this question has no correct answer, but it can open many avenues for scientific discussion and an opportunity to learn.

The nature of education should never be just one-sided in which the teacher delivers knowledge or simply provides "correct" answers. Even in the scientific world, the generally accepted "correct" answers and scientific ideas change over time, from Geocentrism to Heliocentrism, from the phlogiston theory to the chemical elements. Not only did these ideas change, they also changed fast. In only seven years, the historical scientific model of an atom changed from J. J. Thomson's plum pudding model (1904) to the Rutherford model (1911). More importantly, education should inspire students to think and analyze courageously so that they can use more detailed and comprehensive perspectives to view and interpret the world around them. Facing the vastness of knowledge, teachers are as small as students. Teachers should allow their students to realize that when facing boundless knowledge, the teacher is just a pioneer who walks a few steps earlier than the students, not an almighty "encyclopedia." Facing an unknown field of knowledge, teachers should not feel embarrassed or even intend to hide their own shortcomings in front of students. Instead, they should lead by example and use a cautious and rigorous attitude to face these "things that teachers do not know," acting as role models for their students.

Bailey gave a good demonstration of how to respond to a question when she did not know the answer. First, she praised the students for thinking deeply. No matter what the question is, the students themselves are worthy of encouragement. Asking questions can promote students' thinking and learning. If teachers can give immediate compliments to the questioner, it can stimulate students' willingness and intrinsic motivation to learn. Second, when Bailey was unable to answer students' questions, she admitted that she did not know the answer. Students have asked good questions, and it is OK if the teacher does not have an explanation or if the question has no correct answer. In situations like the one in this case, teachers can simply respond with "I wish I had the answer for that" or "I haven't thought about it before" and then do the appropriate research to gain understanding. In this way, teachers can use the question as a learning opportunity for both the teacher and the students.

In addition, teachers can open up the original question to other students by not responding immediately, enabling more students to participate in answering and thinking about the question. This strategy enables more students to participate in the discussion and shifts the focus to the students so that teachers can also have more time to think about questions and think about answers. If the content of the question exceeds the scope of the students' knowledge and life experience, teachers can ask the students to find materials from home or the library and then discuss them together next time in class.

The world is changing unimaginably fast. It is a global education trend to equip children with the competence to face the unknown future, rather than knowledge that will eventually become obsolete. Past knowledge will be outdated, and there is no guarantee that children can use it to solve future problems, but competence will never become obsolete. In this sense, competence refers to the knowledge, ability, and attitude that a person possesses in order to adapt to present life and face future challenges. Only through lifelong learning can we not fall behind the times. Teachers, like all people, are in the same situation and should acknowledge the need for continued learning to maintain and improve professional practice. Therefore, as learners themselves, teachers should not be afraid of the questions that they do not know the answers to, because education is a lifelong journey.

Mu-Yin Lin is a former secondary school science teacher with a Ph.D. in science education from the University of Georgia. She has a master's degree in chemistry. She is currently teaching in a dual-language program in a public middle school in Chapel Hill, NC. Her research interests focus on Indigenous science education and teacher education.

Anthony B. Thompson is a research chemical engineer at RTI International with expertise in inorganic chemistry, catalysis, and clean energy technologies. He has experience in teaching laboratory classes and training numerous students and postdocs in hands-on chemical laboratory techniques.

Case: We Are Now a STEM School with a Summer STEM Program? How Do We Do That?

32

Helen Douglass and Geeta Verma

Abstract

This case occurs in a large suburban district in the American West with over 30,000 students. It addresses the support teachers need as they navigate a complex physical and emotional environment related to STEM education, where anxiety and excitement simultaneously reside. The team of teachers learn how to teach in a summer school environment, wrestle with the expectations related to being designated a STEM school, and come to see opportunities for inclusive and participatory experiences for all students. This case highlights three teachers, Maddie, Janet, and Daniella, as they await their first staff meeting and professional development related to teaching in the new summer STEM program.

A recent trend in elementary education in the USA is the designation of schools as Science, Technology, Engineering, Mathematics (STEM)-focused schools. There are perceived benefits to being a STEM-designated elementary school, including being seen as relevant for college and workforce readiness, having resources for engaging in robust STEM instruction, and possessing the expertise to meet the needs of the community. These perceived benefits of becoming a STEM-designated school also may create pressure for schools to pursue this designation through a school district mandate, parent/guardian community influences, or other extrinsic forces. In addition, these decisions sometimes are made hastily with little or no input from the teachers. Such is the case for Eagle Rock Elementary School.

H. Douglass (✉)
Department of Education, The University of Tulsa, Tulsa, OK, USA
e-mail: hed2054@utulsa.edu

G. Verma
School of Education, University of Colorado Denver, Denver, CO, USA
e-mail: geeta.verma@ucdenver.edu

© The Author(s), under exclusive license to Springer Nature Switzerland AG 2023
S. Jeong et al. (eds.), *Navigating Elementary Science Teaching and Learning*,
Springer Texts in Education, https://doi.org/10.1007/978-3-031-33418-4_32

After a year of preparation, Eagle Rock Elementary School is ready to launch as a newly designated leadership school where students would focus on developing leadership qualities. However, two days before beginning their school-wide leadership program, district administration informed Eagle Rock Elementary's principal, Dr. Graham, that the school is now a designated STEM school and is responsible for hosting a yearly summer STEM program for its students. Eagle Rock Elementary has never hosted a summer program, let alone one focused on STEM learning. After hearing the news of their new STEM designation and summer responsibilities, Eagle Rock Elementary teachers had to act quickly—they have only eight weeks to finish the school year and prepare for the summer program.

Guided by the new framework for science education, teachers choose to focus the summer STEM program on the integration of science and engineering experiences and decided to use a commercial curriculum created for informal summer or after-school contexts. The summer program also includes "maker" experiences in which students engage in design-based activities and learn to define a problem, brainstorm solutions, and then use consumable materials to create prototypes of possible solutions without an explicit curriculum to guide them. Although the Eagle Rock Elementary teachers completed science methods courses during their educator preparation programs, they do not have experience with maker programs or engineering education. They will have to participate in a week-long professional development workshop to learn about integrating science and engineering and using the commercial curriculum they chose. Preparation for the new summer STEM program will involve more than simply planning fun activities!

The three teachers, Janet, Maddy, and Daniella, are a team that will be teaching incoming 3rd- and 4th-grade students during the summer program. Janet is a veteran teacher employed as a 4th-grade teacher at the school full-time during the school year. Maddie is a mid-career teacher, employed as a third-grade teacher at the school full-time during the school year. Daniella is a newly certified teacher who has been a substitute in the building across all grade levels. At the first professional development meeting, teachers arrive a bit early and share with each other their reservations:

> Janet: What do you think about this STEM program we are going to teach? I am excited, but I don't know what to expect. I was really looking forward to the leadership focus we were going to have in the fall, and I didn't plan on teaching summer school."

> Maddy: I know! I spent so much time working on the leadership focus, but I am excited about teaching this summer. It is so different, though, to be told to have a summer STEM program and to be told we are now a STEM school. It happened so fast. Didn't we hear in April we would now be a STEM school and that we had to have a summer STEM program that starts in June? It is pretty overwhelming.

> Daniella: Yeah, I was surprised to get a call about teaching summer school. I am excited, as it will help me learn, but I didn't think Eagle River Elementary had summer school. I also have no idea about what the STEM curriculum might be like. When Dr. Graham called me, she said we would have all the supplies we would need and would be using a new, informal

curriculum that combines science, engineering, and making. I am open to seeing what this is about, but nervous, too.

As the other teachers make their way into the media center, there is pleasant chatter and catching up. There also are expressions of curiosity about the crates of materials, books, supplies, and several technology carts around the perimeter of the room.

Dr. Graham and the new STEM coordinator, Hannah, enter the room. After brief introductions and some get-to-know-you warm up activities Hannah begins to explain the structure of the summer experience and introduces the curriculum and materials for the 6-week summer STEM program. As an example, she shares how the teachers will be facilitating design-based challenges with the students along with teaching a combined science and engineering curriculum. Hannah explains how design-based challenges follow a process to make prototypes as a problem-solving tool: "These challenges can be guided but have many ways to reach a solution. There are a variety of consumable materials such as cups, string, tape, beans, cardboard, and many others available for students to select from as they engage in making their prototypes."

The information portion of the meeting ends, and the professional development begins for the teachers with a modified design challenge. The directions are minimal, and all teachers are included in the activity. Strings and streamers are brought out, and teachers are asked to use these to complete the task of showing a variety of connections among the school staff. The task couldn't be completed on one's own, and when finished, the teachers had formed a web-like creation with many colored streamers connecting the staff in multiple ways. The spirits are high, the anxiety is low, and in a debrief, the three teachers share their thoughts.

Daniella: This activity was fun! Do we get to do things like this with our students?"

Janet: I'm going to. The activity highlighted collaboration and creativity, plus critical thinking. These are things we are building up, right, in this summer program?

Maddie: I think so. This afternoon, we have a design challenge to try with the other teachers. I am still not sure what it means to have a summer program that is STEM-focused or what it means to be a STEM school, but I am willing to keep learning more.

Throughout the week in the professional development, teachers are able to see and experience the engineering curriculum, work with materials for prototyping and making, and collaborate together on how to enact the curriculum and technology experiences. This time, teachers get to work in small collaborative groups to make a prototype of something that would make their morning routine easier. Using consumable materials, groups brainstorm and prototype their ideas to improve a morning routine. Prototypes included devices for automatically turning on the shower, a breakfast-making machine, and an alarm clock attached to a drone. The curriculum's design challenges ask students to use a variety of materials, collaborate using science and engineering concepts, and make prototypes of solutions to a problem. For example, after reading texts about students in Nepal and how communities have to travel to get water, teachers (and later, students in the summer program) explore principles of building bridges. They engage in a design challenge that includes ways for members to cross a deep cavern to reach their water collection site in a safe and efficient manner. Participants test their designs and obtain feedback on how to make improvements.

After the first week of summer school, the three teachers talk again.

> Maddie: You know, I had a lot of tears and arguing when I had the students do the design challenge from the curriculum. Several students wanted to know the 'right' way to do the 'challenge.' I thought they would really like having to design a crossing for the students in Nepal, with the goal of helping them get to school in a quicker, safer manner.

> Daniella: I had something similar happen in my group. I had students argue with me that I was wrong when I suggested working together to find a solution. That was surprising.

> Janet: Well, I had something interesting happen. When we were doing the challenge, I had students who have not been seen or heard from too often in my class during the school year be a lot more engaged and help others with the challenge. They were really putting their ideas out there, too.

The teachers agree that the pace and ambiguity is stressful. They see both constraints and benefits to the informal curriculum and the focus on science, engineering, and making. As the 6-week program continues, the teacher team tries new approaches, using more materials and promoting student engagement, voice, and choice. At the end of the summer program, during the final staff meeting, the teachers reflected on their experience:

> Daniella: I actually like this informal environment, but I really need more time and exposure to the engineering curriculum.

> Janet: I'm glad I get to work on a team, as together we figured out some things and made some adjustments. Do you think we will be doing integrated science and engineering like this during the school year?

> Maddie: I'm still not sure what it means to be a STEM school and what the expectations will be. We still have to meet all the other standards and benchmarks, so I am not sure how to fit STEM into my day.

> Janet: Well, do you think we can do some planning in teams? Or work with Hannah? I am thinking that Dr. Graham will work with us, and we will come up with some common learning goals and expectations for the summer STEM program and how to better communicate our goals and expectations for the program with families.

The teachers agree to keep working together during the upcoming school year to continue learning about being a STEM school. Although there are many more details to work out and uncertainties ahead for the school year, the summer staff has begun to accept the designation as a STEM school and the responsibilities of offering summer experiences using the informal science and engineering curriculum. Nevertheless, they do have lingering questions as to what the summer program will look like, and the professional development, support, and resources they will have to continue their learning.

For Reflection and Discussion

1. With the phenomenon of becoming a STEM school, what challenging environments are created for teachers and students related to resource allocation, management, infrastructure, and preparation?
2. What are possible outcomes along a continuum of excitement through anxieties for teachers? For students? For administrators? For parents and guardians?
3. What are implications for teachers teaching beyond their comfort level for designing and implementing STEM learning experiences?
4. What are the pros and cons of becoming an elementary STEM school?
5. What kind of professional development opportunities should a teacher seek in an elementary STEM school?

Helen Douglass is an assistant professor of STEM education at The University of Tulsa. Her research interests include the intersections of formal and informal STEM learning environments, gender equity in science and engineering, and co-creating inclusive learning and teaching environments. She teaches elementary mathematics and science methods courses, as well as design thinking and introduction to STEM education.

Geeta Verma is a professor of science education at the University of Colorado Denver. Her research interests focus on (in) equity issues in science education, and she strives to create imaginative and innovative science, technology, engineering, and mathematics (STEM) learning environments. Geeta serves as the co-editor-in-chief for the Journal of Science Teacher Education.

Commentary: Supporting Teachers in STEM Instruction

Adronisha T. Frazier

Abstract

This is a commentary to the case narrative, *"We Are Now a STEM School with a Summer STEM Program? How Do We Do THAT?"* written by Helen Douglass and Geeta Verma.

As a graduate student, my interests have centered around secondary and post-secondary education until my recent experience with elementary teachers. I had the opportunity to assist with a teacher professional development (PD) work-shop co-sponsored by Boston's Museum of Science's Engineering is Elementary (EiE) program and three Louisiana universities. This development opportunity was designed to train and support teachers in the implementation of engineering prac-tices and activities in their elementary classrooms. A research study investigated the participating teachers' experiences during and after the workshop. This experi-ence has shaped my insights into elementary education and my responses to the case at hand.

Since the publication of the Next Generation Science Standards (NGSS) in 2013, there has been a shift across the nation to identify approaches to integrate engineering practices with scientific inquiry. In this case, the teachers' sudden reality of working in a STEM school seems to be as swift a transition as were the implementation of the NGSS for some school districts. California, Delaware, District of Columbia, Kansas, Kentucky, Maryland, Rhode Island, Vermont, and Washington implemented NGSS as early as 2013 (Citizens for Objective Public Education, 2020).

A. T. Frazier (✉)
Natural Sciences Department, Northshore Technical Community College, Lacombe, LA, USA
e-mail: adronishafrazier@northshorecollege.edu

Like the teachers in this case, many elementary teachers have had few opportunities to engage with engineering design. Fortunately, schools, post-secondary educational institutions, and organizations have recognized the need for preservice programs and professional development opportunities to educate teachers on ways to integrate engineering in the classroom. Some schools have implemented a commercial curriculum that specializes in engineering education, such as EiE (Professional Development, 2020) or Teach Engineering through the University of Colorado (PD Workshops n.d.). Some educator preparation programs, such as Mary Lou Fulton Teachers College at Arizona State University (Elementary Education n.d.), aim to specifically prepare elementary teachers to teach engineering and STEM concepts in the curriculum.

In becoming a STEM school, it is prudent for administrators to explore discipline integration as curricula development or selection is underway. The multidisciplinary, interdisciplinary, and transdisciplinary teaching of STEM, in contrast to the single disciplinary approach, should be considered as the STEM school cements its objectives, mission, and vision for student success. The multidisciplinary approach focuses on individual learning of "each discipline but within a common theme" (English, 2016). The interdisciplinary approach links two or more concepts with the intent of deepening knowledge and enhancing skills. The transdisciplinary approach uses the "knowledge and skills learned from two or more disciplines" to solve "real-world problems and projects" (English, 2016). In the professional development experiences described in this case, the teachers demonstrated the transdisciplinary integration of science, engineering, math, and technology as they prototyped the solutions to problems with everyday tasks. In the process, teachers realized science and engineering practices should occur in engineering design similar to scientific inquiry. For instance, teachers and students ask questions. Predictions are formed. Constraints are identified to develop an engineering prototype in the engineering design process. Numbers are needed to scale the prototypes. Trials test the prototypes and can lead to improvements. As teachers and students worked through STEM activities in this case, I believe they saw the usefulness of each discipline as an interweaving piece rather than focusing separately on mathematics, technology, or science. They will also discover the differences between the experimental constraints in the classroom versus actual experimentation. In elementary classrooms, students often complete trials in a typical classroom experiment, but authentic science and engineering experiences rely on replicability of the outcome to render consistent results and solutions, respectively. Therefore, the number of trials and tests can quickly enter double digits during the design process.

The authors of this case briefly highlighted each teacher's personal and scientific background. Recall, in Janet's classroom, students were engaged. The kids that were normally reserved during class were more excitable during the summer camp design activities. It would be interesting to see if the student engagement was driven by Janet's experience as a veteran 4th-grade teacher. However, Maddie is a mid-career 3rd-grade teacher. Her students wanted the right answers, which is a common practice seen in the classroom even at the college level. Students want to know what is right and what is needed to "pass." I wonder if the students' orientation to the "right" answer could be impacted by Maddie's understanding of the role of questioning in the classroom. Daniella is recently certified and substituting across all grade levels, but she taught 3rd and 4th graders during the summer school experience. Like Maddie's students, Daniella's students, too, challenged how she encouraged collaboration.

Let us take a moment to look at Maddie's concern from another perspective. The students' desire to do things the "right way" could be a starting point for the investigation and design process because it is vital that the student "make sense of phenomena and design challenges." (National Academies of Science, Engineering, & Medicine, 2019). When Maddie's students questioned her, she could have followed up with "Why do you think I am doing it incorrectly?" This teacher move could have created a discourse in the classroom for students to explain their ideas and compare them to what Maddie proposed. This approach does not imply unpreparedness on the part of the teacher. On the contrary, the use of this teacher move reveals an understanding of more than one way to complete a task in the engineering lesson. In the National Academies of Sciences, Engineering, and Medicine Report (2019), an entire chapter explores "How Teachers Support Investigations and Design." Analyzing this resource can help teachers find tips and strategies to incorporate in the classroom.

At the end of the summer professional learning experience in the case, the teachers were still uncertain about how to proceed in their upcoming school year. It would be ideal for the school administrators to follow up with teachers throughout the year to facilitate their development with the content. It would be advantageous for the STEM coordinator, Hannah, to designate time to co-teach a lesson with teachers. Hannah can observe areas of improvement, things that work in the classroom, and how the teacher can support the students through periods of uncertainty in the process. These areas could be further investigated in one-on-one debriefing sessions or as a teacher group.

Despite the professional development experience, all three teachers seem interested in transforming their current school into a STEM school. Each teacher will need support in the classroom to raise engineering concepts to a significant level of importance. As elementary teachers work to incorporate engineering content into their classrooms, they should keep the following in mind:

(1) It is okay to tell the students there is no right answer. The engineering design process is trial and error. There are multiple ways to reach a conclusion.
(2) Take the support and advice offered by your administration. This is not a normal occurrence for some schools.
(3) You are setting the foundation for how your students perceive engineering. Please be kind to yourself. Be patient. Your students will imitate you.

References

Citizens for Objective Public Education. (2020). *State adoptions of science standards since 2013.* NGSS Science Standards. https://www.copeinc.org/docs/State-Adoptions.pdf

Elementary Education (Science, Technology, Engineering and Mathematics). (n.d.). Mary Lou Fulton Teachers College, Tempe. Retrieved July 15, 2021, from https://education.asu.edu/degree-programs/undergraduate-programs/elementary-education-science-technology-engineering-and-mathematics

English, L. D. (2016). STEM education K-12: Perspectives on integration. *International Journal of STEM Education, 3*(3), 1–8.

National Academies of Sciences, Engineering, and Medicine. (2019). *Science and engineering for grades 6–12: investigation and design at the center.* The National Academies Press, Washington, DC. https://doi.org/10.17226/25216.

PD Workshops. (n.d.). *Teach engineering.* Retrieved July 13, 2021, from https://www.teachengineering.org/

Professional Development. (2020). *EiE.* Retrieved June 4, 2021, from https://www.eie.org/

Adronisha T. Frazier is an assistant professor of biology and the Natural Sciences Department Chair at Northshore Technical Community College. She is also a doctoral candidate in curriculum and instruction specializing in science education at Louisiana State University. She is interested in creating and implementing open educational resources (OERs) in STEM for secondary and higher education, enhancing interest in microbiology education through hands-on learning, and incorporating emerging topics in all biology courses and activities.

Part III
Meeting Science Standards

K. Renae Pullen and Sophia Jeong

Rene Descartes once said that curiosity predisposes us to acquire scientific knowledge. At a very young age, children are curious about the natural world around them. They enjoy making sense of their world through exploration and designing solutions to problems that matter to them. They are capable of learning sophisticated science and engineering concepts as well as engaging in the practices of scientists and engineers. To establish a foundation of science learning in future grades, young children must be given the opportunity to engage in meaningful science learning experiences. A strong foundation in the earlier grades opens opportunities for young children to consider future careers in science and engineering, to develop a well-informed citizenry who are critical consumers of scientific information, and to instill a sense of confidence and enthusiasm for science. Furthermore, providing all young learners with meaningful science learning experiences helps them identify and see themselves as able to learn and do science.

In 2012, the National Academies of Sciences, Engineering, and Medicine (NASEM) published a consensus report called *A framework for K-12 science education*, commonly referred to as The *Framework*. The *Framework* provided a common vision for K-12 science education that was multidimensional; additionally, it conceptualized a set of science standards integrating coherence and progression to support students building on their science knowledge throughout grade levels. An important shift in the *Framework* was the emphasis on students' investigating and designing solutions to real-world phenomena, positioning all students as thinkers, doers, and problem-solvers in science. To this end, the Next Generation Science Standards (NGSS) grounded in the vision of the *Framework*, envisioned anchoring—phenomenon-based science teaching and learning. Many states developed and adopted *Framework*-aligned standards.

The 2022 NASEM's report, *Science and engineering in preschool through elementary grades: The brilliance of children and the strengths of educators*, concluded that young children are capable of engaging in meaningful science and engineering

K. R. Pullen
Department of Teaching and Learning, Caddo Public Schools, Shreveport, LA, USA

S. Jeong
Department of Teaching and Learning, The Ohio State University, Columbus, OH, USA
e-mail: jeong.387@osu.edu

across multiple contexts and settings. However, science and engineering instruction is often under-resourced and not highly prioritized in elementary schools. On average, time to learn science is substantially less compared to that for English language arts and mathematics. These issues often serve as barriers to children receiving high-quality, robust science instruction, especially young learners from historically under-represented communities.

Elementary teachers are content generalists who are knowledgeable of children's learning and teach multiple subjects. However, teachers need time, funding, and professional development opportunities to support them in science teaching. This part includes five cases of elementary science teachers sharing their experiences navigating challenges and barriers to teaching science in their diverse learning settings, while also attempting to meet science standards. For example, the case, *Learning to Be Literate: Navigating the Tensions of Literacy In, About, and for Science*, written by LeeAnna Hooper highlights a classroom teacher negotiating the question of what counts as literacy. In *Growing in Understanding NGSS Science and Engineering Practices*, Akarat Tanak and Debi Hanuscin share the dilemma of a classroom teacher navigating the meaning of student engagement as described in the NGSS. In *Too Loud to Learn*, Megan Lynch and colleagues also raise questions about student engagement and participation in science classes when NGSS-aligned curriculum is implemented. In *When State Standards Change: Dilemmas of Teachers of Color*, Bhaskar Upadhyay and Stefanie Marshall illustrate concerns and potential benefits of state standards for students from immigrant families with respect to fostering cultural relevance of science in their communities. In *Fluffy, Puffy, and Made of What?*, Chelsea Sexton and Jeremy Peacock highlight a teacher's practice, as described by state standards with respect to the science practice of argumentation.

Each case in this part illustrates a diversity of issues that intersect classroom teacher experiences, teaching practices, and standards. To help navigate these challenges, elementary teachers need access to high-quality curriculum resources, effective professional learning experiences, and opportunities for collaborations with science educators. This structured support is necessary for teachers to implement standards-based instruction. Through the cases in this part, elementary teachers not only share their dilemmas, but show how experiences and expertise can be leveraged to improve young students' abilities to learn complex science ideas and build students' science identities as individuals who value and do science.

References

National Research Council. (2012). A framework for K-12 science education: Practices, crosscutting concepts, and core ideas. The National Academies Press. https://doi.org/10.17226/13165.
National Academies of Sciences, Engineering, and Medicine. (2022). Science and engineering in preschool through elementary grades: The brilliance of children and the strengths of educators. The National Academies Press. https://doi.org/10.17226/26215.

Case: Learning to Be Literate: Navigating the Tensions of Literacy In, About, and for Science

34

LeeAnna C. Hooper

Abstract

Sweeping reforms in education over the past decade have created tensions for elementary classroom teachers. An emphasis has been placed on integrating literacy and science, but a roadmap for how to do this work raises questions about what counts as literacy in, about, and for science. This case highlights the tensions teachers face with navigating how to capitalize on science instruction to teach discipline specific literacy. In the case below, a post-observation discussion unfolds between an elementary teacher and her administrator. Their discussion highlights the negotiation of meaning between the two as they define literacy for science and what counts as literacy for science.

Monday morning: Mr. Wolf, the principal at Cedarbrook Elementary, walks into Julia's classroom for a formal observation of Julia's teaching just as she asks her third graders to join her on the carpet for the start of their science lesson: "Scientists, when you join me on the carpet for science, I would like you to bring your science notebook and a pencil." For today's investigation, Julia has set up a small hot plate at the front of the classroom so that the students can observe from a safe distance. On top of the hot plate is a large saucepan. With the students settled, Julia begins, "Scientists, tell me a little bit about what we have been working on in science for the last few days." The students raise their hands without hesitation, and Julia calls on a few to share information about their past investigations. Julia then takes some time to introduce the purpose of today's lesson and has students draw the setup of the materials in their science notebooks.

L. C. Hooper (✉)
Department of Curriculum and Instruction, The Pennsylvania State University, State College, PA, USA
e-mail: lkh5212@psu.edu

As the students shuffle through all of the pages, Julie sees that they had already filled in their science notebooks and is pleased with her decision to use the notebooks with this unit. One student asks, "Wait, is it hot water in the pot?" Julia smiles at this question because she knows the student has started to think about all of the information that should be recorded about their materials set up. Rather than answering the question, Julia instructs the class, "Right now I want you to draw the setup. Then we will talk as a class about how we should label our picture. I promise." Another student promptly asks as she points toward the heater part of the hot plate, "Do we need to draw that boxy thing?" Rather than telling the student, Julia asks "What do you think?" The students all respond in excited agreement indicating that the heater is an important part of the setup.

Next, Julia facilitates an important discussion in which she elicits the students' ideas about what should be included in the drawing of the materials. Julia draws on a whiteboard as the students provide essential information about what should be included in their drawings. She listens carefully to students' words and their connections to prior investigations. Julia wants students to recognize that this setup involves heat in order to make sense of evaporation.

With many of the students done with their initial drawings, Julia transitions to help students make scientific predictions. Reaching for the glass bowl, Julia gains the attention of her students and says, "Ok, now there's one more piece that I am going to add to this today and that is this bowl." Julia picks up the clean glass bowl and holds it up for the students to see, asking, "What do you notice about this bowl right now?" A couple of students share their ideas aloud, but Julia calls on Michael who has raised his hand. Michael says, "It's clear" and Julia repeats that statement. Moving the bowl so that it is right above the saucepan Julia says, "I am going to put this bowl over top of it like this. What do you think is going to happen when I do that? I want you all to use the observations from the previous lessons as evidence to make your predictions" A couple of students raise their hands. Samir says, "It's going to get foggy." Still holding the bowl, Julia looks at Samir and asks, "Did we see something get foggy before?" Samir responds, "yes." Looking out toward the class, Julia asks, "Who can think back to a time when we saw something get foggy before?" Julia facilitates the discussion until it comes to a natural lull. Then Julia asks the students to note in their science notebook the ideas they shared about what they think will happen.

After the students have made their predictions, Julia picks the glass bowl up once more and sets it on top of the saucepan of boiling water. She tells the students, "Scientists, I want you to look closely and think about what you notice." The room soon begins to fill with chatter as the students share what they are noticing. using words like "foggy," "steamy," and "blurry." Julia calls their attention to the water that has started to drip down the inside of the bowl. She asks the students what is happening and a couple of them exclaim that it is like "rainfall." Julia's goal with this lesson was for the students to begin to pull together all the ideas that they had shared over the last few lessons. Some students made naïve references to evaporation last week during the first two investigations, and today, some have noted that the "fog" they have seen in these investigations is water, but she is not convinced that they have put together the piece about temperature change. Knowing this gap in students' ideas, Julia continues to facilitate the discussion to help students make sense of what is really happening with evaporation and condensation.

Pointing to the lines that students have noticed on the inside of the bowl that has been created by drops of running water, Julie asks the students, "Where have you seen water drop like this before?" Maya shares her thinking, "Umm, when it rains. I think I know how it rains!" Julia nods, encouraging Maya to continue sharing her ideas, "because the sun hits the lake and when the lake gets hot it goes up into the clouds and when it's all cloudy and rainy the rain comes back down." Maya's thinking is what Julia was looking for, but she knows she can press the students to say more. Containing her own excitement, Julia poses another question to the class as she points to the glass bowl that is still on top of the pan of boiling water, "Ok, what's happening in here then?" Maya's hand shoots up once again and she says, "Uh, like it's hot and when you put that on there," Maya points to the glass bowl, "that bowl is kind of like a cloud. It's kind of like it's taking water and then it's putting it back down."

Julia turns to the whole class and asks, "Does anybody know the science word for that?" Julia knows that Emma has an idea. Julia has given Emma space to share her thinking and presses her to explain what she means when she uses the word, "evaporation" without defining the word for Emma or the rest of the class. Sure enough, Emma has raised her hand and quickly shouts, "evaporate!" Julia puts her hand up to her lips as if she is thinking, repeating the word, "Hm, evaporate. What does that word mean?" Still making her thinking face, Julia turns to Sam and says, "Sam, what does the word evaporate mean?" "It means when water changes into something else and then that something else goes up into the air." Pleased with this answer, Julia turns the students' attention back to their science notebooks to add their knew ideas.

After giving the students time to add this new information to their science notebooks, Julia gains the students' attention in order to ask them about another new science word. "Ok, so we use that word evaporation and then there's another word about what's happening in here. Think about if there's another science word that you know that would make sense when we think about the saucepan and the glass bowl." Julia points to the picture that has been co-constructed on the whiteboard and motions for the students to look at the pictures that they have drawn in their science notebooks. A shy student, Norah raises her hand, "I think I know what is happening. The water is warming up and turning into steam, so it's rising I think to the top of the bowl and then the gravity is pulling it back down." Intrigued by what Norah has just shared, Julia asks a follow-up question to Norah, "ok, so tell me more about that—why would gravity pull it back down?".

Norah: Because it started turning back into water. Cause it went off like the hot part.

Julia: Ok, so it starts as water and then what happens?

Norah: It warms up and turns into steam. And then it goes up, because it's a gas and then it hits the bowl and then eventually the bowl is not hot as the pot and it starts turning back into water and then the gravity can pull it back down.

Julia: Great, thank you. Norah, Class, can we think for just one moment about one of the words that Norah just used? Norah said the word gas. Have you heard of that word before?

Students (nodding): Yes!

Julia: Hmmm. What does that mean to have something that's a gas?

Norah (jumping in to answer): It's like the heated form of a liquid.

Reaching for another whiteboard from the shelf behind her, Julia holds it up so that she can write on it at she positions it toward the students, "Ok. Let's think about what we like to call states of matter. "Matter is everything that we have on our planet and beyond. So, we have talked about solids and we have talked about liquids. And Norah just talked about another one that we have and that is gas." Julia lifts the glass bowl off of the saucepan and holds it in her hand, "Can you see that?" Miguel shouts out, "Oooh! I need to draw more steam." Julia chuckles and turns toward Miguel and says, "That is the gas, right? So, can you find a point in your picture to label the word gas?" Julia has placed the glass bowl on the floor in front of her. "You decide. Where do you think it is?".

Later that week: Julia sits in Mr. Wolf's office waiting for him to start their post-observation meeting. Julia, a first-year elementary teacher, has participated in post-observation conferences as a student teacher and once with Mr. Wolf. Although participation in an observation cycle has become something with which she is reasonably comfortable, today she cannot shake the nervous feeling in her stomach.

Mr. Wolf has been a principal at Cedarbrook Elementary for ten years, and in this capacity, he has worked with teachers at all levels of their career. In Julia's time at Cedarbrook Elementary, Julia has learned that while Mr. Wolf is a supportive administrator, he places a great deal of value on discussions about student growth through the lens of standardized tests, especially in English Language Arts and Math. Often, he brings conversations about teaching and learning back to the actions that teachers are taking to prepare their students to successfully perform on formal standardized assessments.

Mr. Wolf: Julia, I have been looking forward to our meeting. How do you think your Monday science lesson went?

Julia (after a pause to think before responding): I was really pleased with the student's engagement during the lesson. The students were able to make insightful connections between the observations they made on Monday and the two other condensation and evaporation lessons that we had done before that.

Mr. Wolf: I am curious, Julia. What made you decide to invite me to a formal observation of your science teaching?

Julia (with a smile): I chose science because I wanted you to see the important work the students are doing to make sense of the world around them.

She had been anticipating this question, because she knows that it is rare for teachers at Cedarbrook Elementary to invite Mr. Wolf to observe lessons that are outside of English Language Arts and Math. Mr. Wolf (taking out his observation notes): "Ok, Julia, let's talk about what I observed." And with that, Julia and Mr. Wolf talk through the lesson. The post-observation conference begins to draw to a close. Mr. Wolf has been complimentary of Julia's teaching practices and her ability to have a high level of student engagement in her learning community. Mr. Wolf glances once more at the observation form to ensure he has covered all of the vital components. After a moment raises a final question for Julia. "Julia, as you know at Cedarbrook Elementary we want to be sure that we are keeping in mind the connections that can be made across disciplines. And while your lesson was engaging, I cannot help but notice that you did not use a read aloud or any type of non-fiction text. I know you have chosen not to use a science textbook. Therefore, I am curious, in what ways are you integrating literacy and science?" Mr. Wolf glances as Julia before adding, "In your science teaching, what counts as literacy?".

For Reflection and Discussion

1. How might Julia respond to Mr. Wolf's final question, what counts as literacy in science?
2. In what ways has Julia embedded literacy practices in the evaporation and condensation science lesson?
3. What literacy practices could be considered by teachers when designing science investigations?
4. Julia leverages a literacy for science approach as opposed to literacy in science. In what ways are those approaches different? In what ways are those approaches similar?
5. What are ways in which Julia elicits students' ideas? Why do you think Julia elicited students' ideas and has students construct drawings?

LeeAnna C. Hooper is an elementary teacher in the State College Area School District in State College, Pennsylvania. She obtained her Ph.D. from The Pennsylvania State University in Curriculum & Instruction: Science Education. Her research draws from the fields of science education, literacy education, and anthropology and is centered on the nature of literacy practices embedded in elementary science learning communities and the role that literacy practices play in shaping student sense-making and the epistemic culture of the classroom.

Commentary: Negotiation of Meaningful Literacy

35

Kathryn M. Bateman

Abstract

This is a commentary to the case narrative, *"Learning to Be Literate: Navigating the Tensions of Literacy In, About, and For Science"* written by LeeAnna C. Hooper.

Although the 1960s space race rocketed science into the forefront of education, *No Child Left Behind* in 2001 saw science take a backseat to the math and literacy required to comply with national laws. To ensure states complied with *No Child Left Behind*, they were required to "assess" students early and often in mathematics and literacy. States interpreted this to mean standardized tests that focused on math skills, reading comprehension, and basic writing skills. Another common result of *No Child Left Behind*'s focus on math and literacy was the push to have other content areas, like science, teach and reinforce math and literacy. Science classrooms became places where vocabulary was pre-taught, memorized, and quizzed and found in non-fiction readings that asked students as much about the main ideas as the content of science. Although the 2015 *Every Student Succeeds Act* replaced *No Child Left Behind*, the standardized tests and definitions of literacy remained. However, literacy in science can mean much more than reading non-fiction texts and writing essays or lab reports—i.e., literacy *about* science. Literacy *for* science means that students will develop the skills to participate in the social practices of science. This includes reading background information and writing to communicate findings with peers and the public, but it also means using a variety of texts, making sense of science in the norms of the practice, and negotiating shared meaning of language in practice (Hooper & Zembal-Saul, 2019).

K. M. Bateman (✉)
Department of Psychology, Temple University, Philadelphia, PA, USA
e-mail: kmb1182@gmail.com

We see in Julia's story that she does not open a book, read aloud, or list off vocabulary words, and yet her classroom is quite rich in literacy *for* science. She engages students in creating text in their science journals in the form of diagrams. These diagrams have become a norm for Julia's third-grade classroom, but the students further develop literacy in their negotiation of what counts, what belongs in the diagram, and what is important to document. This allows Julia and her students to talk together, make decisions, and come to a consensus, all things practicing scientists need to determine as a community. Science vocabulary still has a prominent place in Julia's classroom, but you will not find a list of terms on a pre-printed worksheet. The science words are generated from the conversation and used in context—evaporation, gas, and matter. Instead of being abstract terms to define and then use later, Julia introduces new words into the lesson as they arise, and students have some observable phenomenon on which to hang the word.

Finding the right time to introduce science words is critical, as some words have multiple meanings, i.e., scientific and everyday language. To help avoid confusion between scientific and everyday language, in some classrooms, students will make nonsense phrases for phenomena they haven't been able to label yet. For example, another classroom studying energy transformations through bouncy balls does not use the word "energy" until the class can come to an agreement on what they are observing, instead opting to call it "BAM!" Although these teachers could provide students with a formal, scientifically normative definition of the word "energy," it could easily be brushed off as already known because of its commonplace, every-day use. Students might say something like, "I didn't get much sleep last night so I have no energy" or look at a puppy chasing its tail and say, "wow, he has a lot of energy." Students already think they know what words such as energy, matter, gas, and weight mean. They use these words in non-scientific conversation all the time. Yet, many third graders' definitions of gas are more likely to be of the bodily function variety than the state of matter. In Julia's lesson, she negotiates the mean-ing of words like gas with the students and associates the meaning of words with an observable phenomenon which students then turn into text in the form of their science journal diagrams. They do not use "gas" until they have observed steam in the demonstration and can connect the term with the phenomena or characteristics they have observed.

To support her students in literacy *for* science, Julia needed to carefully plan out her science classroom norms and practices. However, just like Julia's third-graders practice science in Julia's classroom culture, Julia practices *teaching* science in Principal Wolf's school culture. Like Julia has set up goals and objectives for her students' science lessons based on her values and beliefs, Mr. Wolf has goals and objectives for Cedarbrook Elementary based on his values and beliefs. Principals and teachers do not always share the same value and beliefs, which we can see in his comment to Julia, "in your classroom what counts as literacy?" Julia and Mr. Wolf appear to share a common value in helping students develop literacy, but what that literacy looks like in development differs. For Julia, her literacy is *for* science, which does not mimic the literacy practices of language arts classes. Julia recognizes this mismatch and decides to showcase her students' flourishing skills in negotiating meaning, creating non-traditional texts, and engaging in scientific dialog with the class. Julia wants Mr. Wolf to see that literacy *for* science is just as valuable as literacy *about* science; she is preparing her students to be scientifically literate citizens and future scientists.

Convincing principals and other administrators to shift their thinking about what counts as literacy can be challenging, particularly for novice teachers like Julia. Administrators hold a certain amount of power over teachers—they make policies, conduct observations to evaluate teachers, determine who teaches what, where, when, and whom. Policies like standardized testing come from higher than school-level administrators like principals (national and state level) but how that becomes part of the school culture depends on how principals interpret the policy and translate it for their teachers. For Mr. Wolf, this meant a more traditional definition of literacy, one upon which Julia wished to push back. To do so, she invites him to see her classroom do things differently and do them well. She has leveraged her learning outside Cedarbrook Elementary to create a classroom culture that values literacy *for* science. She has not forcefully or blatantly told Mr. Wolf he is wrong about literacy. Instead, she's demonstrating another possibility for him to see.

Supporting policy pushback like Julia's requires development of certain cultural features (Bateman, 2019). One cultural feature is establishing a collegial relationship with the principal. Given Julia's invitation to see her science classroom, she opened her doors to a conversation with Mr. Wolf in which they, like her students in the lesson, can negotiate the meaning of what counts as literacy. In contrast, Julia could have viewed her principal as managerial, someone with authority over her and knowledge she does not possess. This stance would have resulted in Julia expecting Mr. Wolf to show or tell her what is expected of literacy in her classroom and little to no negotiation of meaning. The type of relationship Julia has with her principal is neither pre-determined nor fixed. Seeking out a collegial relationship with the principal provides a teacher with opportunities to make sense of policies and negotiate what counts as good teaching to meet the values of everyone involved. Creating a school culture that has a shared vision, values, and beliefs about what counts as good science teaching starts with establishing those relationships to enable collaborative negotiation of meaning.

References

Bateman, K. (2019). *Assembling policy dilemmas: science teacher responses to educational policy* (Doctoral dissertation, Pennsylvania State University)

Hooper, L., & Zembal-Saul, C. (2019). Literacy practices for sensemaking in science that promote epistemic alignment. In E.A. Davis, C. Zembal-Saul, & S. M. Kademian (Eds.), *Sensemaking in elementary science: Supporting teacher learning* (pp. 64–77). https://doi.org/10.4324/978 0429426513

Kathryn M. Bateman is a research and professional learning manager at the Museum of Science, Boston. She obtained her PhD from The Pennsylvania State University in Curriculum & Instruction: Science Education. Katie is a former Philadelphia middle school science teacher and Marine Biologist. Her research is embedded in the everyday practices of teachers through a lens of postmodern philosophy with a goal of advocacy for equitable education.

Case: Growing in Understanding NGSS Science and Engineering Practices

Akarat Tanak

Abstract

This case centers on the challenges teachers face in supporting student engagement in the science practices described in the Next Generation Science Standards. Ms. Taylor, an experienced primary teacher, is reviewing her second-grade students' work at the end of an inquiry about plant growth. Disappointed, she tries to figure out why her students are struggling to form explanations about what plants need to grow. Some thoughts on Ms. Taylor's dilemma are provided after the case.

Ms. Taylor sat at her desk, her disappointment growing as she reviewed student work. Her second graders had just completed an investigation she had planned to help students meet NGSS Performance Expectation *2-LS2-1: Plan and conduct an investigation to determine if plants need sunlight and water to grow.* She had asked them to write a scientific explanation, but that wasn't what she was reading now. Most of her students had written some variation of "plants grow best on the windowsill" supported by evidence that "the plants on the windowsill were taller." She was encouraged by one team's answer that "plants grow best where there is sunlight" but even that didn't go as far as she had hoped.

Ms. Taylor's class had spent weeks on the investigation. She thought back to the day she introduced the unit and how eager they were to answer the question she had posed, "where do plants grow best in our classroom?" Many students had offered up conjectures immediately, relating these to where they saw plants in their homes and in other rooms in the school.

When she told them that they would be working collaboratively as teams of scientists to test their ideas, they had actually cheered! It was clear they were

A. Tanak (✉)
Division of Education, Department of Education, Kasetsart University, Bangkok, Thailand
e-mail: akarat.t@ku.th

excited and motivated by the thought of caring for their own plants. They examined the seeds she provided them with a hand lens, expressing wonder at what type of plants they would become. She was careful to provide lots of time for planning how to carry out the investigation, and she had been impressed with how quickly students had grasped the idea of a fair test. Two teams were growing plants on the windowsill, two in the far corner of the room, and finally, two placed theirs in the storage closet. They had decided as a class that they needed to keep the soil they planted the seeds in and the amount of water all teams gave their plants the same for this investigation so that only the location varied.

Her students could barely contain their excitement when the first sprouts emerged! She smiled, remembering how each morning they had eagerly retrieved their iPads to photograph their greenhouses and record the height appearance of their plants. In the scientist meetings that followed, they were curious to compare their plants to those of other groups—and surprised to note the differences that emerged between them. She had focused a lot on helping them make good observations. Students had noticed differences in when the seeds sprouted, how tall they grew, the color of the plants, and how many leaves they had. She had encouraged them to keep track of these observations and keep thinking back to the investigative question.

It was clear the investigation was a huge success, if judged by students' enthusiasm alone. One of the parents had even commented on how their child said it had made them felt like they were a "real scientist." Students were so proud of their investigations! Given that, it made her growing disappointment as she reviewed their work quite difficult. Though they had clearly enjoyed the experience, it wasn't clear they had met the learning goals.

She had been so sure that the students would be able to provide an explanation that plants needed sunlight to grow! When she looked over their responses to the investigative question, however, what she read were reports of the outcomes and data—not scientific explanations of why some plants grew better than others based on what plants need to grow. For example, the statement "plants grow best on the windowsill," told what happened, but didn't extend to *why* that might be the case. Had she overestimated the abilities of her second graders? Did they just need more practice with scientific explanations? What should her next steps be? She had been planning for them to vary the amount of water in their next investigation while keeping the amount of light the same, but she wasn't sure now if that would help them meet the standard, either.

For Reflection and Discussion

1. How does the way in which Ms. Taylor framed the investigative question has implications for the kinds of answers her students provided in the end? How could Ms. Taylor has (re)phrased the investigative question in order to encourage students to make claims or offer explanations, as opposed to merely reporting outcomes?

2. Lessons do not always work out as intended. How might Ms. Taylor proceed, now that she has identified this issue? How can she do so without negatively impacting their enthusiasm and the pride her students feel in their accomplishments?
3. How might Ms. Taylor approaches her next investigation with students, to better support their engagement in the science practices?

Akarat Tanak is an Associate Professor in the Faculty of Education at Kasetsart University in Bangkok, Thailand. She teaches undergraduate and graduate courses in science education including elementary science learning, scientific thinking, and instructional media and innovation. Her research focuses on the development of teachers' practice with an emphasis on thinking skills and technology integration.

Commentary: Science Starts with a Question

37

Deborah Hanuscin

Abstract

This is a commentary to the case narrative, *"Growing in Understanding NGSS Science and Engineering Practices"* written by Akarat Tanak.

While Ms. Taylor is focused on what's happening at the *end* of the investigation, she is missing an important factor in students' difficulty that goes back to the beginning of the investigation and starts with the question. The problem encountered by Ms. Taylor is one described by McNeill and Berland (2017) as "data as answers." That is, students' answers to the investigation question are a statement of the outcome of the investigation—rather than an explanation. One potential reason for this is how the investigative question was phrased. That is, "plants grow best on the windowsill" is a perfectly reasonable answer to give response to the question posed by Ms. Taylor, *"where do plants grow best in our classroom?"* In contrast, the idea that Ms. Taylor was expecting students to grasp, *"plants need sunlight to grow,"* does not follow logically in response to the question.

Focusing on the practice of constructing explanations is unlikely to help the situation, if this is the question that students are expected to answer. Rather, by phrasing the investigative question differently (to align with the practice of asking questions as envisioned by the NGSS), Ms. Taylor could actually help improve students' ability to engage in the practice of using evidence to construct explanations. For example, the unit she is teaching focuses on how plants obtain the materials they need for growth. This learning goal of the unit can be rephrased as an explanatory focus question for unit: How do plants obtain the materials they need for growth? The investigations students conduct should provide them with the

D. Hanuscin (✉)
Science, Math, and Technology Education, Western Washington University, Bellingham, WA, USA
e-mail: hanuscd@wwu.edu

Table 37.1 Science practices organized in a sequence of investigating, sensemaking, and critiquing (McNeill et al., 2018)

	Investigating practices	Sensemaking practices	Critiquing practices
Science practices	1. Asking questions	2. Developing and using models	7. Engaging in argument from evidence
	3. Planning and carrying out investigations	4. Analyzing and interpreting data	8. Obtaining, evaluating, and communicating information
	5. Using mathematical and computational thinking	6. Constructing explanations	

building blocks for this larger explanation. Investigative questions such as *how does the amount of sunlight affect plant growth?* and *how does the amount of water affect plant growth?* can't be answered simply by providing "data as answers"—students would need to use evidence to develop explanations of the cause-and-effect relationships.

A fundamental shift in the NGSS is toward seeking answers to how or why phenomena occur—not just descriptions of the phenomena themselves (McNeill et al., 2017). This shift is reflected in the practice of *asking questions* that are explanatory in nature. Explanatory questions—how and why types of questions (e.g., "how do plants absorb water and nutrients?")—help students go beyond gathering observations of phenomena to building explanations. Explanatory questions can help students participate in science in more authentic ways, rather than merely following steps or directions to demonstrate a concept.

This case illustrates also that the science practices do not operate in isolation but are interrelated. Students' engagement of science involves an overlapping sequence or *cascade of practices* (Bell et al., 2012). The scientific practice of asking questions is closely linked to the practice of constructing explanations. In this case, how the initial question was phrased (practice 1 in Table 37.1) had consequences for how answers were formulated (practice 6 in Table 37.1). As shown in Table 37.1, students engage in multiple practices as they are investigating, sensemaking, and critiquing. As teachers plan for students to engage in the science practices, they must also consider these interrelationships. Starting with asking a scientific question to investigate, students have continuing opportunities to plan and carry out investigations, develop models, interpret data, and argue about explanations from data as the way to make sense of the science they are engaged in.

Teachers need to be able to identify investigable questions that lead students to collect evidence to support explanations and to generate claims. A challenge is that teachers ask a variety of questions during instruction for a variety of purposes. For example, asking "in what kinds of places do plants grow?" would help a teacher elicit students' prior knowledge. While appropriate as a *pedagogical* question, it is not an appropriate *scientific* question (Tanak & Hanuscin, 2020). Thus, teachers also need to be able to help students develop investigable questions.

Finally, this case calls attention to the question of *who should be asking questions in science classrooms*? Questions posed by the teacher can be a model for the kinds of questions that scientists investigate. However, we also emphasize that in order to meet the vision of the NGSS, *students* should learn to pose their own investigative questions in science classrooms. While teachers need to be able to pose appropriate scientific questions for investigation, they also need to be able to support students in asking questions (Hanuscin and Tanak forthcoming). As students gain confidence in their ability to pose questions and pursue the answers on their own, this can further contribute to the kind of ownership and enthusiasm for science Ms. Taylor wants to foster.

References

Bell, P., Bricker, L., Tzou, C., Lee, T., & Van Horne, K. (2012). Exploring the science framework. *Science and Children, 50*(3), 11.

Hanuscin, D. L. & Tanak, A. (forthcoming). When is asking questions a scientific practice? *Science and Children.*

McNeill, K. L., & Berland, L. (2017). What is (or should be) scientific evidence use in k-12 classrooms? *Journal of Research in Science Teaching, 54*(5), 672–689.

McNeill, K. L., Lowenhaupt, R., & Katsh-Singer, R. (2018). Instructional leadership and the implementation of the NGSS: principals' understandings of science practices. *Science Education,* 102(3), 452-473

McNeill, K. L., Berland, L. K., & Pelletier, P. (2017). Constructing explanations. In *Helping students make sense of the world using next generation science and engineering practices,* pp. 205–227.

NGSS Lead States. (2013). *Next generation science standards: For states, by states.* The National Academies Press.

Tanak, A. & Hanuscin, D. L. (2020). An exploratory study of preservice 'teachers' framing questions for classroom science lessons. In *Paper presented at the annual meeting of the Association for Science Teacher Education, San Antonio, Texas*

Deborah Hanuscin is a former elementary teacher and informal science educator. She is currently a Professor in Elementary Education and Science, Math, and Technology Education (SMATE) at Western Washington University. Her research focuses on elementary science teacher learning and the design of professional development and curricula to support that. She is a past president of the Association for Science Teacher Education, and recipient of the ASTE Outstanding Science Teacher Educator award.

Case: Too Loud to Learn

38

Megan E. Lynch, Jennifer L. Cody, and May Lee

Abstract

This case raises questions about expectations for student engagement and productive participation in science classes. Veteran kindergarten teacher, Melissa Perry, is new to phenomena-anchored, NGSS-aligned science teaching. During a sustained investigation on cockroaches as part of a unit on the characteristics of living things, Ms. Perry's expectations for her students and her already established norms of traditional classroom discourse (turn-taking, raising hands, waiting quietly) came into conflict with the ways in which students were participating in the science investigation. As a result of this contradiction, tensions emerge when Ms. Perry co-teaches with an instructional coach during a science lesson. Questions about school culture, school-wide behavior plans, access to high-quality teaching/curricula in schools with majority-minority populations, and the character and quality of coaching are all present within the case.

"Show me S.T.A.R. behavior!" announces Ms. Perry, loudly emphasizing the word "star." Within two seconds, her 26 kindergarten students are sitting straight in their

M. E. Lynch (✉) · M. Lee
College of Education and Human Services, University of North Florida, Jacksonville, FL 32224, USA
e-mail: m.lynch@unf.edu

M. Lee
e-mail: mhl11@psu.edu

J. L. Cody
State College Area School District, State College, PA, USA
e-mail: jlc479@psu.edu

J. L. Cody · M. Lee
Curriculum and Instruction, College of Education, University Park, PA, USA

chairs, stiff as a board, hands clasped and on the table, looking straight, and silent. Beaming proudly, Ms. Perry congratulated the students for quickly demonstrating such good classroom behaviors, "we are getting better each day at showing S.T.A.R. behavior! I think you all have earned time at the end of the day for a few minutes of play time on the carpet!" The students began to smile and nod their heads. A few started to tap their fingers on the table or kick their legs quietly under the table. "Whoa! Let's not lose that S.T.A.R. behavior just yet! Today, we are starting a new investigation with Mr. Hightower's help, and I need you on your best behavior."

Mr. Hightower is a new instructional coach for the district, offering a pilot program in the school that integrates science content with language learning in the in the early grades. The program requires teachers, such as Ms. Perry, to co-plan and co-teach science lessons with Mr. Hightower once a week. Ms. Perry was a bit nervous about teaching science since the school did not have a curriculum for science through third grade, and she had not been required to teach science in her 12 years of teaching.

Ms. Perry teaches in the Mid-Atlantic region of the US in an "urban" Title I school, Robinwood Elementary. The student population is characterized as "majority-minority," with over 50 percent of the students at Robinwood classified as English learners. The students are predominately part of Spanish-speaking families from Latin American countries. For the past five years, the school, and the district at large, has underperformed on all state measures, ranking near the bottom against schools across the state. School-wide initiatives have focused almost exclusively on ELA and math instruction to improve state measures, with no attention to science and social studies curriculum development. Additionally, there are only two ESL teachers at Robinwood to serve more than one hundred emergent bilingual students. In fact, this is why Mr. Hightower was newly hired as an instructional coach and asked to work primarily with the teachers at Robinwood Elementary.

Mr. Hightower enters the room and walks over to the "empty" aquarium that has been sitting on a table near the chalkboard for a couple of weeks. Over the past two weeks, the students have been mildly curious about what was inside. "Why isn't there any water?" "Will it have fish?" "Why is there some dirt and a tree in it?" "Why does it have a lid if nothing's inside?" However, none of the children spent more than a few minutes inspecting the aquarium and asking a question here and there.

Now that Mr. Hightower and Ms. Perry are in close proximity to the aquarium, and the students are displaying a renewed interest. Ms. Perry reminds them that they are continuing their year-long investigation of living and non-living things. "There is something in the aquarium that I would like to share with you all. I am going to call you up by your tables one at a time based on which table is showing the best S.T.A.R. behavior." In a flash, students sit up straight, clasp their hands together, and hold their breath in an effort to stop moving or make a single sound.

One by one, the students are called up by their tables until they are all hovering over the aquarium that just seems to have "some dirt and a tree in it." Mr.

Hightower asks the students to look carefully for something they had not noticed before. "I see a little bit of water in the corner." "There is something like broken potato chips over there." Mr. Hightower asks the students if any of them have looked under the log or picked up the log. Their eyes widen. Ms. Perry, not a huge fan of creepy, crawling things, picks up the log, and shows the bottom to the students. The students jump back. Some squeal with excitement and terror. "Eww!" "What is that thing?!" "I can't see! I want to see it!" "Ahh!" "Get that away from me!" Gross! "What is it?" "Can I touch it?" "Bugs!!"

Without revealing what they are, Mr. Hightower picks up one of the dozen-plus Madagascar Hissing Cockroaches and holds it carefully in his hands. He asks the students if they would like a closer view. Some students are jumping and bouncing up and down trying to see in the palm of Mr. Hightower's hand; others are running back to their chairs. Some students lurk nearby, curious, but not sure of how close they want to be. Mr. Hightower and Ms. Perry shift to putting four of the Madagascar Hissing Cockroaches into petri dishes. Students continue to move around the room until Ms. Perry asks for S.T.A.R. behavior so that all students move quickly and quietly to their tables. She whispers to Mr. Hightower, "I'm so sorry they are acting like this. We knew the cockroaches would get them excited and engaged, but they are not acting well-behaved. I'm going to move them to the carpet and then we can continue."

Ms. Perry begins naming students who are exhibiting S.T.A.R. behavior, "I see Manuel is being a S.T.A.R. student." "Ana is also showing S.T.A.R. behavior." Other students quickly follow suit. Ms. Perry and Mr. Hightower bring them over to the carpet sitting in a large circle. Mr. Hightower explains that he wanted the students to notice and wonder about the cockroaches. Ms. Perry pulls up a large post-it note paper and writes "notice" on one side with large cartoon eyeballs and "wonder" on the other with a large question mark. She draws a line down the middle. At the same time, Mr. Hightower is distributing the four petri dishes of Madagascar Hissing Cockroaches around the circle, one for every 6–7 students. The students are highly engaged in the notice and wonder. Their talk do not stop. They share with each other as they passed around the cockroaches. Students are turning and talking to one another. Their ideas are clever and insightful. Ms. Perry calls on students one at a time to share what they are noticing:

"This one has lines on the back."

"I counted six legs."

"Does it bite or have poison?"

"Where does it come from?"

"Are these his eyes?"

Mr. Hightower listens to the students and lets Ms. Perry know what they say: "Ms. Perry, Emmanuel is noticing that the cockroaches have something that's maybe like hair on their legs." Ms. Perry continues to write what students are noticing and sharing.

One of the students, Paula, passes the petri dish to Sophia, who is sitting next to her. Paula asks, "Do you think it's a baby or grown?" Sophia responds, "I don't know. It's bigger than any bug I've seen." Paula nods in agreement. "It's not moving. Is it dead?" Sophia wonders out loud. Paula lifts up the petri dish and looks under: "No, I can see things in the front move." Paula and Sophia continue noticing and wondering about the Madagascar Hissing Cockroach in their petri dish. The classroom volume increases as more students turn to talk to one another.

Ms. Perry hears one of her emergent bilingual students, Isabel, share something in Spanish and asks her to repeat it. Ms. Perry knows that Isabel probably won't understand what she says, and she is a little uncomfortable not knowing how to respond to Isabel, but hopes that another student will offer a translation. As Ms. Perry listens to Isabel share her thoughts in Spanish, she reminds the class to use Level 1 voices. At the same time, Ms. Perry reminds herself that the science lessons tend to be louder because students are so excited and engaged and that can be okay. She reassures herself that the student movement and volume on the carpet are okay. It is not a time of direct instruction so students can talk freely to one another. Yes, it goes against the dominant school culture—if her colleagues were to walk in or pass by the hallway, they might think things like "Ms. Perry can't manage her class," "her students are out of control," or "she must bribe them to be so interested," but she tells herself that what she is hearing from her students is productive talk.

As Ms. Perry observes the room, she notices that her students seem to be engaged in so many small conversations that they might not be able to hear one another. She asks them to lower their voices, to keep their eyes on the teacher and the notice and wonder chart to stay on task. Ms. Perry asks a student, Mateo, to share his idea. As he starts, "I wonder about what these things might eat because…" another student gives out a blood-curdling scream. Jack had dropped one of the petri dishes, causing it to fall open and leave the cockroach exposed. The scream startled many of the students who were now talking at an unbearable volume. Ms. Perry once again asks for S.T.A.R. behavior and sends the students back to their desks.

Mr. Hightower collects the petri dishes to put the Madagascar Hissing Cockroaches away and suggests to Ms. Perry that since the students are seated, this would be a nice time for them to write in their Scientist Notebooks on one of the "notice and wonder" pages. Ms. Perry sees the value in this suggestion and directs students to pick up their notebooks and begin writing about what they noticed and are wondering. With about 10 min remaining, Mr. Hightower and Ms. Perry circulate the room, read what the students are writing, and comment on their ideas. While circulating, they stop and confer with one another about a good time to debrief. Diego, an emergent bilingual student whose family is from the Dominican Republic, walks over to Mr. Hightower and Ms. Perry to share his writing. "Ms. Perry, how do you spell "body"?" Ms. Perry responds, "Did you do three before coming to me? Think about it, ask another student, look at the Tricky Words Wall?" "Yes, I did." Ms. Perry offers the correct spelling and asks Diego to return to his seat. She looks over to Mr. Hightower with exasperation, "I am so sorry he interrupted you. Let me get them ready to wrap up and then we can sit and debrief."

As soon as the students leave, Ms. Perry sits down next to Mr. Hightower, puts her head in her hands, and just starts crying. Through the tears, she says, "I am just so disappointed in my students. At times, it felt like it was the first day of school all over again. We spent so many days and weeks practicing S.T.A.R. behavior again and again, taking turns to speak, sitting still. Just look around! I have posters on all my walls reminding them of S.T.A.R. behavior. It felt like all of the progress for the year was completely undone. I honestly don't even know what the students got out of the lesson. I just feel like they are so excited and are genuinely interested in science and the phenomena we have, but they treat it like gym or recess instead of learning. The expectations are just so different in other parts of the day, sometimes I feel like science is just too much for them."

For Reflection and Discussion

1. What are Ms. Perry's expectations for her students' classroom behaviors and what informs those expectations?
2. In what ways is equity relevant to this case?
3. How are the students engaging in science practices like observing real-world phenomena, science discourse, and scientific thinking?
4. Instructional coaches often work with classroom teachers in ways that can push teachers in new directions, sometimes uncomfortable directions. What are some of the potential benefits and drawbacks for Ms. Perry working with an instructional coach?
5. What's next for Ms. Perry and Mr. Hightower? Specifically, how do you think Mr. Hightower could respond? Why?

Megan E. Lynch is a Postdoctoral Fellow at the University of North Florida in the College of Education and Human Services. She engages in practice-based research with clinical faculty and school partners in Professional Development Schools and partnership schools and supports research initiatives in the Urban Education program. Her research interest is in the development of socially just pedagogy and political activism with teacher candidates, in-service teachers, and P-12 students within school-university-community partnerships.

Jennifer L. Cody is a Ph.D. candidate in Curriculum and Instruction at Penn State University and works as a professional development associate and research assistant on the Science 20/20 Grant Project: Bringing Language Learners into Focus through Community, School, University Partnership. She is currently a full-time fifth-grade teacher and has taught for 15 years. Her research interest is centered on the ways in which one's communities of practice influence equitable and inclusive pedagogical practices in science instruction.

May Lee is an Instructor in the Department of Curriculum and Instruction at The Pennsylvania State University. She facilitates courses on teaching English learners and serves as the Coordinator of the World Campus M.Ed. Program in Curriculum and Instruction. As a former ESL teacher, her work examines ways in which in-service and preservice teachers support and challenge their emergent bilingual students' learning.

Commentary: Too Loud or Too Controlled to Learn?

Maria Varelas

Abstract

This is a commentary to the case narrative, *"Too Loud to Learn"* written by Megan E. Lynch, Jennifer L. Cody, and May Lee.

So many conundrums, tensions, and questions puzzled me as a teacher, a teacher educator, and a science education researcher as I imagined Ms. Perry and her class while reading this case. What kept coming to my mind is grappling with issues of power and choice and the many dialectical relationships that teachers and instructional coaches have to consider all the time—dialectical relationships like freedom and authority, transformation and reproduction, and body and mind (Varelas, 2018).

I first focused on the emphasized "S.T.A.R." behavior and wondered what being a "star" (someone who shines through, who is bright) necessitates for a young person, a child, a student. Does it necessitate taming and controlling the body so that the mind can be led to a set of ideas to be discovered? Or does it mean being able to choose to create together with body and mind something that is new and meaningful? Or does it mean something in between, something that is constantly negotiated by both students and teacher as they recognize, and attend to, the need for distributed power and choice, for both freedom and authority, and for both transformation and reproduction of ideas and meanings, engaging both their bodies and minds?

I pictured Ms. Perry being tormented by the idea that her students were misbehaving, not only when she was alone with the children, but more so when Mr. Hightower, a male instructional coach who presumably was comfortable teaching

M. Varelas (✉)
Department of Curriculum and Instruction, University of Illinois Chicago, Chicago, IL, USA
e-mail: mvarelas@uic.edu

© The Author(s), under exclusive license to Springer Nature Switzerland AG 2023
S. Jeong et al. (eds.), *Navigating Elementary Science Teaching and Learning*,
Springer Texts in Education, https://doi.org/10.1007/978-3-031-33418-4_39

science (all identities that she did not seem to share), was present in her classroom and co-teaching with her. Maybe Ms. Perry was seeing her students' obedience as reflecting her expertise as a teacher, which could show Mr. Hightower that she was in charge of her class and would also make her students more ready to learn from him. Ms. Perry seemed to be positioning herself relative to Mr. Hightower in similar ways in which she was positioning her students relative to her: those with more power deserve respect and obedience by those with less power. And these have been the most prevalent, dominant ideologies and discourses in classrooms across the US, especially in classrooms with students of color, emergent bilinguals or multilinguals, and students whose ethnoracial and linguistic identities have been minoritized, discriminated upon, and marginalized. In my own science education research and educational research in general across school subjects including science, Black and Brown children are more subjected to behavioral management to be "good," which some construct as part of being good in science or being good scientists (Varelas et al., 2011). This is one of the dangers I worry about for Ms. Perry's class that seems to comply with a school-wide practice.

Ms. Perry seems to feel the need to conform with larger school structures that assume quietness (level 1 voices) and limited mobility as necessities for learning. In some ways, she shows us in this case what Maria Rivera Maulucci et al. (2015) called structurally reproductive teacher agency—namely, her agency as a teacher that maintains the structures put in place to maintain control. Even play time seems to be a privilege rather than a right for kindergartners in Ms. Perry's school, as children earn more play time on the carpet only with "S.T.A.R" behavior, when we know that, especially for young children, play is necessary and actually part of learning (Hirsch-Pasek & Golinkoff, 2008). In social systems, like the larger system of schooling, power, and control is not constructs that are created, owned, and maintained at the individual level but are systemic, sociohistorical, and sociopolitical schemas. Structures that systems are based on, and function with, regulate what degrees of power and control are available to different individuals, including teachers and students. However, this does not mean that structures cannot be changed or transformed. Structure and agency are in a dialectical relationship.

Ms. Perry also realizes that her students were engaged, making observations and asking questions when the "bugs" nesting under the log of the seemingly empty aquarium were revealed. Her students' "noticing" and "wondering" seemed to temporarily give her permission to break school rules about level 1 voices and accept her students' loudness without trying to totally suppress it. At the same time, she was recording her students' thinking on the large post-it note paper by listening to what they were contributing as Mr. Hightower was also sharing with her what he was hearing from other students. The real "bugs" in the petri dishes the students were passing around, the children's English and Spanish speech, and the written text that Ms. Perry was composing to capture their meaning-making allowed for multimodal ways of thinking and communicating, each with complementary affordances. Ms. Perry's nurturing of multimodality (with Mr. Hightower's support for the hands-on component, including his touching and holding the Madagascar

Hissing Cockroaches for students to first see outside of petri dishes) offered her students multiple entry points to science learning (Jewitt, 2009).

However, despite the beauty and brilliance of her students' engagement with science, Ms. Perry did not come to recognize their observations and wonderments as important signs of learning—namely, making meaning both of science ideas and of themselves as producers of science knowledge. Her goal of knowing exactly what students learned at the end of a lesson (and I wonder whether this is feasible in every lesson or even desirable) seemed unmet. The children's behavior (being excited, loud, interrupting her conversation with the instructional coach) overshadowed for her the opportunities for learning that the children were attempting to co-construct with their teachers (Ms. Perry and Mr. Hightower), and from my viewpoint, succeeded in, despite all of Ms. Perry's top-down control.

The conundrums in which I imagined Ms. Perry being caught seemed to be experienced not only in her mind but in her body too, as her whole self was showing her disappointment, her sadness, her anxiety, and her assessment of having failed relative to what she was identifying as being a good teacher. Her crying at the end of the lesson encapsulated and expressed the powerful emotions she was feeling, along with the meaning she was developing of her children's actions and learning. Ms. Perry's teacher identity and her knowledge and practice of teaching (including science teaching) were shaped by her school's structures (which, like all structures, include both schemas and resources) and configured in profound ways both in her body and her mind as she was engaging in teacher learning, in further developing her teaching, with Mr. Hightower.

However, Ms. Perry could not see how her students' bodies, their emotions, the loudness of their voices, and their movements were assets that could have been celebrated and built upon in learning science. Ms. Perry was not finding joy in her kindergartners' casual questioning about the "empty" aquarium that they were noticing for days in their classroom; in their intense reactions to realizing that there were bugs living in the aquarium; in their willingness to question and reason about the bugs' body parts and age; in having another colleague, a fellow teacher, in her classroom providing resources that she and her students could not have access to otherwise; and in engaging in conversations with him about teaching and learning. Instead, she appeared to experience excessive stress that many teachers have been led to frequently feel.

So, I am wondering how teacher education efforts (preservice and in-service/ professional development) could support teachers like Ms. Perry to resist many schools' inequitable and unjust privileging of authority over freedom, control over choice, containment over exploration, certainty over questioning, correctness over failure, rational thought over emotions, and thinking over sensing and acting with the physical body. Reflexivity coupled with explicit interrogation of knowledge, beliefs, and practice is necessary dimensions of equity-minded teachers. As teachers, we need to be working with others to (a) explore our own ways of (teacher) learning and (b) examine how our teaching practice attends or not to our students' learning. Constructing our own ways of balancing the dialectical relationships noted above may offer us opportunities to develop our own transformative (instead of reproductive) teacher agency that could lead to collectively changing oppressive structures.

References

Hirsch-Pasek, K. & Golinkoff, R. M. (2008). Why play = learning. In R. E. Tremblay, M. Boivin, & R. V. De Peters (Eds.), *Encyclopedia on early childhood development* (pp. 1–6). Centre of Excellence for Early Childhood Development and Strategic Knowledge Cluster on Early Child Development.

Jewitt, C. (2009). An introduction to multimodality. In C. Jewitt (Ed.), *The Routledge handbook of multimodal analysis* (pp. 14–27). Routledge.

Maulucci, M. S. R., Brotman, J. S., & Fain, S. S. (2015). Fostering structurally transformative teacher agency through science professional development. *Journal of Research in Science Teaching, 52*(4), 545–559.

Varelas, M. (2018). Dialectical relationships and how they shape (in)equitable science learning spaces and places. In L. Bryan & K. Tobin (Eds.), *13 questions: Reframing education's conversation: Science* (pp. 183–191). Peter Lang.

Varelas, M., Kane, J. M., & Wylie, C. D. (2011). Young African American children's representations of self, science, and school: Making sense of difference. *Science Education, 95*(5), 824–851.

Maria Varelas is a professor of science education and chair of the Department of Curriculum and Instruction at the University of Illinois Chicago (UIC). She co-coordinates the Mathematics and Science Education PhD program and the Science Education MEd program, and was the Director of the UIC Center for the Advancement of Teaching-Learning Communities. She is a Fellow of the American Association for the Advancement of Science and her research focuses on possibilities and challenges related to student learning when practicing and preservice teachers consider and practice science education aiming at equity and social justice.

Case: When State Standards Change: Dilemmas of Teachers of Color

40

Bhaskar Upadhyay and Stefanie L. Marshall

Abstract

In this case narrative, Bhaskar and Stefanie share a discussion among teachers during a school staff meeting. Mrs. Ali, Ms. Blue, and Mr. Vue each belong to underrepresented racial/ethnic groups. Teachers of marginalized communities often recognize the white-normed ways teaching, instructional coaching, and that these ways often also inform decisions in their schools. The teachers share some of the consequences the new state science standards may yield for students, and teachers, specifically concerning curriculum and assessment decisions. They share their concerns and the potential benefits of new state standards for students from immigrant families in learning content and the cultural relevancy of science in their communities.

Rookwood is a high-poverty urban city school located in the Midwest. Forty percent of the student population are of color, a large percentage of whom are immigrants or children of immigrants. Of the fifteen teachers on staff, there are five teachers of color, three of whom are also either immigrants or children of immigrants. Rookwood has been deemed a *Turnaround school*—an academically low-performing school that now must employ extensive and comprehensive interventions. The case below highlights a discussion that takes place during a staff meeting led by Principal Nolan.

B. Upadhyay (✉)
Comparative and International Development Education, Department of Organizational Leadership, Policy and Development, University of Minnesota, Twin Cities, MN, USA
e-mail: bhaskar@umn.edu

S. L. Marshall
Department of Teacher Education, Michigan State University, Twin Cities, MN, USA
e-mail: marsh413@msu.edu

The school administrators and teachers are both curious and concerned about the effects of new state science standards on curriculum, pedagogy, and assessment. One of the concerns of the administration seems to be about losing the status of Turnaround school and related extra district funding. Therefore, the principal wants teachers to strategize about how to support student success.

Principal Nolan (White Female): Next on the agenda is the results from our state assessments. Overall, we have some work to do, especially in science. Our scores from 2018 to 2019 improved in reading and math, which is great. In math, we went from 14.2% of students demonstrating proficiency in Math to 20.1% in 2019. In reading, we went from 18.8 % proficiency to 22.8%. However, our science proficiency percentage declined from 19.6 to 7.8%—a pretty significant drop. With the new Science Standards having just been adopted by the state, we need to think about how we provide our students with hands-on experiences that align with the standards. You should have received a copy of the standards a few weeks ago in your email. The new standards will be fully implemented in two years. Ms. Arnold will share some essential information from the State Science Standards Committee meeting.

Ms. Arnold (White Female): I think that the main idea to focus on is that the science and engineering practices ground the standards rather than the content. The new State Standards are based on the Next Generation Science Standards (NGSS) framework and are highly influenced by dimensions of learning—disciplinary core ideas, science and engineering practices, and crosscutting concepts. So, the new standards are about skills development as the basis for how science knowledge is gained, generated, and reaffirmed. The committee discussed whether the standards should be written with content as the guide to continue the previous practice of being content-led. What led to the decision to center on the practices is that they help students understand how scientific knowledge is produced. Also, science and engineering practices embody science and engineering cultural practices, providing more realistic ways to do and engage in science. Given that 40% of the student body are immigrants or are children of immigrants, the practices should engage students in authentic ways. The focus on how the standards translate into high-poverty urban schools and Native schools was of concern. Still, the committee decided that a critical factor in enacting the standards is the socio-cultural considerations that are part of teachers' everyday decisions. That's all I have.

Principal Nolan: Ok, everyone. It sounds like the new standards may be a real opportunity for us to think outside the box. I was hoping you could take a few minutes at your table to reflect on the Standards. What does this mean for our students? What do we need to do to prepare to employ these new standards? I have the questions up for you all. Let's take a few minutes and be prepared to share out.

A humming of chatter takes over the library. Mr. Vue (Second generation Hmong-American who has been working for 8 years in the district), Ms. Blue (Native American who has been teaching for 10 years), and Mrs. Ali (Somali immigrant who is a novice teacher in her second year) begin to talk at their table.

Mr. Vue: So, how are you all feeling about the new Science Standards? Honestly, I'm struggling with what they mean, especially for our immigrant students.

Mrs. Ali: Me too. I get that it is thought that the practices allow for socio-cultural perspectives. However, from what I read, the practices of understanding science and engineering are already prescribed.

Ms. Blue: As a Native teacher, my practices of understanding nature and the local environment don't get reflected in science standards. Native contributions get reflected in appendices and footnotes, if that. I am concerned with how non-Native teachers engage Native students in science and engineering practices. I'm worried that Eurocentric views of science will frame the practices if teachers do not know how to engage with the science and engineering practices with a socio-cultural lens.

Mrs. Ali: Yes. Social and cultural are needed.

Ms. Blue: Storytelling is deeply embedded in my culture, and I do not know if our peers will engage our students in making sense of ideas without valuing this part of our culture.

Mrs. Ali: Somali elders or adults share different things about land and culture through stories with their children. Stories are essential for us, too, especially with our children.

Mr. Vue: I think most of our—Hmong, Somali, Native—culture and knowledge is passed on through stories.

[Pause]

Mr. Vue: The prescribed practices could allow for other ways of knowing, but most teachers have not been supported to engage with student ideas in ways where they can see the value and ingenuity that comes through their rich ideas. I guess one thing that stands out is that we just heard that many of our students are not proficient in science. I'm just not sure how to prepare students for an assessment that is not culturally relevant.

Mrs. Ali: I feel like they just care about the test results. I see people focusing on the assessments rather than valuing student sense-making. The assessments don't match what my students know about the phenomenon. There are so many opportunities to support learning based on the practices, but they are also limiting.

Mr. Vue: We serve a large immigrant population. We must consider that some students have had substantial amounts of science, while others may have had limited exposure. Our stories as immigrants are not monolithic! Our Hmong students tell me they don't see the diversity of their culture in what is being taught in many classrooms.

Mrs. Ali: Somali children are telling me the same thing. I'm in just my second year, but students find me to talk because they know I will listen to them. When we share out in a large group, it might be essential to ask how teachers will be supported to genuinely *see* and *hear* our students. Our children are brilliant—but I feel like so many miss out on their brilliance.

Ms. Blue: One thing I noticed in the new standards in our state is that they really focus on practices.

Mr. Vue: Mostly engineering and science practices. Do we have to know which are engineering and which are science? I'm confused about how content and practices link.

Mrs. Ali: You know practices are important, like knowing how to get an answer, but how do I assess? I hope the district or the state gives us some guidance.

Ms. Blue: I like the practices part because it might fit better with Native ways of finding solutions. Practices will be good for Native students because it's like telling stories.

[Pause]

Mr. Vue: I didn't think about that. Yes. It's like storytelling. Practices focus on how (process and skills) to get what (knowledge).

Ms. Blue: We have to be aware of the content, though. This means new resources, maybe fewer topics, so we can do more of the "how" (practices), not just the "what" (in science and engineering). The focus on the practices means more time to do inquiry activities.

Mr. Vue: I'm glad you mentioned resources. I'm worried about resources and when we will get them along with professional development. I honestly hope we don't have to make too many changes in our curriculum and lessons.

Mrs. Ali: I agree.

[Pause]

Mrs. Ali: You know I may have to translate these engineering and science practices into Somali, and Mr. Vue, you also translate materials into Hmong for our students, right?

Mr. Vue: I do. I hope there is a consideration for resources that will support our students.

Ms. Blue: I'm anxious.

Mr. Vue: Me too. Very anxious…I think we need to keep thinking about immigrants and other underrepresented groups.

As the time for small group discussion ended, Mrs. Ali suggested finding some necessary information on immigrants' cultural, linguistic, and social nature in the US. This information helps everyone develop a better understanding of their students and parents and aids teachers in employing pedagogy that better suits student learning needs.

Mrs. Ali shared information that she had compiled for teaching purposes:

- Immigrants and migrants are born in a different country or territory other than where they now live permanently.
- Based on the most recent data from the United Nations (2019), the United States host the largest number of migrants in the world, totaling 50.7 million.
- Immigrants make up 13.7% of the US population and represent almost every country globally (Budiman, 2020) and the languages and dialects they speak.
- The education of immigrants and migrants varies widely based on their social, political, cultural, religious, and economic circumstances and natural and human-created disasters in their country of birth. Therefore, some immigrants may have received different amounts of formal education.

Diversity among immigrant families creates a multitude of opportunities for teachers and school administrators in science classrooms.

For Reflection and Discussion

1. Why do you think this group happened to be sitting together during the staff meeting?
   a. How might the makeup of the group of teachers impact their conversation?
   b. What issues concerning the new Science Standards are these teachers worried about?
   c. What concerns do they have for the students in this school? Why?
2. How have you noticed white-normed/Western ways of knowing prioritized in schools?
3. How might race (anti-blackness, racism, caste, colonialism) impact a teacher's instruction?
4. As a first-year science teacher, how would you decide on your science teaching methods (pedagogies) for your students in this school?
5. What could the principal have done to support science teachers feel more comfortable and prepared to teach with the new Science Standards?

References

Budiman, A. (2020). *Key findings about US immigrants*. The Pew Center. https://www.pewresearch.org/fact-tank/2020/08/20/key-findings-about-u-s-immigrants

United Nations. (2019). *International migrant stock 2019 [electronic resource*. Department of Economic and Social Affairs. https://www.un.org/en/development/desa/population/migration/data/estimates2/estimates19.asp

Bhaskar Upadhyay is a professor of STEM education at the University of Minnesota. His scholarly work focuses on exploring issues of STEM education surrounding race, equity, social justice, indigeneity, and global STEM education that affects mostly teachers, students, and schools of underrepresented communities.

Stefanie L. Marshall is an assistant professor of science education at Michigan State University.

Commentary: The Unfortunate Realities of Elementary Science Instruction

41

Terrance Burgess

Abstract

This is a commentary to the case narrative, *"When State Standards Change: Dilemmas of Teachers of Color"* written by Bhaskar Upadhyay and Stefanie L. Marshall.

The case of science teaching at Rockwood Elementary School demonstrates an unfortunate and all too frequent dilemma for teachers and administrators across the USA. According to a 2018 report by the National Survey of Science and Mathematics Education (NSSME), elementary schools in the USA deliver to students approximately 20 min of science instruction per day on an infrequent weekly or monthly basis. We can use the positioning of science instruction at Rockwood Elementary to understand the nuances surrounding its common absence from the greater elementary curricula.

First, though, is the question of *"why science?"* Rockwood Elementary was identified as a "turnaround" school; with this designation comes several accountability metrics—including several in science—to meet the state's testing expectations. Because of this, Rockwood Elementary School's Principal Nolan identified science as the area of improvement to maintain state compliance. New science standards are also being adopted, including three-dimensional learning goals, described by Ms. Arnold as "embody[ing] science and engineering cultural practices [that] provide realistic ways to do and engage in science." This adds an additional layer: the word *culture* has been invoked to suggest that teaching science in this new way will meet the cultural needs of the diverse student population. However, with no further description, this loaded claim becomes misleading.

T. Burgess (✉)
Department of Teacher Education, Michigan State University, East Lansing, MI, USA
e-mail: tburgess@msu.edu

Historically, the culture of Western science has marginalized the identities and experiences of persons of color. In many cases, it has been used to justify upholding such marginalizations. Thus, Ms. Arnold's claim, in which she conflates cultural practices of science and engineering with racial and ethnic diversity, becomes problematic. This is affirmed by Mrs. Ali, a Somali immigrant, who noted in her discussion that, "from what I read, the practices of understanding science and engineering are already prescribed." The NGSS framework lists eight science and engineering practices, which have no "cultural" connections. Additionally, the NGSS Framework includes a set of appendices designed to provide concrete tools for implementing the standards. *Appendix D*, titled "All standards, all students: Making the next generation science standards accessible to all students" (National Research Council, 2012), includes several case studies, which speaks to the "footnote" from Ms. Blue's observation. Without proper centering, such footnotes become lost, resulting in the perpetuation of the instruction the standards were intended to disrupt.

Therefore, we should ask: What would happen if the core of science instruction meaningfully uplifted various cultural ways of knowing within the elementary classroom? Might this improve test scores *across* subjects? Ms. Blue noted that while the practices can evoke storytelling—an Indigenous way of teaching and sharing knowledge—the assessments are unlikely to authentically capture this learning, thus negating its purpose. Additionally, we must consider what *accessibility* means for students whose linguistic needs are often overlooked within curricula, and the invisible labor of teachers is trying to meet these needs. For example, Mrs. Ali and Mr. Vue regularly translate materials into the home languages of their students, a service which is typically outsourced and takes several weeks for a school to complete, resulting in missed learning opportunities. While this would be unfathomable for subjects such as mathematics and literacy, this is the unfortunate reality for elementary science instruction.

The truth is, we do not experience the world in silos. As we venture out and make sense of our surroundings, we leverage these experiences in ways that allow us to apply these understandings to new situations. Science teaching and learning should be no different. Elementary teachers are uniquely positioned to teach *across* disciplines, making the opportunities for meaningful and interdisciplinary science instruction boundless. For instance, we can consider the Flint, Michigan water crisis.

Flint, MI is a financially dispossessed city predominantly populated by Black-identifying people. The Flint Water crisis was spawned when the former state governor decided to cut costs and route water into Flint through the local Flint River, rather than importing treated water from neighboring Detroit, MI. This rerouting went through the city's original infrastructure, which was constructed using lead pipes. The lead leached into the city's water supply and ultimately wreaked havoc on the lives of the citizens, predominantly affecting the neurological development of preadolescent children.

The Flint water crisis is an exemplar case for deep exploration within and across grade levels and where we can address several state standards. Within the classroom, this social justice issue can be examined from science, social studies, literacy, and mathematics perspectives. Within the fifth-grade standards, students could engage content in the earth and physical sciences while also incorporating elements of engineering design to explore the idea of water as a resource. They could then leverage the engineering design standards to propose methods for wastewater and resource management, allowing for collaboration with community members to propose feasible solutions to this local problem. As students begin investigating water as a resource and their obligation for its responsible maintenance, social studies standards that encourage public discourse, decision-making, and civic participation could be incorporated to illustrate the interconnected nature of this issue. Finally, after making these connections, the class could then engage in a public service campaign informing their school and local community of the dangers associated with poor resource management. This would allow them to engage in civic participation beyond the health-related discourse around the effects of water contamination to broader discussions of the socioeconomic implications for a community subjected to such water exposure. Interdisciplinary science instruction like this allows students to be agents of their learning in ways that allow teachers, as Mrs. Ali so aptly noted, "to see and hear the brilliance of our students."

The students of Rockwood Elementary deserve culturally relevant science instruction because it is their right (Tate, 2001), not simply because it helps the school stay off the state's radar. Further, we must support teachers like Mr. Vue, Ms. Blue, and Mrs. Ali, whose roles are much deeper than simply a classroom teacher. Their efforts to see the humanity and brilliance of their students allow them to be critical of state science standards that are positioned as promoting equity for those of historically marginalized identities in STEM. Their expertise allows them to question the infrastructure required to support such an initiative within their school in ways that their White colleagues miss. It is only when voices such as these are heard and privileged that elementary science instruction and schools truly "turn around."

References

National Research Council. (2012). *A framework for K-12 science education: Practices, crosscutting concepts, and core ideas.* National Academies Press.

Tate, W. (2001). Science education as a civil right: Urban schools and opportunity-to-learn considerations. *Journal of Research in Science Teaching, 38*(9), 1015–1028. https://doi.org/10.1002/tea.1045

Terrance Burgess is an assistant professor of science education in the Department of Teacher Education at Michigan State University. His research focuses on how increasing equitable science learning opportunities for elementary youth of color influence their multiple identities within the urban school setting. Additional areas of his research explore how teachers' positionalities and their implementation of standards-driven curricula tend to youth's multiple identities.

Case: Fluffy, Puffy, and Made of What?

42

Chelsea M. Sexton and Jeremy Peacock

Abstract

Jeremy is a veteran teacher with a dozen years of experience in his classroom in rural Georgia. Understanding the importance of mentoring preservice and induction-stage teachers, he invited Matt to student teach in his fourth-grade classroom. Argumentation is Jeremy's favorite science practice, so he created a lesson with a focus on student discussions around explanations of a phenomenon and the evidence that supports the explanation.

Jeremy was anxiously excited about the upcoming science lesson. Today was the first day that his student teacher, Matt, would be taking over the fourth-grade classroom for science. Jeremy had been discussing the water cycle and had organized his lesson around standards related to this topic. Matt would focus on *Georgia Performance Standard S4E3e: Students will differentiate between the states of water and how they relate to the water cycle and weather. Investigate different forms of precipitation and sky conditions (rain, snow, sleet, hail, clouds, and fog).* Since Matt had been observing the class for the past few weeks, he knew about some of the different resources that Jeremy used to start his science lessons. One of Jeremy's favorites was a YouTube channel featuring elementary-aged students asking questions about phenomena.

C. M. Sexton (✉)
Department of Mathematics, Science, and Social Studies Education, Mary Frances Early College of Education, University of Georgia, Athens, GA, USA
e-mail: cmsexton@uga.edu

J. Peacock
Director of Secondary Education, Jackson County School System, Jefferson, GA, USA
e-mail: jpeacock@jcss.us

Matt had chosen to start the inquiry process in the same way, having the class watch a video in which a student-produced question is asked about a specific phenomenon. After students were introduced to the YouTube question, he planned for them to conduct an investigation and construct a possible explanation for the question. Matt had learned about this inquiry cycle in his preservice teacher courses and was ready to try something new. While Jeremy had seen the overall plans for the lesson, he had not seen all the specifics because Matt was tweaking it until the end, wanting it to be perfect. The students were coming back from music class in five minutes and Jeremy asked, "Hey Matt, have any last questions or want to run through anything before the students get back?"

Matt replied, "No thanks, Jeremy. I think I got this. Today's gonna be a great day!"

Jeremy said, "Alright, I am ready to see it."

The students excitedly filed back into the classroom, still singing songs from their previous class. Omar was even practicing his air drumming skills as Jeremy tried to shush them through the hallway. Matt was ready to begin and had the opening credits to the video "What makes up clouds?" on the smartboard. Students, recognizing the familiar YouTube channel, excitedly wiggled in their seats. Matt called the class to attention.

"All right, everybody. In continuing our learning about the water cycle, we're going to do an investigation today to learn about clouds. So, does anyone here have any questions about clouds?"

Amy piped up, "Yeah, are they made of pillows? In movies, I always see angels sleeping on clouds!"

Andy jumped in and said, "Oh, Amy, they're not made of pillows. They're full of rain."

Amy replied, "Well they don't look like water, they look more like fluffy pillows."

Matt interrupted their squabble and prompted, "Does anyone else have questions about clouds?"

Steve answered, "I wonder how high they are."

And Jay chimed in with, "Why do they move across the sky?"

Matt continued, "Those are all really great questions! We're going to jump on a little bit of what Amy originally asked and think about what clouds are made of and how they form."

"Yes," whispered Amy as Matt pressed play on the video. The child on the screen started his monologue: "Hi, my name is Jamar, and I'm from Montgomery, Alabama. My question is, 'What makes up clouds?'" The screen shifted away from Jamar and began showing a montage of different fluffy structures across the screen: cumulus clouds, nimbus clouds, stratus clouds, cumulonimbus clouds, cirrus clouds. While the students could not name them yet, they saw examples of many different cloud structures.

At this point, Matt paused the video and said, "All right, let's divide into think-pair-share groups with our elbow partners, and let's come up with some ideas and explanations for the question, 'what makes up clouds?' Remember, we will get to revise our explanations after the investigation we do."

Jeremy sat in the back of the room quietly taking notes and smiling as he listened to the prompt. His students were murmuring curiously, creating explanations about the composition of clouds. Matt was cycling throughout the room, whispering to elbow partners as they quietly conversed about what they thought made up clouds.

Benjamin shared his idea with Amy that clouds were cotton balls, and since she thought they were pillows, they both agreed that it had to be something solid like that. Andy was sticking to his idea that they were full of rain. His elbow partner, Omar, kept asking him why the water didn't fall if it was full of rain. He didn't understand how Andy thought that rain could just float in the sky. "It definitely doesn't float when we see it. Or when it falls on our heads."

As Matt finished his circle around the room, he asked elbow partners to write out their explanation in response to the question "What makes up clouds?" on their whiteboards. Matt really liked using whiteboards because they were easy to erase, which was helpful for students who wanted to change their claims after completing an investigation.

"All right. Would anyone like to share?" Matt excitedly asked from the front of the room. Amy and her partner excitedly raised their hands and shared. "Well, I thought clouds were like pillows, and Benjamin thinks they are like cotton balls. So, we think that different clouds might be different things, but they are all made out of solid things."

Matt replied, "All right, I think it's a pretty good explanation that different clouds may be made of different things, but they are all solid. Does anyone else want to share?"

Omar and Andy shared the idea that clouds are made of rain, but "There has to be strong wind that always pushes it back up into the sky."

"Or gravity doesn't work up there," interjects Jay. "Patty and I think they are sponges that hold water, and the atmosphere holds up the sponges."

"What if they are made of wool?" Steve asked apprehensively.

Some groups agreed with each other, and some groups had very different ideas about the composition of clouds and their formation. Jeremy laughed to himself in the back of the room as he listened to the creative ideas his students imagined. He wished Matt had also asked students to draw pictures of their cloud explanations because their ideas were so diverse.

Matt was pleased with the way the conversations were progressing and decided it was time for students to begin an investigation. He hurriedly pulled a cart around from behind the bookshelf, where it had been hidden from view. It had laundry tubs full of water, shaving cream, and blue food coloring. Everyone looked on with excitement, and the students started to wiggle in their seats again. Jeremy was somewhat perplexed at the array of materials on the cart, but he was also excited to see what Matt had in store.

Matt passed out materials to students so that each table received a laundry tub full of water and spoons. He then walked around the room and put two drops of food coloring in each tub and added two generous pumps of shaving cream, so that they floated on top of the water. "Alright class, what do we have in front of us?" Amy shouted, "Our clouds look like pillows!"

Matt quickly diverted her excitement into something a bit more scholarly (he hoped) and said, "No, no, what are our observations? Like, what does it look like? Do they look wispy? Do they look like any of the types of clouds that we saw before?"

Steve said, "I mean, yeah kinda? And the blue food coloring does look like the sky! So, they look like the big puffy clouds." Matt replied, "Awesome observation, Steve. Okay. Now, everyone, use your spoons to scoop up a little bit of the water and pour it over your shaving cream. What do you see?"

Omar said, "Oh hey, our cloud looks like it's separating and starting to break apart. Now, it kind of looks like those wispy ones." Amy agreed, "Yeah, the ones that look like cotton balls after you tear them apart."

Matt continued, "Think about the idea you originally wrote on your whiteboard—your explanation for the question, 'What makes up clouds?' What evidence did you see in the investigation we just conducted to support your explanation? On your whiteboard, write down two examples of the evidence which you think supports your explanation. Remember, we want to be able to revise our explanations based on the evidence we see."

For the most part, the students ignored Matt's directions and continued investigating properties of the water and shaving cream. Patty started mixing her simulation very quickly. "Look, a tornado just came through my clouds!"

They were excited to see all of the different ways they could create interactions between their materials. Jay said, "Wow, it's so cool that we get to play with actual clouds like this!"

Matt heard Jay and gave him a bit of a puzzled look—this was a simulation of clouds, not an actual cloud. But, given Jay's excitement, Matt ignored the comment and continued onto the pair at the next table, not wanting to quelch his enthusiasm.

Jeremy heard Jay's outburst too and glanced up to see if or how Matt would redirect him. Seeing Matt's confused face, Jeremy put his grading to the side and gave his full attention to the classroom.

After making a few rounds, Matt asked group to share what evidence they have collected from the investigation to support their initial explanations.

Omar suggested, "I observed that clouds are very fluffy."

Steve continued, "I can break apart clouds with water."

Patty exclaimed, "The clouds dissolve, and the whole sky turns light blue! I guess that what happens that's what happens after a rainstorm or a tornado."

Andy sullenly said, "I thought clouds were made of rain, but I guess they are actually made out of shaving cream."

Matt was now visibly confused and stammered to get through his next thoughts. "Wait... wh- wha- why? Bu- but the clouds...?"

Jeremy wanted to step in, but he waited to see how Matt might recover from accidentally leading his students to believe that clouds were made of floating shaving cream in the sky.

For Reflection and Discussion

1. Why might have students gotten the idea that clouds were made of shaving cream? Which parts of Matt's lesson were helpful in learning about clouds as one piece of the water cycle?
2. Was Matt using the inquiry cycle in a productive way? Why or why not?
3. How could Matt have used the standards to design or choose an investigation that might help students develop a more sophisticated understanding of the composition of clouds in relation to the water cycle?
4. At this point in the lesson, what would you do if you were Matt?
5. In what ways did Matt's lesson address or fail to address this standard?
6. The water cycle is a complex process unlike the simplistic diagrams often found in textbooks to depict it. Similarly, an understanding of different forms of precipitation and their role in the water cycle can be challenging for students. Do you think that the standard for this lesson, as written, is appropriate for fourth-grade students? Why or why not?

Chelsea M. Sexton is a former high school environmental science and research methods teacher with a background in marine ecology. She currently teaches elementary preservice teachers while working on her doctoral degree in science education at the University of Georgia. Her research interests center on education for sustainability, justice-centered and case-based pedagogies, and preservice science teacher preparation.

Jeremy Peacock is the K-12 Director of Secondary Education for the Jackson County School System, where he supports innovative teaching and learning that engages students in figuring out the world around them and solving real-world problems. A past president of the Georgia Science Teachers Association, Georgia high school science teacher of the year, and having received his Ed.D. at the University of Georgia, Jeremy also consults and provides workshops around the state and beyond focusing on phenomenal 3-D learning and assessment.

Commentary: Incorporating Models in Science Instruction

43

Lynn A. Bryan

Abstract

This is a commentary to the case narrative, *"Fluffy, Puffy, and Made of What?"* written by Chelsea M. Sexton and Jeremy Peacock.

Matt's experience in *Fluffy, Puffy, and Made of What?* illustrates a teaching situation that is not uncommon, especially for new science teachers. What Matt experienced in implementing his cloud activity happens in one way or another to just about every science teacher—you choose an activity that you believe will interest and engage students in learning about core scientific ideas or phenomena. However well-intentioned, the activity does not go as planned, and students seem to construct ideas that are scientifically non-normative and so perplexing that you do not know how to respond!

From the case description, we know that Matt intended to use an inquiry-based approach to teaching in which students were encouraged to ask questions, gather evidence, and propose scientific explanations based on their observations of shaving cream "clouds." A notable feature of Matt's activity for his instruction is *modeling*—the activity was intended to model clouds. Developing and using models is one of the eight core scientific and engineering practices identified in the *Next Generation Science Standards* (NGSS Lead States, 2013). In fact, modeling is a core scientific practice emphasized in every science education reform initiative for more than three decades and for good reason. Science is essentially the process of articulating, testing, evaluating, and refining/revising models of the world. The ability to construct and reason with models is widely regarded as the core of "doing" science (Giere, 1990, 2004) and, therefore, should take center

L. A. Bryan (✉)
Center for Advancing the Teaching and Learning of STEM, Purdue University, West Lafayette, IN, USA
e-mail: labryan@purdue.edu

stage in K-12 science instruction. Nonetheless, modeling—specifically *developing and using models*—has "too often been underemphasized in the context of science education" (National Research Council, 2012, p. 44).

As Matt experienced in this case, the models themselves do not do the teaching. What students observe and interpret from interacting with models does not always align with what the teacher intends for students to learn. Matt's case illustrates what happens when a model is not appropriate for achieving the learning goal(s) of a lesson. While we do not have a copy of the activity or the lesson plan that Matt used, the case suggests that he aimed for children to accurately describe what makes up clouds. So, how do teachers choose a model or modeling activity that is appropriate for helping students learn what is intended to be learned?

First, teachers need to be thoughtful and purposeful in choosing the models for science instruction. With access to an unlimited number of activities and ideas for teaching available on the internet, the judicious teacher will look beyond the "fun" factor of a model or a modeling activity. Evaluating a model to determine its appropriateness for instruction entails asking questions such as: What is the purpose of the model in the activity—e.g., to learn about what things are made of and how the different parts that make up something relate to each other (structure)? Or how something works, what it can be used for, or how it behaves (function)? Or how a phenomenon happens and to predict what will happen in the future (mechanism)? What aspects of the phenomenon does the model represent? How will the features of the model and interactions with the model help students build the understandings intended to achieve the learning goals? What are some possible ideas that students may generate from the model—both intended and unintended ideas?

Evaluating models and modeling activities for instruction also requires teachers to consider how they will guide students with feedback, prompts, and questions as students interact with and interpret models. The role of a teacher's discourse in scaffolding students' learning experiences with models is crucial but often underestimated. Creating opportunities for students to verbalize their thinking with peers and the teacher is critical to helping them make sense of scientific ideas. As we read in this case, students were left to explore and interact with the model without much direction or scaffolding. It appears that Matt took a "discovery approach" to teaching, which assumes that students will learn by engaging and manipulating experimental materials as they "discover" scientific principles on their own. Matt observed, asked open-ended questions, and encouraged exploration, but he did not engage students in discourse in a way that purposefully scaffolded concept development or challenged students' claims. Ideally, Matt would have *planned* for engaging students in a discourse-rich learning experience, for example, thinking through possible ideas that students may generate from the activity so that he could prepare for what he might say to direct students' attention to features of an appropriate model or what questions he might ask to scaffold students' use of evidence to support their emerging ideas. Engaging students in productive discourse as they make sense of science aligns with an inquiry-based approach to learning science.

In the case of *Fluffy, Puffy, and Made of What?*, it is hard to imagine how the activity with shaving cream "clouds" in water would lead to students understanding and articulating a basic description that clouds are made of an accumulation of water droplets or ice crystals and an explanation of how clouds form. If a teacher cannot determine how students will make sense of the science as they progress through a modeling activity, then the model or modeling activity likely is not suitable for students to achieve the learning goal(s) set for them.

Another consideration in choosing appropriate models and modeling activities is derived from the academic standards that the teacher aims to address. As with many academic standards, the one in this case, *Georgia Performance Standard S4E3e*, is complex and includes several related core concepts. Teachers must "unpack" a standard to determine the progression of ideas that students will need to learn to understand the core concepts. The process of unpacking a standard and mapping the progression of ideas will help inform the choice of models for instruction. For example, where in the progress of ideas does "what makes up clouds?" fit, and what science concepts will students need to understand in order to build an understanding of cloud formation? As standard S4E3e suggests, learning about "what makes up clouds?" requires a basic knowledge of the particle nature of matter (i.e., water can exist as a solid [ice], a liquid, or a gas [vapor]) and phase changes (i.e., freezing, melting, condensing, and evaporating). Considering these underlying concepts for understanding what makes up a cloud, it seems prudent to design a lesson, particularly in the context of the water cycle, in which the students engage with a cloud model that incorporates the evaporation and condensation of water.

Despite the outcome of the shaving cream cloud activity, Matt seems to be on the right track for providing meaningful science learning experiences for his students. He has a commitment to teaching inquiry-based science; he values students' generation of knowledge and opportunities for students to create and articulate explanations; he is learning to use models and modeling in his science instruction. Matt is fortunate to be working with a veteran teacher who shares the same values for science teaching and will be able to provide feedback and guidance to help Matt reflect on planning and enacting meaningful, inquiry-based science instruction.

References

Giere, R. N. (1990). *Explaining science. A cognitive approach.* University of Chicago Press.

Giere, R. N. (2004). How models are used to represent reality. *Philosophy of Science, 71*, 742–752.

National Research Council. (2012). *A framework for K-12 science education: Practices, cross-cutting concepts, and core ideas.* The National Academies Press. https://doi.org/10.17226/13165

NGSS Lead States. (2013). *Next generation science standards: For states, by states.* The National Academies Press.

Lynn A. Bryan is the inaugural Director for the Center for Advancing the Teaching and Learning of STEM (CATALYST) at Purdue University. She is professor of science education and holds a joint appointment in the Department of Curriculum and Instruction and Department of Physics and Astronomy. Her research focuses on science teacher education, particularly teachers' development and enhancement of knowledge and skills for teaching science through the integration of STEM disciplines and modeling-based inquiry approaches.

Part IV

Fostering Science and Engineering Practices

Lynn A. Bryan

Children are natural-born designers. They build towers with Legos®; construct forts with couch cushions and blankets; design costumes with paper bags, scissors, and markers; play school, house, or grocery store with elaborate rituals, routines, and rules. As children participate in these kinds of activities, they explore ideas, solve problems, and make sense of the world around them—much like scientists and engineers. At the core of "doing" science and engineering are *practices*. The term *practices* describes the actions and behaviors in which scientists engage as they investigate, evaluate, build, extend, and apply models of the natural world to new problems. Practice refers to the actions and behaviors in which engineers are involved as they use science, technology, and mathematics to design, build, and improve models and systems to solve problems and meet needs. Thus, learning science concepts such as plants and animals, weather and climate, or states of matter involves more than understanding concepts alone. Learning science also necessitates engaging children in what scientists and engineers do—the practices of science and engineering.

The most recent framework for science education, *A Framework for K-12 Science Education: Practices, Crosscutting Concepts, and Core Ideas* (National Research Council [NRC], 2012), emphasizes eight essential practices for students of all ages to learn. These eight practices relate to both science and engineering because the work of scientists and engineers involves similar and complementary actions and behaviors:

- Asking questions (for science) and defining problems (for engineering)
- Developing and using models
- Planning and carrying out investigations
- Analyzing and interpreting data

L. A. Bryan
Center for Advancing the Teaching and Learning of STEM, Purdue University, West Lafayette, IN, USA
e-mail: labryan@purdue.edu

- Using mathematics and computational thinking
- Constructing explanations (for science) and designing solutions (for engineering)
- Engaging in argument from evidence
- Obtaining, evaluating, and communicating information (NRC, 2012, p.3)

These eight practices do not suggest a specific linear sequence—they are not a set of "steps." However, neither are they independent of one another. Science and engineering practices are interconnected; they work in tandem. Scientists investigate and discover new things about the natural world and how it works. Engineers translate those discoveries into solutions to practical problems.

Few would argue that engaging children from a young age in *doing* science and engineering has tremendous benefits. Children who are allowed to ask questions, investigate, build models, and explain the world around them become active in their own learning. They begin to develop foundational knowledge and practices of not only science and engineering but also language, literacy, mathematics, and other subjects, as we see in the case, *Demystifying Magic with STEM*. In this case, first-grade teacher Michelle integrated science and engineering practices into a literacy unit about fairytales to demystify the magic of spinning straw into gold in Rumpelstiltskin. As this case exemplifies, children learn to articulate, clarify, and negotiate ideas as they work with each other to build scientifically normative explanations and design, test, and optimize a solution. In doing so, they may acquire new vocabulary, apply mathematics, and learn about the historical or geographical context of a question they are investigating or a problem they are solving. Children also will realize that there is not always one solution or outcome. Moreover, they may begin to learn how to handle failure, revise, and optimize when their investigation or design does not work as expected.

Teaching children to develop and use science and engineering practices *in the context of what they are learning in the classroom* is crucial. As opposed to stand-alone lessons in which children participate in activities to illustrate an uncontextualized practice, opportunities for children to engage in science and engineering in compelling, content-rich, and relevant contexts will pique children's interest, wonder, and imagination. For example, *But Miss, There are Six Oceans, Not Five?* presents the case of Victoria, who utilizes a "Wonder Wall" to inspire her instruction of science content. The Wonder Wall is where her children place all of their wonders and questions about the science topics they are studying. At the beginning of an oceans unit, Victoria's students observed and interpreted models of the Earth (maps and globes) as they identified the world's oceans. When some confusion arose about the Pacific Ocean, Victoria considered ways to help students make sense of how two different models represent the Pacific Ocean. She pivoted to her lesson to address this confusion by having children design their own globes and honored their subsequent wonders by letting them explore the reverse—making maps out of the deflated globes. The case, *The Bible and the Beast*, provides another example of children developing and using science and engineering practices in the context of the science content they are learning in the classroom.

Darcy's third-grade students were conducting a series of investigations in which they were learning how to construct scientific explanations in the process of understanding geologic time and prehistoric organisms. Throughout the inquiry, children asked questions, analyzed models and artifacts, constructed explanations based on scientific evidence, and evaluated and communicated information. However, Darcy unexpectedly found herself contemplating how to respond to the evidence that one student used to support his claim.

By engaging in the practices of science and engineering in a meaningful context of science content, children become active participants in making sense of the content while learning to think critically, communicate, and collaborate as part of a community or team. The role of the teacher becomes one of facilitator and moderator, scaffolding questions to help children realize patterns and connections; asking open-ended questions that prompt their participation and elicit ideas; creating a learning environment in which all children are afforded the chance to contribute to co-construct understanding. As the case, *Is Only Sticky Important? Sensemaking through Equitable Discussion in a First-Grade Engineering Lesson*, illustrates, this approach to teaching can be "messy," but the payoff is worth the challenge. In the end, children will develop a rich understanding of how the work of scientists and engineers contributes to the betterment of society. Children may begin to see themselves as innovators and become the ones who design solutions for securing cyberspace, eliminating world hunger, providing access to clean water, addressing climate change or climate change, or designing medical interventions.

Reference

National Research Council. (2012). A framework for K-12 science education: Practices, crosscutting concepts, and core ideas. The National Academies Press. https://doi.org/10.17226/13165.

Case: Demystifying Magic with STEM

44

Thomas Meagher, Michelle Simon, and Gillian Roehrig

Abstract

Fairy tales, folk tales, or fables are integral components of the elementary literacy curriculum and the events in these stories often rely on the use of "magic" to overcome obstacles or solve problems. But what is magic? By definition, magic is: "he power to influence events through mysterious and/or supernatural forces." Children easily accept that magic is the cause of many events and phenomena they observe and experience in their lives. Magic is also an easy way to explain science and engineering concepts that we as adults may not fully understand. However, as educators, we have a responsibility to improve our student's knowledge and understanding of the world around them and integrating science and engineering standards with literacy standards can be one of the most effective ways to teach about real life and address common misconceptions. This case explores how Michelle, a first-grade teacher at a STEM school in rural Minnesota, uses fairy tales that the children are required to read as a jumping-off point for teaching essential science and engineering standards, as required by the state-mandated curriculum. Magic is demystified as students learn to use scientific and engineering concepts and practices to solve real-world

T. Meagher (✉)
Owatonna Public Schools, Owatonna, MN, USA
e-mail: tmeagher@isd761.org; meagh014@umn.edu

Department of Curriculum and Instruction, University of Minnesota, Minneapolis, MN 55455, USA

M. Simon · G. Roehrig
McKinley Elementary STEAM School, Owatonna, MN 55060, USA
e-mail: msimon@isd761.org

G. Roehrig
e-mail: roehr013@umn.edu

Department of Curriculum and Instruction, University of Minnesota, St. Paul, MN, USA problems that arise in the stories. It is through Michelle's creative and engaging ways of using both the constraints of families adverse to their children reading stories with magic and the required learning targets that we can see firsthand how effective STEM integration can work in an elementary classroom as she teaches her students to become readers, engineers, and scientists.

Michelle has spent most of her teaching career in the lower elementary grades and a short time teaching sixth grade. When McKinley Elementary transitioned to a STEM school, Michelle transitioned from teaching traditional sixth grade into teaching first-grade STEM, in an integrated fashion. Michelle requested to switch to the primary grades because she wanted to ensure younger students had meaningful learning opportunities in science and mathematics, rather than waiting until the upper grades when children must take the state-mandated science test. She also loved first-graders' creativity, wonder, and enthusiasm for learning that propelled them into the upper grades. Michelle believed that all students, no matter their age, can engage in scientific and engineering practices; Michelle and her first-grade team embraced the idea that STEM is not what to teach, it is *how* to teach.

When Michelle began planning for first grade, she was surprised to learn that fairytales were an integral part of the language arts curriculum. Fairytale storylines can be cruel or gruesome, but often "sugar-coated" in how they are presented to students. For example, in the story of Rumpelstiltskin, there are several cruel actions made by each character in the story, such as when the king kidnaps the miller's daughter and locks her in the castle tower. This old European story can be effective to teach moral and ethical lessons in human behavior; however, it also poses severe misconceptions about the properties of materials and that "magic" is how to change one thing into something else, for example, Rumpelstiltskin spinning straw into gold.

Michelle and her team saw this story as an opportunity to *STEMify* (i.e., integrate science, technology, engineering, and mathematics into) the language arts curriculum and have students use science and engineering practices to demystify the magic of turning straw into gold. She considered that it would help students to understand the story through the lens of science and that materials have many properties that make them unique, not just color. She thought, "I can have students learn how to sort materials and identify their properties and realize that color isn't always the most important property of matter." However, she was not prepared for the negative reactions of several of her students' families when they learned she was going to teach the fairytale Rumpelstiltskin.

McKinley Elementary STEM School is situated in rural Minnesota; however, the school's student population has shifted over the past two decades due to immigration from east African and Latin American countries since the year 2000. This shift in the composition of the student body has left the district's language arts curriculum out of date with the cultural background of many of its students.

Rumpelstiltskin is a Brothers Grimm fairytale from Western Europe and is often included in the reading curriculum in Minnesota school districts where community populations are historically descended from German, Norwegian, and Swedish immigrants. Therefore, the social and moral references made within the story of Rumpelstiltskin do not match well with the cultural backgrounds of current students whose families immigrated from Somalia, Sudan, Mexico, or Honduras.

Michelle found that families would not want their students to read stories that had instances of magic or characters who engaged in practicing magic. For many families that hold conservative religious and cultural beliefs, "magic" of any kind is considered taboo for their children to learn about in a public school. Michelle felt caught between needing to teach the required curriculum and meeting the needs of families who objected to the fairytales, plus, wanting her students to see how science and engineering can help people understand the reality behind what at first appears to be magic.

The first-grade team approached the dilemma of parent objection by planning how science and engineering could provide students with a scientific understanding of fairy tales that have instances of "magic." The team designed lessons where students could explore the properties of materials such as straw, cotton, nylon, and metals that were similar in color to gold. Students would also investigate how plant and animal fibers could be spun into yarn using hand spinners they design and create in coordination with the art teacher. Michelle wanted her students to observe how materials have unique properties and that a substance does not change into a different substance when you change its shape. A "physical change" may change a material's appearance but not its fundamental properties.

For student teams to observe the phenomena more carefully, Michelle planned to have them engineer methods of spinning different kinds of fibers together to create yarn to weave into a small cloth. During this unit, the students would record observations about the properties of the fibers they used and whether any of these materials changed into different substances or just took on new shapes of the original fibers. The STEM lessons would follow each section of the reading assignment of the fairytale Rumpelstiltskin. The class discussed how their experiences with science and engineering compared to what the story called the "magic" of turning straw into gold.

The first-grade team felt positive that their newly **STEMified** fairytale unit would reduce the negative reactions of parents who objected to fairytales or to exposing their children to "magic." Michelle was excited to get started, her team bought different color samples of plant and animal fibers, coordinated with the art teacher to have students design and make hand spinners with a 3D printer, and sent emails to the families of her students preparing them for the new STEM fairytale unit.

Teamwork and collaboration among students are essential for effective STEM lessons. Each unit needs to include crucial social and emotional skills components to engage students in science and engineering practices. Michelle decided she needed to build a culture in her classroom that would help students apply the skills of science and engineering to explore fairytales and demystify the magic in

Fig. 44.1 Michelle uses questioning strategies to help her students explore the properties of materials

the story. By building a classroom culture where science and engineering practices were valued and understood, students could explore fairytales and learn that magic is not something to fear but processes we can figure out together. This is the essence of the science investigation or engineering design process: identify a problem, design a solution, try it out, assess whether it is effective, and then improve on the design. When Michelle approaches instruction this way, it means embedding trust as an integral component of learning, trust in the teacher's directions, trust in each other, and most importantly, trust in themselves as they explore new ideas (Fig. 44.1). Michelle explained her ideas:

> I begin the lessons by setting expectations of kindness, access to materials, and accountability. We [students and teachers] specifically talk about the words we can use if we get frustrated, how we can take a break, and how we can be a team player. We talk about how the materials and resources we have may not be ideal, they will make us frustrated but we need to solve problems as we use them together and use them in different ways. We may use the resources we have in different ways than anyone has ever thought of before! Sometimes we even touch on things such as biomimicry in the process and making connections to solve problems. This foundational work is probably the most important piece.

> I try to be a resource more than a problem-solver, I want them to go to their neighbor to learn from them. We also take a lot of breaks. During our breaks we celebrate "breakthroughs" and review strategies, discuss needs, and words for solving problems.

> I often have them work with "week-long" partners. I want them to figure out and internalize strategies for getting along with another person. There are often a few kids in the class that have trouble working with others, unless there is a severe need they have to work with a partner, for at least part of the project.

On the morning of the first STEM lesson, Michelle set up stations for students to explore all the different types of plant and animal fibers that they could use to spin the yarn. Michelle began the day by reading the first section of the fairytale Rumpelstiltskin, and the students were fascinated with the idea that the little man could weave straw into gold. One boy said, "We've got a barn at my grandma's house that's full of straw, we're gonna be rich!" Michelle thought this was a perfect transition to observing different kinds of natural fibers and comparing plant and

Fig. 44.2 Team of McKinley first-grade students investigating how materials can be combined together to form new shapes and designs

animal fibers from straw, grasses, cotton, and wool. The students took out their STEM journals and began to draw and describe the materials that were set out at each fiber material station, which included straw and gold-colored metal wire (Fig. 44.2).

Each day the students became more interested in what would happen to the miller's daughter and if she would figure out Rumpelstiltskin's name and what the king would do when he found out she really could not weave straw into gold. Michelle observed that students grew increasingly doubtful of the story as they experimented and engineered ways of spinning fibers and making the yarn that they were weaving into cloth. The students began to realize that just changing the shape of the fibers didn't change the properties of the fibers into different kinds of materials. They began to see that the magic in this story was not real and that gold is not made out of yellow straw, it is just similar in color.

After several days of exploring and spinning with the different kinds of fibers to make yarn, followed by weaving the yarn into cloth, Michelle felt that the unit was working and students were recognizing that magic can be investigated and explained with STEM. The boy who thought he would become rich came over to Michelle and said, "I know I'm not gonna be rich now and that king was just dumb." Michelle felt sad and disillusioned by this comment and thought, "Now I've squashed their imagination. How can I help them keep their creativity and imagination but still see the wonder of the world through STEM? Aren't all the amazing technologies that seem like magic today really the results of people following their dreams and imagination? How can I help children think like scientists but still have the imagination of a child?"

Michelle approached her team with these questions to redesign the new STEM fairytale unit: "How can we teach students to see a world where they can explore, investigate, invent new things or solve problems, but still keep their dreams and imagination?" Michelle's team began redesigning the STEM fairytale unit determined to keep the science and engineering aspects of the unit intact but solve the new issue of where students can practice observation, experimentation, and explanation and while engaging their creativity and imagination.

* * *

Magic is a common theme found across all cultures in the world when characters in a story are unable to explain how something happens, such as turning straw into gold in Rumpelstiltskin, beans growing into the clouds in Jack and the Beanstalk, or the cunning trickster coyote who is found in folktales around the world. When we do not fully understand how something happens, it is easy to believe in the power of magic to explain these outcomes. Believing in magic is especially strong in children.

In elementary schools, science is often taught with the perspective of building students' scientific literacy skills, which translates into mostly "reading" about science and rarely "doing" science. Teaching through the lens of STEM where scientific thinking, technology use and creation, engineering design process and mathematical analysis of data and numerical patterns relies on students engaging in science and engineering practices. Students of any age and background can develop scientific mindsets when they participate in asking questions, making observations, collecting and analyzing data, and forming conclusions about what they experienced. By utilizing an integrated approach, STEM teaching and learning is a pedagogy and way of learning more than content to be memorized by students. Teachers who use STEM as a way of teaching and learning do not think "This is one more thing I have to fit into an already crowded day"; they view STEM as HOW to teach the different content they need to fit into a school day.

So how can elementary teachers engage in STEM teaching and learning when they are under pressure to have students perform well on literacy and math assessments? The case dilemma about Michelle, with her first-grade team, approached teaching by having students explore an authentic and motivating challenge. When students are engaged in a fascinating story about magic, teachers have the opportunity to address common misconceptions and introduce how the properties of a substance remain intact when that substance goes through a physical change. In other words, a piece of straw remains a piece of straw regardless of how the fibers are chopped, broken, spun, or sliced together. This experience helps students understand more accurately how materials have specific properties and that "magic" cannot simply change these properties by changing something's shape, or physical structure.

Without the experience of handling materials, recording their observations, and making sense of the results of their experiments, young students could easily accept the magic in a story to be "real." By engaging students in authentic inquiry or engineering challenges, they can begin to master science and engineering practices and observe how this mindset and way of looking at the world can help them "demystify the magic" of what we do not fully understand.

However, teaching students to see everything through a lens of science and engineering can have pitfalls, for example, dampening the imagination and joy that children bring to their lives through what they see, think, and feel. For many elementary teachers, having close relationships with their students and engaging them in learning is most important to bring out their students' potential and find their gifts. Thus, this balance of providing deep learning, mastering key literacy and math skills, while developing their creativity, often makes teachers feel stretched and out of their comfort zone to teach far beyond reading and math. This is when integrating STEM teaching and learning with language arts, art, social studies, or math lessons can help students have real-world experiences from which they can draw meaningful connections and form deeper ideas and concepts.

The question remains: How can we as educators balance the joy a child experiences from learning new ideas while helping them build a creative and imaginative way of understanding the world they live in? Creativity and imagination are essential for someone to succeed at problem-solving and seeing the world with endless possibilities. To accept things that we do not understand as "magic" may sustain our minds for a short time; however, when we develop ways of asking questions and finding answers, we begin to see that the universe is a series of endless questions, just waiting to be answered.

For Reflection and Discussion

1. What specific connections between STEM practices and concepts and literacy skills does Michelle create while teaching students about fairy tales?
2. How can scientific concepts or engineering principles help students explain real-world phenomena without the mysterious forces of magic?
3. Fairy tales, folk tales, or fables often try to relate lessons in relationships among characters that revolve around specific events. How could characters in a fairy tale benefit from having accurate knowledge of science to overcome problems faced in the story?
4. Discuss instances when you learned about new scientific information that seems like "magic" to you sometime in your life? For example, figuring out how magnets attract and repel objects, or how chemical reactions work, or how genes express as dominant or recessive traits in an individual.
5. Fairytales and folktales from around the world often have a "trickster" who uses magic to fool the other main characters of the story. What are examples of other stories that would engage students to figure out how the magic or the tricks can be explained with STEM? Be sure to discuss both the plot of the story and how science and engineering practices could demystify the "magic" of the trickster.

Thomas Meagher is District STEM Education Coordinator for Owatonna Public Schools providing professional development, instructional coaching, and curricular support for over 200 teachers grades K-8. He is Frequent Presenter at national and international conferences on Science and

STEM Education sharing new ideas in designing and team-teaching engaging, STEM integrated lessons for teachers working in grades K-8.

Michelle Simon is First-grade Teacher at McKinley Science Technology Engineering and Math school in Owatonna, Minnesota. She earned a STEM Teaching Certificate from St. Catherine University and a master's degree in the science of teaching from Walden University. She was Fellow with the Transatlantic Outreach Program, supported by the Goethe-Institute in Germany. In 2020, she was honored as Minnesota Elementary Social Studies Teacher of the year.

Gillian Roehrig is Professor of STEM Education at the University of Minnesota. Her research focuses on induction support for beginning secondary science teachers and professional development and classroom implementation of integrated STEM in K-12 settings. She currently serves as President of NARST: A global organization for improving science teaching and learning through research.

Commentary: Early Science and ELA: A Mutual Enrichment

45

Joseph W. Spurlock

Abstract

This is a commentary to the case narrative, *"Demystifying Magic with STEM"* written by Thomas Meagher, Michelle Simon, and Gillian Roehrig.

"STEM is not what to teach, it's *how* to teach" (Meagher et al., in this book, emphasis in original). This sentence represents what I have tried to communicate to my elementary science methods students at The Ohio State University. While my work is specifically about early science, the perspective of science as an active process of observation and exploration applies broadly to STEM fields. Science is more than test tubes, glass-bound insects frozen in time, and solar system models with hotly debated planet status. Science is a mindset of investigation and critical questioning that equips learners to make sense of their natural and cultural worlds, and STEM can create a space within this where children learn to create, innovate, and empower.

Michelle's use of STEM as an entry point into literature mirrors my own work teaching undergraduates in early science methods, although we experience this interdisciplinary connection in the reverse order. We explore how to use picture books to introduce concepts and engage students as we prepare them to investigate a topic. We find ways to use young people's writing as a method for them to explain their process to their peers, drawing on Miller and Calfee's assertion that "writing makes thinking visible" (2004). I encourage my students, as they work with their placement class, to make a wide variety of fictional and informational texts available to elaborate concepts collaboratively and to study students' various literacies, which can allow them to evaluate what and how the children are learning

J. W. Spurlock (✉)
Department of Teaching and Learning, The Ohio State University, Columbus, OH, USA
e-mail: spurlock.107@osu.edu

(for those familiar, you will notice my very intentional interpretation of Bybee et al.'s (2006) 5E framework).

The philosophy of our methods class is that bridging literature and literacy into science is a necessity in contemporary early childhood classrooms, in large part because it may be the only way to make science happen in your class consistently. Language arts is an essential component in K-12 education, often afforded the majority of available daily instructional time. When we add on mathematics, lunch, recess, and transitions, time gets terribly short for anything else. It follows, then, that science is pushed to the margins, often delivered in pre-planned, teacher-led activities (often referred to as "experiments") that happen on rare days where all other business has been completed. Even when they are very engaging, these activities tend to look more like carefully prescribed, step-based lessons, in which there is little actual exploration or experimentation. My students and I develop a richer approach to science by co-constructing, through readings and discussion, a definition of what it means to "do science." For us, science is really about investigation and critical thinking, and we discuss it as a way of understanding not only the natural world, but also cultural and social matters. When we think of science as a world view that reaches beyond our classroom, language and literacy become scientific tools in and of themselves. The very act of reading comprehension can be argued as a scientific process where readers collect and represent information, make inferences about its implications, and synthesize new conclusions based on what they have learned.

Michelle's pedagogy reflects this view of "doing science" at multiple levels. The activities she designs create a venue for exploration where students are encouraged to make their own interpretations of their data. At the interdisciplinary level, she creates an experience that translates fairy tales and magic into investigative experiences for her students. At the broader discursive level, when the inherent crossover of classroom and culture leads to religiously conservative parents pushing back against magical fairy tales, she purposefully and iteratively problem-solves to simultaneously quell family concerns, fulfill her district-mandated duties, and create a rich experience for her students.

While Michelle's response is uniquely innovative, the cultural clash between classroom and home cultures is very familiar. As a science methods teacher, I have facilitated a number of challenging discussions that result from this interactional tension. Some students struggle to negotiate their religious beliefs with their responsibility to teach evolution, while others work to overcome their perception of science as mechanical and exclusionary. Other small, yet highly vocal, groups of students take great exception to our department's contributions to contemporary efforts toward social justice and equity. Efforts to press students to look honestly at the history of science in Europe and America have, at times, been roundly opposed and filed under the scapegoat banner of the much-maligned and rarely understood critical race theory (CRT) (Ladson-Billings, 2021). As you are reading this, it may seem odd to implicate CRT in a commentary about STEM education, but I can hardly find a more appropriate arena to discuss inequities than science, given the long history of European weaponization of "the scientific process" for the benefit of colonization and imperialism. In my class, we discuss things like the history of eugenics and phrenology, and I encourage these future educators to speak frankly and honestly with young children about exactly why we do not see more women and people of color in our science materials.

These discussions do not happen without language and literacy. When teachers do the work to connect science practices with culture and lived experience, they make it a real and enduring topic for their students' discourse, creating yet another avenue for critical thinking and reflection. Just as STEM is about *how* we teach more than *what* we teach, allowing students to truthfully face issues of science and cultural values is not about teaching them what they should feel about them. In the same way that Michelle allows students to determine for themselves the line between magic and reality, STEM is about allowing them to decide, iteratively and meaningfully, how they feel and where those feelings will take them.

References

Bybee, R., Taylor, J. A., Gardner, A., Van Scotter, P., Carlson, J., Westbrook, A., & Landes, N. (2006). *The BSCS 5E instructional model: Origins and effectiveness.* BSCS.

Calfee, R. C., & Miller, R. G. (2004). Making thinking visible: A method to encourage science writing in upper elementary grades. *National Science Teachers Association. University of California, Riverside, 42*(3), 20–25.

Ladson-Billings, G. (2021). Critical race theory—What it is not! In *Handbook of critical race theory in education* (pp. 32–43). Routledge.

Joseph W. Spurlock is a Ph.D. student at The Ohio State University in Columbus, Ohio. He received his early childhood education license from Capital University. His research is rooted in the connections between early childhood literacies, literature, and science learning, specifically how they co-define one another in classrooms and how matters of social justice and equity can manifest and be supported by these pedagogies.

Case: But Miss, There are Six Oceans, Not Five?

46

Andrew Gilbert and Valery Erickson

Abstract

This case highlights an effort to support science content learning in a second-grade classroom by fostering children's sense of wonder. Victoria Moon was in her second year of teaching and demonstrated considerable potential through both her teacher preparation program and her first year of teaching. She saw engaging students with wonder as a viable and compelling approach to science practice. In particular, she framed most science content approaches through student-derived questions surrounding their wonders via a classroom "Wonder Wall." This sometimes led to challenges with regard to planning, assessment, and staying on track with grade-level teammates. This case depicts a new teacher's process as she incorporated wonder and worked to connect it to science standards.

46.1 A New Teacher's Goal to Engage Children's Sense of Wonder

Victoria is working at Elm Street Elementary School (ESES) in a metropolitan area just outside Washington, DC. ESES has just over 700 students representing a broad range of ethnicities: 5% Asian, 15% Black, 64% Latino, 12% White, and 4% multiple ethnicities. In addition, just over 77% of the school is eligible for the free and reduced lunch program. There are a number of English language learners

A. Gilbert (✉)
College of Education and Human Development, George Mason University, Fairfax, VA, USA
e-mail: agilbe14@gmu.edu

V. Erickson
Fairfax County Public Schools, Fairfax, VA, USA

© The Author(s), under exclusive license to Springer Nature Switzerland AG 2023
S. Jeong et al. (eds.), *Navigating Elementary Science Teaching and Learning*,
Springer Texts in Education, https://doi.org/10.1007/978-3-031-33418-4_46

in this context, and Victoria also happens to speak Spanish, which serves her well in communicating with many of her Latinx students. There is a palpable sense of joy within the school community of ESES across both the committed staff and enthusiastic children who are eager learners.

Victoria's second-grade class is no exception. She leads her class with a gentleness that permeates all of her actions but also frames her high expectations for the children in her class. However, Victoria is actively working to solve a problem that faces many new teachers... how to construct her own vision of inquiry in her classroom. She has a burgeoning vision of inquiry as built on children's thinking and is working to engage their sense of wonder in the world around them. In this respect, she makes clear efforts to link children's wonderings to the standards she is expected to teach as part of the district goals. She is still making sense of when to engage in inquiry approaches, which type of inquiry approach to use (open vs. guided inquiry), and how she could build a sense of wonder in her students with regard to science. She knew she wanted to build a sense of wonder as a means to engage her young learners in science content and practices that were both connected to their lives and interests while also meeting the institutional goals of the district and school. But she wondered how she could make that desire a reality in the "rough and tumble" of practice, which is no small task for a new teacher.

However, Victoria is excited to work at ESES because of her clear belief in the abilities of children and the fact that her desire to utilize inquiry-based approaches resonated with the overall culture at ESES. The school's interest in inquiry teaching allows each grade-level team some flexibility regarding their approaches, assessments, and timing of lessons throughout the year. Victoria also feels she has a greater degree of flexibility since second grade is not tested in terms of content like the higher grades. One tool that she uses throughout the year is a "Wonder Wall" where children are encouraged to place all of their questions and wonders with regard to the science topics that the class is studying. She then uses the questions as a means to start or facilitate discussion or design experiments and encourages children to bring up their ideas (or their classmates' questions) from the wall during class discussions. This frames the context for the classroom and how Victoria's science-related pedagogical and content goals came into focus regarding a teachable moment during a unit on oceans with her inquisitive second-grade children.

46.2   How Do the Animals Get from One Pacific Ocean to the Other?

Victoria could see it written across the faces of her students, the realization that her explanations were not making any sense to them. In fairness, this happens to every teacher at some point. The situation arose as Victoria was beginning what she deemed "one of her more traditional lessons" during the ocean unit, where she had the children gather on the carpet to move through interactive smartboard slides.

Ms. Moon: So when we go through the 5 Oceans, we have the Southern Ocean, Arctic Ocean, Indian Ocean, Atlantic and Pacific…

Matías: But Miss, there are six oceans, not five!

Ms. Moon: Matías, why do you say there are six Oceans?

Matías: See, look at the map (pointing to map hanging at front of the room), it has two Pacific Oceans.

Patricia: Is that why the Pacific is the biggest because there are two?

Matias: (Pointing toward the wonder wall) That was my wonder too… How do the animals get from one Pacific Ocean to the other?

Ms. Moon: Okay, let's talk about this for a minute.

It is important to note that the foundation of the class is built on a desire to create a questioning community where being a risk-taker is valued. This foundation includes a willingness to speak up when things do not make sense. Victoria was happy to engage in the conversation with her students and confidently jumps into an explanation for the group as she sees this as a teachable moment. She provides examples to the second-grade students, first, by picking up the globe and comparing it to the map.

Ms. Moon: See if you wanted to make the map in the shape of the globe, it would wrap around, and the Pacific would be connected together. So, it is the same ocean; it was just cut to make the map.

Patricia: See, that is why it is the biggest because there are two of them.

Ms. Moon: Let's try this (wraps sentence strip it around the globe). See how when you wrap the sentence around the globe, it connects on the other side?

Intuitively, Victoria knows that her explanations are not connecting with the class as students keep harkening back to how the Pacific is actually two discrete oceans. She makes one last attempt by taking down a map from the front of the room, wrapping it around the globe to demonstrate how the map connects. This only seems to add to students' confusion as Victoria looks into the blank looks on their faces, knowing they are not understanding what she is trying to demonstrate. She finishes the lesson that day and gives her students the promise that they will revisit the question and work to explain it.

46.3   Changing Plans to Investigate the Globe

It is no secret why many elementary teachers can be intimidated by opening up spaces for inquiry through wonder-based questions since they can most certainly take classrooms in unanticipated directions. These new directions can follow circuitous paths that may or may not be directly aligned with content goals, unit plans or might move the class out of step within their grade-level teams. This unit started with an open-ended engagement meant to hook Ms. Moon's students on the content by investigating shells for the first month of school. This approach was meant to drive home ideas related to the five senses, noticing, observing, sketching, and learning to craft questions, which were all directly related to standards associated with scientific practices involved in inquiry-based learning. When pivoting toward maps, Victoria started with an exploration by providing a wide range of maps and had the children consider the representation of oceans. She encouraged them to think through their ideas and write some wonders in their journals and postings to the wonder wall if they desired. Victoria would often go to the Wonder Wall asking, "Have we answered this? Do we still need to answer this? What else do we need to investigate?"

In this case, the children were stuck and not making sense of her explanations for their question about the Pacific Ocean. Victoria wanted to find a method to engage the children in such a way that would provide long-lasting insights into their question. She spent the afternoon and evening considering several ideas before settling on an approach that had children reconstruct globes using balloons. That led into the next day's activity of creating the balloon world, which involved children developing a process to recreate the globe using a blown-up balloon. The children got to watch the balloon morph from being flat into its inflated shape as the air was added. Children then devised ways to recreate the continent shapes and map them onto their own balloons from the class globes (see Fig. 46.1).

In this globe-making activity, students could tangibly take each continent and place it in the right spot. Some students started to make sense of the "two" ocean problem as they constructed and saw the globe taking shape, noting, "So a map is just a cutup globe." Interestingly, a few days after this activity, another chance arose to drive home the connection between maps and globes when the air in the balloons seeped out, and the students were left with deflated globes. A few students asked if they could cut up their former globe "into a map." The students chose different places to cut their maps and noticed that there was only one Pacific Ocean or if they cut it into two, they were once joined together. Although this process might seem inefficient, it cannot be understated that the children were highly motivated to answer this wonder, and it drove them toward deeper understandings regarding the representations of Earth via maps and globes.

Fig. 46.1 Construction of globe balloon models

46.4 A New Teacher Reflects

The following is an excerpt from a conversation with a colleague that took place shortly after this lesson. Victoria describes her vision of this event and frames considerations raised in the process.

Things like map lessons are what I love. I just like the flexibility that it allows, and it was totally their driving question. It's just so much more ownership than if it's just, "Okay, check that off. Now onto the next thing that we are supposed to learn." It was also better than leaving the question unanswered because that's often what happens, too: if you're just pushed to move on. Students following their wonder is so much more what we want than just what they put on the assessment. This can be a battle with standards, and I think that's the hard part of inquiry.

I think the hardest thing is that we have all these amazing lessons, ideas, and things that drive what students are thinking, and the students question them in insightful ways. However, the assessments are the problem: we still have to do these traditional assessments. Even in the primary grades, we have benchmarks on assessments that we have to reach. A lot of times, it feels like we are telling our students, "Oh, what you were just learned was awesome, and you walked away with so much, and I guarantee you'll remember it in 10 years, but now I have to give you this." I believe in teaching this way, but then we still assess that way. I think the biggest battle is with that disconnect.

At the beginning of the year, my collaborative team discussed how inquiry is not just this open-ended free-for-all. Inquiry can be guided compared to the idea of completely open inquiry. I think that's the biggest thing that my team and I have learned this year. I think that idea helped us a lot, even having the structure of guided inquiry has helped us incorporate it more often than other lessons. There are times where we do explicitly teach, but knowing that you can still explicitly teach and then move into an inquiry lesson is comforting as we strive for excellent teaching practices. Not everything has to be, "Oh, throw maps all over the room and let them explore." That's a piece of it, but it's not all of it. I think that's what we learned as a team.

46.5 Considerations and Questions

Victoria was open to the possibilities as she changed a traditional lesson into one of inquiry, with hands-on, minds-on involvement that brought science content to life for children. The administration of ESES provided school faculty with autonomy for the individual grade-level teams to construct their own paths toward addressing state standards and not be directly tethered to the district pacing guide. This flexibility gave Victoria the chance to wrestle with how she approached this teaching challenge. It lessened the distance between the teacher she feels the system wants her to be compared to the teacher she wishes to be. She was able to balance a sense of wonder that was steeped in the questions that children brought to the classroom while also enacting the expectations for the standards-based content. However, she still struggled to make the link between wonder-based pedagogy and the district expectation for standardized assessments.

For Reflection and Discussion

1. In what ways can teachers navigate the tensions of enacting inquiry-based science pedagogy versus the reality of district expectations, standardized assessments, and/or pacing guides?
2. What factors may need to be addressed in order to foster a sense of wonder with children in public school contexts?
3. Consider the activities that Victoria enacted with the children, and devise a possible assessment that would adhere to the content expectations as well as the spirit of wonder.

Andrew Gilbert is an associate professor at George Mason University in Fairfax County, VA. He has taught in a variety of K-12 settings in both urban and suburban contexts in the Washington, DC metro area. In addition, he has two decades experience as a science teacher educator at universities across the USA and Australia. One of his main research areas focuses on bringing inquiry-based approaches to elementary classrooms through engagement with wonder.

Valery Erickson is a second-grade classroom teacher in Fairfax County Public Schools. She has taught for three years and works primarily with language learners. Her intentionality with creating a relationship-centered classroom fosters a space where students feel seen, known and loved. Her goal is to build this foundation so that students can freely and confidently explore and engage in the process of being a lifelong learner.

Commentary: The Wonder of Childhood and the Sacred Act of Wondering

Adam Johnston

Abstract

This is a commentary to the case narrative, *"But Miss, There are Six Oceans, Not Five?"* written by Andrew Gilbert and Valery Erickson.

When my oldest kid, Anna, was in second grade, she came home one day to ask me, pointedly and accusingly, "Where did Pluto go?" She knew I taught physics and astronomy, and she thought I might be able to explain or at least represent the community that she and her distraught classmates could accuse: What did you do to a beloved planet? Her teacher, accurately and innocently, had told them that in their list of planets, they would not need to include Pluto any longer. Anna and surely others interpreted this as some destruction of the entire astronomical body. Though it felt a little like breaking the news that someone's beloved pet had died, I explained that Pluto was just in a special recategorized group. It still existed, but in a better place, a better designation than what we'd first thought when we'd labeled it as a "planet."

In retrospect, I so wish that Anna would have had a Wonder Wall in her class. This isn't because I didn't want to answer her question, but because I think that the beauty of the learning process starts with the engaged wonder of children (or adults). I can only imagine that her experience could have been richer if she'd been in a context where a child could be curious about where Pluto went, imagine why dirt is brown, deliberate how caterpillars stick to trees, and wonder why there might only be five oceans when, quite obviously, there are six presented to us on a map split at the international date line. It just so happens that two of those seem to have the same name, truly a strange way to label oceans.

A. Johnston (✉)
Center for Science and Mathematics Education, Weber State University, Ogden, UT, USA
e-mail: ajohnston@weber.edu

© The Author(s), under exclusive license to Springer Nature Switzerland AG 2023
S. Jeong et al. (eds.), *Navigating Elementary Science Teaching and Learning*,
Springer Texts in Education, https://doi.org/10.1007/978-3-031-33418-4_47

The wonder of children and the use of their questions brings to the surface lots of provocative notions for me. To start, I think it highlights why many of us have endeavored into this beautiful, tortured gig of teaching in the first place. True, there are some versions of education that robots could do: present a few pieces of information and let others recite back to us what they've been told. Or we might simply direct and listen to how fast a 7-year-old spits out a line of oral reading, over and over again. Depending on how we think about "learning," our view of "teaching" can become any number of things, ranging from presenting information we ask our children to memorize to—and dare I hope for such a world—a view of learning where students are co-constructors of their understandings alongside their co-learners and teacher, where they can learn in a context that highlights them as individuals while working alongside others. The classroom led by Victoria—Ms. Moon—inspires faith that this is possible.

I don't necessarily want to proselytize and try to bring others into some cult of wonder, yet I think that this case should fundamentally challenge and inspire us as teachers. It shouldn't just make us think differently about what we might *do* to teach a class but instead make us rethink what the purpose of education is in the first place. An orientation to wonder makes me reconsider what I think learning means, what the role of the student is, and what the responsibility of the teacher should be. Wonder, to me, is a tenet of what it means to do science, as well as what it means to be a person. There isn't much that distinguishes us from other organisms on this perplexingly round planet, but the ability to question, reconsider, reflect, etc. are facets of humanity that we should develop and celebrate. We could stand to do more of this.

Many of the best scientists I know are second graders, and the students in Ms. Moon's class live up to this. Consider this: beings who've existed on the planet for less than a decade are confronted with a two-dimensional projection of this three-dimensional planet and they, quite rightly, start to question what's being shown to them—the idea that there is a certain number of oceans. But this fact is not the real hang-up of these scholars. They are trying to reconcile what it means for Earth to be round, an extraordinary mental feat. In fact, research is clear that the shape of Earth is fundamentally misconstrued by young people from every corner of the globe.[1]

The wonder, and with it an advocacy for learning, radiates from these young scientists. They saw a problem and they challenged the idea before them, out of curiosity and humanity rather than requirement or request from someone else. I love to imagine the delight (infused with small amounts of panic) that Ms. Moon must have felt as she witnessed this community begin to tag-team on each other's ideas, noting the supposed split of the Pacific Ocean on the map and then start to problematize this idea: "Is that why the Pacific is the biggest because there are two?"

---

[1] Thanks to the editors for allowing me to keep this sentence.

An orientation to wonder gives our learners agency. It provides them with the means to pursue learning on their own terms. It creates community that builds upon one another's constructions. The Wonder Wall, rather than a simple display of cute ideas from children, becomes a shared community display of the problems that a group of learners is actively engaged in. In this context, the teacher is an advocate rather than a gatekeeper, a co-learner instead of a simple vault of knowledge. Children become comfortable with the supposed risk of asking questions that are not simply about clarification, but are at the core of who they are in that moment. Seeing two edges of a map, each labeled with a Pacific Ocean on different sides of the page, of *course* they should wonder how the animals get from one side to the other. After all, we've done things to our planet that would seem at least as harmful.

At the same time that this could be a freeing opportunity for these scientists, for the elementary teacher it could create the tenuous feeling of a lack of control. For similar reasons, we're inclined to push back against reforms in upper grades and adult learning. We are afraid of what might happen if teaching isn't a simple pouring of knowledge from the well of a teacher to the bucket of the learner. Victoria reflects this tension by pointing out that standard assessments don't call on learners to be wonderers. Summative tests ask for answers rather than questions and reflection. Maybe we won't see an educational world in which such assessments are reformed—but at the same time I have faith that the students who are reconciling two-dimensional continents on three-dimensional balloons are going to be better off not just on a standardized test but for the rest of their lives. They get to experience learning in an authentic manner, constructing understandings for themselves and recognizing that they can figure things out.

Through it all, it's from their questions that the learners find direction. They know where this is all going because they were the ones who set it into action. Victoria gets to go home and figure out how to address this, an act that is one of my favorite features of teaching. There was no lesson plan or preconceived activity designed to teach about a set number of oceans, but I suspect that, as Victoria continues as a second grade teacher, she will find that this question and similar ones come up again. How she is addressing these questions is unique to these children; but other children will come to this on their own terms as well, and this work of the teacher will get revisited in other classrooms in future years.

In this case about second graders, multidimensional representations of Earth, addressing questions, honoring wonder, and all the rest, there's one facet that I think we should especially honor: Ms. Moon is a collaborator with these students, not as they might yet become in some future adult state, but as they are as children. It's easy to imagine that even without this episode these kids would grow to see the Pacific Ocean as one coherent expanse of water once they are 33-year-old engineers. But the number of oceans and the way that we construct a map of Earth *is* vexing to these children in their present. We need spaces where 7-year-olds can exist and learn for themselves, honoring the humans that they are in that beautiful moment. I hope that we can continue to remember that our children are fully formed individuals, capable of asking questions and directing their own learning. And I hope that others can spend time in the classroom spaces—bounded only by their Wonder Walls—that celebrate this.

Adam Johnston is a professor of physics and director of the Center for Science and Mathematics Education at Weber State University in Ogden, Utah. His scholarly interests originated with conceptual change theory and the nature of science, evolving into professional learning and science teacher preparation. He currently coordinates efforts to promote and support science and math teaching at all levels, in classrooms as well as in backyards, parks, and the outdoor settings beyond.

Case: The Bible and the Beast

48

Jaclyn Kuspiel Murray

Abstract

Darcy (Ms. Kirk) teaches in the suburb of the state's largest city in the southeast. She's been teaching third grade for seven years and prides herself on using the best science education practices to plan, implement, and reflect upon the inquiry-based lessons she prepares for third graders. In the scenario below, Darcy asks students to observe a set of fossils. Her purpose is to uncover prior knowledge before students engage more fully in the science unit. Initially, students participate in a series of investigations about geologic time and prehistoric organisms. They learn how to construct explanations using the Claim–Evidence–Reasoning (CER) framework. Throughout the initial investigations, Darcy presses for evidence and reasoning when students posed scientific claims. However, in the following set of narratives that chronicle a series of events, she is not prepared to respond to the evidence Sam uses to support his claim for how he knows dinosaurs existed a long time ago.

48.1 Episode 1

Darcy: Okay, everything is set up in the back of the room. Is everyone ready?

Class: Yes, Ms. Kirk!

Darcy: Before we begin, can somebody remind us all about what we are going to do? Yes, Molly, what are we doing?

Molly: We're going to write down what we notice when we look at what's on the table.

J. K. Murray (✉)
College of Education, Mercer University, Macon, GA, USA
e-mail: murray.jaclyn.k@gmail.com

Darcy: "And after that, what will we do?"

Sam: We will return to our seats and write down our wonderings.

Darcy: Did everyone hear that?

Class: Yes.

Darcy: What will we bring with us to the back table?

Chandra: I will take my pencil and science notebook to write what I notice.

Darcy: Great, let's get started and remember we notice with our eyes, not with our hands or mouths.

The children crowd around the table and record what they observe. Then Devon leaves the table in a hurry; Darcy questions his actions. Devon replies, "I'm getting a ruler." Upon returning from the supply cubby, he begins to measure the fossils. After the children return to their seats and record their wonderings in their science notebooks, Darcy directs the class to discuss with their tablemates what they noticed about the objects on the back table. As she moves around the room, Darcy listens to students talk about the color, size, and roughness of the items they viewed. Then, from the other side of the room, she hears Sabina say, "I wonder what these objects have in common? Ms. Kirk is always asking us to compare and contrast." "Remember, everyone; we are only discussing our noticing now. We will get to wonderings in a minute," Darcy reminded her students.

James: I see teeth, a lot of dirt, and smooth orange marbles with insects inside.

Andrea: Yes, also the rock has an animal tooth buried inside.

James: Oh yeah, now I see.

Devon: The larger tooth in the rock is a lot bigger than the teeth around the rock. The one in the rock is about 3 cm long, while the others are half the size or smaller.

The lights blink for a moment—the signal Darcy uses to gather the children's attention.

Darcy: Wrap up your conversation about noticings and prepare to discuss wonderings with your tablemates. What is a "wondering?"

Sam: Like, what questions we have, and what ideas we might have about what things are and where they come from.

Darcy: Chandra, I see a hand. Did you have something to contribute?

Chandra: Yes, ma'am. Do you mean, like, where they come from outside the classroom?

Darcy: Yes! You observed the objects on the back table, but that is because I placed them there. Where in nature might we see these objects? Some of you may have seen these before. What might they be made of? There are lots of things you could wonder about.

48.1.1 Chandra's Table

Cam: I wonder why some of these shapes look familiar while others don't.

Chandra: Right, I see insects, teeth, sponges, and snail shells, but I don't know what the other things are.

Devon: Also, I wonder why the teeth have different shapes.

Cam: The snails and clams are so tiny.

Chandra: I know what these are! They are fossils. My family and I found some on our vacation at the beach.

Cam: What is a fossil?

Chandra: Dead animals.

Cam: Oh, so these little clams died when they were babies.

Devon: Maybe, but remember our field trip to Colonial Williamsburg? All of the doors and ceilings were low because people were shorter back then. Maybe snails and clams were a lot smaller years ago than they are now.

Chandra: Huh, I didn't think about that.

48.1.2 Hannah's Table

Ben: Why do the sponges look so hard?

Tara: Probably because they are dry. They soften when you add water. Because Ms. Kirk said not to touch the objects, we couldn't add water.

Sam: I don't think that's a sponge; it's too tiny to be a sponge.

Hannah: "Then what is it, Sam?"

Sam: I don't know.

Hannah: What about the bugs? They look like they are frozen.

Ben: The bugs are inside some orange glass.

Tara: I'm not so sure.

Sam: The objects are all different. Some are in something, or only a little bit in something, while others are not stuck to anything. There is a lot of dirt, but then the insects are in something smooth, not rough. Let's group the objects according to their amount of buriedness.

Tara: We could do that, or we could organize them by size or shape. Many are round and either small or large. The teeth are different; they'll go in the "other" pile.

Ben: Well, which way should we do it?

48.2 Narration 1

Darcy listens in on all of the discussions that take place at each table. She notes what students say and does not interject to correct inaccurate statements. She uses this information to plan a set of science explorations for the following week.

After reflecting upon the children's discussion, Darcy chooses to introduce the students to a roadcut, or layers of exposed earth like one would see in natural settings such as the Grand Canyon. A roadcut results from drilling and blasting the land to make way for a road. The layers of exposed earth reveal evidence of the past. She wants children to make connections across fossil characteristics and where they were found. To aid students in understanding how layers of earth form over time due to deposition, she draws upon their knowledge of falling leaves. It's Fall, and the children see how the dead leaves create a layer above the grass, soil, or other earth material that lies beneath. The students are disinterested at times, and they don't seem to recognize the analogy, so she redirects the lesson toward the idea of dinosaur extinction. Darcy poses an anchoring phenomenon with an image of a partial dinosaur skeleton lying in the dirt.

The first investigative exploration is designed to support students' understanding of how weather, erosion, land movement, and other phenomena disrupt the land, causing the land to build or wear away. The children interact with the weathering and erosion simulations available from the National Geographic website (https://www.nationalgeographic.org/interactive/walters-travels-weathering-and-erosion/). The following day, Darcy has the students examine how fossils form under different conditions with a simulation (https://www.edumedia-sciences.com/en/media/136-fossilization). She wants her students to recognize that a landfall or volcanic eruption can bury the living instantly, causing objects to fossilize, frozen in time in the pose they took their last breath. Despite the abrupt formation of some types of remains, most of the students appear to understand that fossilization typically occurs over time and can manifest as molds, casts, and true fossils.

As Darcy listens to student conversations, she realizes that some students have difficulty understanding the difference between cast and true fossils, because they both appear as true fossils. Additionally, it seems like Chase and Susan are conflating mold and trace fossils. Darcy interjects, "Mold fossils are impressions left behind from the dead object after it has disintegrated, while trace fossils are impressions created by the organism when it was living. Footprints are one example of a trace fossil."

The class had been working with the Claim–Evidence–Reasoning (CER) framework since the first week of school. The students' interest is sparked with the question, "How do we know that dinosaurs lived years ago?" Darcy reminds the class that to support their claim, they must cite evidence and reasoning to connect the claim's evidence. The third graders excitedly attempt to use the CER framework to explain how they know dinosaurs lived a long time ago.

48.3   Episode 2

Darcy instructs the class to clear their desks of the math manipulatives they were working with and prepare for a transition. "It's science time!" We will share and discuss our responses to the question, "How do we know that dinosaurs lived a long time ago?" The class pulls out their science notebooks and turns to the last page, where they had begun to form Claim–Evidence–Reasoning explanations to answer the question. "I'll give you a few minutes to look over your CER statements one last time before we share them with the class," said Darcy. "You might want to talk with a partner at your table."

48.3.1 Hannah's Table

Tara: Well, I think that dinosaurs existed because I see their bones in rocks found beneath the ground.

Hannah: The fossils are below the ground because it took a lot of time for the dirt to pile up, creating the pressure to save the bone.

Ben: There are other ways to make fossils.

Tara: What do you mean? Ms. Kirk showed us that video of the clams dying when the sea and the landslide, wind, and rain poured dirt over them.

Hannah: Yes, she showed that video, but we also know that a volcanic eruption can cause a fossil without years of dirt piling up.

Sam: Yeah, there are other ways of forming fossils and different types of evidence to prove their existence.

Ben: What if we say, "We know that dinosaurs lived a long time ago. There are different types of evidence to prove that they were on the earth. Our evidence includes fossils located throughout the world, at different depths in the ground, and by different processes. Because they were found all over the place and we can see bones and fossils, they must have existed."

Sam: What about the "a long time ago" part?

48.3.2 Chandra's Table

Cam: No one has ever seen a living dinosaur, so I guess we can say that dinosaurs lived before people. People have lived for a long time, so that means dinosaurs lived an even longer time ago.

Devon: We also have evidence to say that because dinosaurs are found further down in the ground than people, they must have lived before humans.

Kelly: Let's add that.

Chandra: I think we should use the sentence helpers Ms. Kirk gave us; she likes when we use what's on the anchor chart.

All children at the table agree and they continue to create a consensus-based explanation to share with the class:

Claim: I know dinosaurs lived a long time ago because they were around before people.

Evidence: I know that fossils found deeper in the ground are older because weather and erosion add layers of dirt to the surface.

Reasoning: The most logical conclusion we can draw from the evidence is that dinosaurs lived a long time ago, before humans, because their bones were found deeper in the ground than human bones. Because humans have been around for a long time, and dinosaurs lived before humans, dinosaurs must have lived a long time ago.

48.4   Narration 2

Darcy was pleased to see that the students at each table were engaged in creating individual or consensus-based explanations of how we know dinosaurs lived a long time ago. Once again, Darcy listened to what each student said as she walked around the room. When the conversation diverged from the task at hand, she asked, "How many are still working?" Initially, the question prompted students to return to the task. Over time, the students had completed their explanations, and Darcy prepared to address the whole class.

48.5   Episode 3

Darcy: It looks like everyone is finished. Let's hear some explanations for why we know that dinosaurs lived a long time ago. Do I have a volunteer? Yes, Sabina! Go ahead. Tell us your claim.

Sabina: A lot of pressure on bone found deeper in the ground causes the bone to harden, just like what we see at the dinosaur museum.

Darcy: And, next, share your evidence and reasoning.

Sabina: The evidence that supports this claim is that we know that the weight of the added ground on top of bones creates pressure causing the bone to fossilize and become harder. Since dinosaur bones are harder, they must be found deep in the ground where the pressure is high, and the age is old. It takes hundreds of years for a bone to become a fossil.

Darcy: Thank you, Sabina. In your science notebooks, write down what Sabina said and state whether you agree or disagree with her claim, evidence, or reasoning. I will give you all a few minutes to write.

Darcy weaves around the tables as the children agree and disagree in written form. When pencils are down on the desk, she continues.

Darcy: Let's get ready for the next claim. Who wants to volunteer?

Many students raise their hands; however, she rarely hears from Sam, so Darcy calls on him to share.

Sam: I know that dinosaurs lived a long time ago because people from the past saw them. My evidence comes from the Bible and fossils. In the book of Job—the large beast—is the story of Behemoth. Since all animals were created on the sixth day, dinosaurs must have existed a long time ago but now are extinct. The fossil pieces found in the ground are evidence that dinosaurs existed a long time ago.

Darcy was stunned. She didn't know what to say. After a moment of hesitation, she directed students to write what Sam had stated in their science notebooks so they could agree or disagree with his explanation. Darcy continued to ask others to share their explanation while thinking about what Sam had said. Darcy often reflected-in-action; however, today, she thought she would not immediately respond and reflect on Sam's response at the end of the day (Fig. 48.1).

Fig. 48.1 One set of fossils the third graders viewed before a lesson about the age of fossils

For Reflection and Discussion

1. Describe the instructional model that Darcy uses to engage elementary students with fossils. Which practice was highlighted in the case? What instructional practice(s) did she use to support student learning?
2. How is science teaching, as depicted in the case above, different from the instructional practices utilized in English and language arts (ELA)?
3. In Episode 2, Chandra's table constructs a set of statements to form an explanation with the CER framework. The evidence and reasoning align with one another; however, together they do not support the claim. How would you assist students in modifying their CER explanation to align the components of the framework?
4. The evidence Sam used to support his claim to the question, "How do we know that dinosaurs lived a long time ago?" was problematic from Darcy's perspective. Do you consider Sam's explanation problematic? Explain why or why not.
5. In what ways could Darcy respond to Sam's explanation, either in the moment or at a later time? Where could she find resources to help her?
6. The students at Chandra's table constructed a scientific explanation with the Claim–Evidence–Reasoning framework. Rewrite the claim to present a coherent, evidence-based statement about dinosaur age.

Jaclyn Kuspiel Murray is an assistant professor of science and engineering education at Mercer University in Macon, Georgia. She is a former biomedical engineer and high school physics teacher. Her research centers on engaging prospective teachers in the sensemaking practices as science learners and facilitators of science learning to create coherence across the science content and science education methods courses in the program.

Commentary: Creating a Space for Students' Understanding of Science and Religion in the Elementary Classroom

49

Ian C. Binns

Abstract

This is a commentary to the case narrative, *"The Bible and the Beast"* written by Jaclyn Kuspiel Murray.

My first thought when I finished reading *The Bible and the Beast* was "I don't think I would've handled Sam's response the way Darcy did." Initially, I thought that Darcy should've addressed Sam's comment orally and not had students record his response in their notebooks. As someone who's focused on the interactions between science and religion for more than a decade, I was certain that this interaction would be a problem. After some time, I realized that it was important for me to take a step back and remind myself that I am an outside observer. Of course, it would be easy for me to come up with a "better" way to handle this situation. Instead, I need to put myself in Darcy's shoes. I should carefully reflect on the entire case instead of just Sam's response and Darcy's reaction.

Looking over the entire case revealed many important aspects of teaching that can make it easier for Darcy to address Sam's response. First, it's clear that Darcy's students understand her expectations when they work with their tablemates. In episode 1, we see students start working when they get back to their tables and have productive discussions. Additionally, Darcy is able to redirect students when necessary. While walking around the room, Darcy overhears Sabina say, "I wonder what these objects have in common?" Darcy uses this to remind the class that their current task is to discuss what they notice, not what they wonder. Finally, when Darcy asks students to explain wondering, Sam responds without raising his hand. Instead of directly correcting Sam, Darcy sees that Chandra has her hand up, points

I. C. Binns (✉)
Department of Reading and Elementary Education, Cato College of Education, University of North Carolina at Charlotte, 9201 University City Blvd, Charlotte, NC 28223, USA
e-mail: ian.binns@charlotte.edu

© The Author(s), under exclusive license to Springer Nature Switzerland AG 2023   263
S. Jeong et al. (eds.), *Navigating Elementary Science Teaching and Learning*,
Springer Texts in Education, https://doi.org/10.1007/978-3-031-33418-4_49

this out in her response, and asks Chandra to share her thoughts with the rest of the class.

Second, Darcy and her students are comfortable with each other. In episode 1, we see students crowd around the table to record their observations. After getting to the table, one student, Devon, leaves the table without any explanation. Darcy questions his actions and when he responds that he's getting a ruler, she lets him do this. This interaction displays a clear level of comfort between Darcy and Devon. Additionally, Darcy walks around the room constantly monitoring students' progress. There's no indication that students negatively react to Darcy's actions. They recognize Darcy's walking around the room as a common tactic and continue working. Finally, several times throughout the case Darcy recognizes times when students are either confused or make inaccurate statements. Darcy purposefully picks when to address these inaccuracies and when to let them drive her next lesson. We see her use both of these tactics in Narration 1.

Third, Darcy's students understand how to use the CER Framework and her expectations. At the end of episode 1, we learn that Darcy has used this strategy since the beginning of the school year. After posing the question, "how do we know that dinosaurs lived years ago," Darcy reminds her class how to use this framework and they get to work. We then see two different groups' conversations at their respective tables. The first conversation includes Tara, Hannah, Ben, and Sam. In this conversation, the students seem to have a productive dialogue about their CER statement, but they don't seem to reach complete consensus. This is evident when Sam asks if their CER statement accounts for "a long time ago." The second conversation includes Cam, Devon, Kelly, and Chandra. Unlike the first group, they reach a consensus on their CER statement after some dialogue. Finally, students understand what's expected of them when other groups share their CER statements. After Sabina shares her CER statement, Darcy directs the class to write down Sabina's statements and whether or not they agree or disagree. Their actions indicate that this is something that Darcy regularly expects.

49.1 Conclusion

The above three aspects of teaching that Darcy utilized in her classroom can be helpful when it comes to addressing Sam's response. As I looked over the entire case, I realized that Darcy has already laid at least some of a foundation for addressing science and religion in her classroom. It is clear from this case that Darcy has established a connection with her students. Establishing a connection with your students is one of the most important topics that I emphasize with my preservice elementary teachers. They need to make an effort to ensure students understand their views are welcome in the classroom. Students need to know that it's ok to give the "wrong" answer. By setting up this type of supportive and welcoming environment, teachers have the space to address unexpected responses.

As I look at how Darcy handled Sam's response, I now think Darcy's response was a successful way to address this situation. Darcy chooses to follow through with her instructional practice instead of bringing unnecessary attention to Sam. We know from Darcy's internal dialogue that she rarely hears from Sam, so having students follow the same procedure they used for Sabina's response avoids possible embarrassment. This course of action can help Sam, as well as the rest of the class, continue to feel comfortable sharing. This also gives Darcy time to reflect on the situation and come up with a positive way to respond.

I think Darcy could first approach Sam about his response prior to addressing the entire class. She can use this opportunity to help Sam remember what distinguishes science from other ways of knowing, like history, mathematics, the arts, and religion. She can remind Sam that science relies on evidence, is testable, and focuses exclusively on the natural world. Additionally, Darcy can help Sam remember that science involves a scientific community, reinforcing the idea that scientific research is collaborative, reproducible, and subject to peer review. Darcy should also let Sam know that she intends to have this conversation with the full class. Again, this is a good opportunity for her to dialogue with the class about what distinguishes science from other ways of knowing.

Ian C. Binns is an associate professor of elementary science education at the University of North Carolina at Charlotte. His research focuses on the interaction between science and religion with the goal of helping people understand what makes science and religion unique and how they both benefit society. Ian is also a host of the podcast Down the Wormhole, which explores the "strange and fascinating relationship between science and religion."

Case: Is Only Sticky Important? Sensemaking Through Equitable Discussion in a First-Grade Engineering Lesson

50

Dearing Blankmann, Alison K. Mercier, and Heidi Carlone

Abstract

This story focuses on Ms. Wallingdale ("Ms. W" as her students call her), a white, female, elementary teacher, and her fifteen Latinx and African American students in an urban emergent elementary school in the Southeast. In this case, we give a glimpse of how Ms. W guides everyday sensemaking in their learning community and strives to create genuine, equitable opportunities for her students to co-construct understanding. This kind of sensemaking is messy, leading to uncertainty and improvisation on the part of the teacher. This story raises questions about what exactly opening up space for equitable sensemaking looks like and the kinds of pedagogical strategies that can leverage students' everyday ideas, experiences, and cultural resources while disrupting power structures.

The industrious hum of children problem-solving fills Ms. W's first-grade classroom. The Title One school in the heart of a small southern city strives toward a STEM focus in the curriculum. Earlier in the week, Ms. W posed an engineering design challenge to her group of first graders. They were tasked with designing a mortar to use in building a model wall. The mortar materials were limited to sand,

D. Blankmann (✉)
Stout School of Education, High Point Univertsity, High Point, NC, USA
e-mail: dblankma@highpoint.edu

A. K. Mercier
School of Teacher Education, College of Education, University of Wyoming, Laramie, WY, USA
e-mail: amercier@uwyo.edu

H. Carlone
Katherine Johnson Chair in Science Education, Department of Teaching and Learning, Peabody College of Education and Human Development, Vanderbilt University, Nashville, TN, USA
e-mail: heidi.carlone@vanderbilt.edu

© The Author(s), under exclusive license to Springer Nature Switzerland AG 2023
S. Jeong et al. (eds.), *Navigating Elementary Science Teaching and Learning*,
Springer Texts in Education, https://doi.org/10.1007/978-3-031-33418-4_50

clay, and/or soil in some combination with water, and the wall composed of little stones could only be 8 inches tall and 12 inches wide.

On this particular afternoon, the class is engrossed in a dynamic discussion about what properties and materials constitute a good mortar. Ms. W is intentional about where and how she assembles her students for productive discourse. The meeting rug, a cheery kaleidoscope of color, is exactly the right size for the group to sit on in a circle. Sitting in a circle encourages Ms. W's students to speak to each other, rather than just their teacher. Being seated closely together and on the floor also brings an intimacy to the talk that lends itself to nurturing a learning community, one in which mutual trust is at the core. Ms. W standing up, while still part of the circle, faces her students with her back to the chart covered whiteboard. Ms. W's goal for this lesson is to facilitate a robust discussion that guides students to engage in collective sensemaking about the properties of the materials offered as options in creating mortar to build their model walls. In response to a student's suggestion that the substance needs to be sticky, Ms. W questions the class about the merits of stickiness in the design. By doing this, she sends the unspoken message to her students that their peer has an idea of value worth thinking over and building on.

"Is only sticky important? What if you only used something sticky?" she asks. Ms. W knows asking a question in two different ways helps clarify what she is asking her students to think about.

"That's the way to make it stronger," someone asserts.

Keera jumps in. "Because if it's not sticky, how is it going to stick, or how is it going to go together?" Her question is directed at her classmates, not her teacher.

"We're curious!" another student chimes in.

Keera continues, "It's just going to rub off on the other materials."

Nivia's hand shoots up as she simultaneously begins to speak. Not wanting to shut her down for interrupting, Ms. W acknowledges Nivia's attempt to adhere to discussion norms. Ms. W realizes when a discussion really starts to gear up, students become excited, and the expectations of a traditional classroom are not necessarily conducive to spirited discussion. She is not afraid to share the floor with her students; in fact, she encourages it because she appreciates her students as knowledge holders and producers. However, recognizing the value of the flow of conversation introduces a moment of dissonance as a teacher. The potential exists for the discussion to take a turn that is unhelpful to both the group and individual students. Ms. W has faith in this group and the direction of this discussion. She has encouraged the class from the beginning of the year to see each other as resources and collaborators in problem-solving. With a subtle nod, she signals to Nivia to continue.

"I think sticky's good for materials," Nivia says. "I remember one time, uh, I made a little wall and then it was snowing, I saw that my sap that it froze a little, and I felt it, and I said, hmmm, I wonder how it feels. I'm going to wait until summer, and then I said, I see, I saw that it melted. And it got really sticky and stuff."

Ms. W sees an opportunity to leverage Nivia's idea in a way that will help her peers start to connect her experience to the problem of designing a durable mortar. "So, your wall and your mortar behaved differently in different temperatures. is that what you're saying?" By interpreting Nivia's thinking, Ms. W works to make her idea accessible to the rest of the community to build on. Nivia nods her head.

Ms. W sums up Nivia's thought, "So it has to hold up in all weather. Should I add temperatures to our chart?"

Ms. W points to the chart where she is capturing the essence of the conversation. The documented student thinking will be a helpful reference for the young engineers as they wrangle with their design challenge.

Nivia nods again while Keera supportively calls out, "Yes, yes, yes, yes!"

Nivia, lost in a stream of thought, continues to muse out loud, "It's like they, the bricks, the bricks, they stayed together."

Ms. W pushes Nivia and the group to dig deeper into thinking through the properties of an effective mortar. She wants her students to consider alternative properties of the materials that might contribute to optimal mortar. She asks the same question in two different ways. Beginning with "Can you only use sticky materials?" followed with a clarifying, "What other properties - other than sticky - could be useful?" She then adds, "I'd like to hear from some other voices today." Ms. W has noticed that Keera and Nivia have been the dominant voices of the conversation. This comment is intended to be a quiet signal to the two girls to allow space for others to share their ideas and a gentle nudge to friends in the learning community who, due to the fast pace of the discussion, may not feel comfortable jumping in. Ms. W routinely juggles creating space for all voices while not discouraging her more vocal students. Orchestration of productive and equitable discourse can be challenging. It is a goal Ms. W reaches for each day.

Keera, very excited about a connection she just made, can't hold in her thoughts and blurts out, "Ms. W there's something about the tree sap!"

While appreciating all of Keera's contributions, Ms. W still wants to leave the invitation open to other thinkers. "Hold that thought, Keera. I want to hear some other voices," she says gently and then without missing a beat continues, "So the question is, are there properties other than sticky that we need in a mortar? Is it sticky to touch when we go outside and touch the brick wall? Is it sticky in the library?"

These references are familiar to her students. Earlier in this unit, Ms. W created posters of different types of walls built with a variety of materials and mortars. She walked with her students around the school hunting for examples in their setting. Such scaffolding provides her students with relatable experiences to connect ideas and vocabulary with these multiple representations of walls. It also provides students with relatable, common experiences which, in turn, help them connect everyday experiences to vocabulary. She has found this strategy particularly helpful to engage her language learners.

Keera can't help herself and answers promptly, "No, because ..."

She is cut off by Ms. W, "Hold on, hold on." Ms. W does not mean to diminish Keera's enthusiasm or dismiss her expertise. This may not have been a move she would have made with other students. However, Ms. W especially wants her quieter and more reserved students to contribute their ideas to the class's collective knowledge building. By drawing on a breadth of experiences and ideas, Ms. W intentionally validates and encourages the notion that everyone sitting in the circle has something valuable to contribute. She scans the circle of students. Nivia's hand goes up. Keera also takes up the cue and raises her hand. Ms. W waits for input from elsewhere in the circle. The silence is almost more than Keera can bear as she struggles to contain her enthusiasm. She is like a kettle that is about to boil over. Eventually, Ms. W wistfully concedes, "I guess we are not getting to hear other voices yet. Nivia, are there properties of mortar other than stickiness that you'd like to share?" Ms. W repeats the original question again.

Ms. W recognizes that while Nivia has taken a while to make her point, knowledge production is often not linear. Students try out ideas in this forum, playing with, analyzing, and synthesizing together as a group. Ms. W remembers a time in her early days of teaching when providing adequate wait time seemed unbearable. In those days, it took every ounce of self-control not to finish a student's thought or move onto the next student too quickly. Keeping the tempo of the conversation going presented a dilemma. "It, it, it was, I think at first it was hard," Nivia begins. "I think at first it was soft, and then it got very hard. I remember every time I go to my cousin's house, I see, like all these kinds of footprints and things."

Ms. W recognizes the insight as an essential piece to the collective puzzling about helpful mortar properties other than stickiness. While Ms. W comprehends where Nivia is going with this idea, she wants to be sure the entire group also understands what Nivia is trying to convey. "In the concrete? You see their footprints in the sidewalk?"

"Yeah, and they be putting their footprints and their names and things in it, and then it gets hard."

The lid has come off the kettle. Keera can't keep it in. "Ooohhhh," she exclaims. "I know what you're talking about! You write your name in the sidewalk when it's wet!"

A quiet voice joins in. "Then it dries," Willow contributes. This was the moment in this discussion Ms. W has been working for. As the discourse evolves, a fresh voice now finds a comfortable space to join in.

Ms. W attempts to revoice these important realizations, "So, I hear…" The discussion has now taken on a life of its own. Ms. W knows this is her students' time now. She ignited the conversation, but now the young engineers are on fire with their collective thinking. Reflecting a trust in her students as collective knowledge builders and partners in meaning making, she lets the conversation unfold dynamically.

Sophia now joins in. She confirms the experience of her peers. "I've seen that."

"It's hard like the wall," says Mia, participating for the first time.

Ms. W is about to rejoin the conversation, but Keera participates first, "It's like a property."

"Yeah, it's good," Nivia responds.

Ms. W has been studying the circle. Dayanara, an emergent English speaker, looks like she wants to jump in but doesn't have an entry point with the rapid pace of the conversation. "Dayanara, we haven't heard from you in a while." Ms. W invites her in so that the prospect of sharing her good thinking isn't lost. Dayanara takes this opportunity. "It's like if you use sap, and then the sap could get hard."

Ms. W knows Dayanara is connecting the class's observations about cement to both a text about trees from guided reading in which sap is discussed and her personal experiences playing underneath the tall pine trees in the school's playground.

"But a bug eats the sap," Dayanara adds.

This observation elicits an eruption of responses from the group as there is a collective "aha" moment. There is a cacophony of affirmations. "Oh!" "Yeah!" "I know what you're saying."

Ms. W smiles. She knows how affirming an experience like this can be for Dayanara and also for the collective learning community, as they grasp that a peer's thinking both confirms their own experiences and also contributes another piece to the puzzle.

Keera speaks directly to her peers: "It's like when Ms. W told us about – Remember? When we were running away from the beetle? The beetle ate wood!" Ms. W concedes the floor to Keera, recognizing her team of good thinkers are connecting the similarities in the properties of cement and sap in addition to the realization that a criterion for mortar may need to include suitability to not just the elements, but also the inhabitants of the natural world.

Emilia, also an emergent bilingual student chimes in, "Or some insects or animals might want to live in there, so, so..." she directs her question to Keera "You saw a beetle?"

Delighted to share the experience, Keera answers, "Yes. That beetle was chewing on the tree."

Ms. W would like to get the discussion to shift back to the properties of materials of mortar. However, she recognizes that this seemingly tangential discussion drew into the conversation students who would have otherwise remained silent. She decides to let the thread continue a moment longer.

Emilia's interest is piqued. "So, it just chewed right through," she muses.

Ms. W, wanting to leverage the fascination and expertise around the beetle experience while connecting the discussion back to the properties of the materials, says, "Hmmm, well, I'm wondering, if your mortar had something in it that tasted good, if that would be a problem."

Keera is on a roll. "Yeah! Wait, what was that special word that we called a problem? I just forgot it."

Ms. W responds, "Umm, hold that thought. I'm not sure which word we're talking about." She doesn't want anything to interfere with the momentum of the discussion, but she also wants to be careful not to shut down Keera's interest in expanding her science vocabulary. Ms. W will circle back to the words after the train of thinking about properties is complete.

Fig. 50.1 Example of student's wall constructed with earth material mortar

Kendra, listening intently all this while, pulls the collective thinking of the last few minutes together.

"If you put too…you could put too much sugar in the honey, ants like sugar, so the ants would probably eat the sugar, and you would just have bricks left." Together, Ms. W and the learning community she guides came to a number of important conclusions about the properties of the materials used in an efficient mortar. The materials of mortar must hold up across temperature, weather, and wildlife.

While the model walls her class will build to test their mortar designs will not have to endure any of these situations, the walls will have to hold up to the force of a "wrecking ball," a golf ball fastened to a string and dropped from four different heights. Ms. W looks forward to the conversations to come. This discussion will play a pivotal role in the decision making of the design teams. The dialogue today will inform how her students create, test, and revise their mortar designs in the days to come (Fig. 50.1).

For Reflection and Discussion

1. Ms. W had a goal of opening up space for equitable, collective sensemaking. Do you think that she achieved this goal in this case? Why or why not? What strategies did she use to try to achieve her goal?

2. How does Ms. W acknowledge and celebrate students' ideas and contributions while working toward facilitating their understanding of properties of mortar to inform their engineering design?
3. How does a discussion of sap as a possible mortar relate to their eventual engineering design challenge, which will use sand, soil, and/or clay and water for mortar? Discuss why you view this discussion of sap as productive or unproductive in helping students make sense of the science and engineering ideas.
4. How would you, as a teacher, pick up the thread of Kendra's idea? What is the teacher's next move?

Dearing Blankmann is an assistant professor of educator preparation at High Point University. Her work concentrates on teacher development in the integration of literacy and science/engineering. Additional areas of focus include questions of equity and working with teachers to create integrated STEM experiences in outdoor spaces.

Alison K. Mercier is an assistant professor of elementary science education at the University of Wyoming. Her work focuses on co-creating responsive STEM learning for both elementary students and elementary teachers.

Heidi Carlone is the Katherine Johnson Chair in science education in the Department of Teaching and Learning at Vanderbilt University's Peabody College of Education and Human Development. She is interested in: (1) connections between the culture of science and engineering learning settings and youths' identity development in and beyond those settings and (2) co-designing teacher professional learning and leadership networks to support and nurture equitable science and engineering instruction.

Commentary: When Culturally Sustaining Science Pedagogy Becomes "Sticky"

51

Randy K. Yerrick

Abstract

This is a commentary to the case narrative, *"Is Only Sticky Important? Sensemaking through Equitable Discussion in a First-Grade Engineering Lesson"* written by Dearing Blankmann, Alison Mercier, and Heidi Carlone.

We have been presented with a rich first-hand account of culturally sustaining science pedagogy which I find extremely valuable for pushing our thinking about elementary science instruction in at least three different directions. These include but are not limited to: (1) the exploration of science as a discourse, (2) the extraordinary capacity of dialogue and narrative to make science inclusive for an increasingly diverse population, and (3) the continued refinement of the role of engineering in elementary STEM education.

Ms. W has not constrained her thinking about elementary science teaching to explaining scientific concepts, exploring their usefulness, and applying them to the world. Rather, Ms. W has committed herself to the evolution and exploration of children's ideas as applied to co-construction of science in children's lives. As Hawkins (1974) has argued eloquently, teachers need to realize their role in the "I, Thou, and It" relationship children have with their surroundings. Ms. W has constructed an environment with her young children where evidence, observations, and experience have weight in and amongst constructs children have access to. As children recall instances of things which "stick" they are encouraged to explore the conditions under which these constructs have limits (e.g., temperature). Ms. W treats science as discourse—the ways children speak, think, and act within a given cultural context. In much the same way that Gallas (1995) conducts her "science

R. K. Yerrick (✉)
Kremen School for Education and Human Development, Fresno, CA, USA
e-mail: yerrick@mail.fresnostate.edu

© The Author(s), under exclusive license to Springer Nature Switzerland AG 2023
S. Jeong et al. (eds.), *Navigating Elementary Science Teaching and Learning*,
Springer Texts in Education, https://doi.org/10.1007/978-3-031-33418-4_51

circles," Ms. W removes herself as authority over content knowledge, but not from the authority of judging what is best for the progression of inquiry. Many have described science as it naturally occurs in professional contexts as competitive and with the gaining of status at the cost of being exclusive. In contrast, Ms. W challenges any notions of teaching as a collection of knowledge to be shared or a set of skills to be learned. She demonstrates teaching is a craft, an artform, not a skill to be learned. Through the pen of Blankmann, Mercier, and Carlone, we see first-hand Ms. W's ability to elevate different ideas, weave experiences together, while conducting an orchestra of voices to pause with one hand and wave in a crescendo of another. There is space for all voices in this symphony—from the loud and dominant trombone of confidence to the meek suggestive tone of the lone violin of Nivia asking a sincere and humble questioning about how substances can transform to another.

I encourage the reader to explore more completely the notion of science as discourse to understand more about the goals teachers may hold for students which stretch beyond the content. Carlone et al. (2021) explain elsewhere more completely what Ms. W is aiming for above all other destinations. They call it epistemic heterogeneity, or in common terms, how Ms. W privileges the ways of knowing to have equal value with what is known. The ways in which children come to know things are unique to each child. In their own ways, through their own language, at their own pace, children express what they want to know. For some, this process is verbal, and for others, it is experiential, visual, or relational. When science teaching is done in such ways as to value multiple ways of knowing, children can make contributions from a wealth of assets they bring with them to the classroom. And what we know from a variety of other researchers is that this process of managing agendas, worldviews, and ways of thinking is as much the science as teaching the content itself (Gallas, 1995; Ballenger, 2004; Moje, 2001).

I value these explications of science as discourse because we know that most preservice elementary teachers have limited experiences with authentic science teaching. In my 25 years of preparing elementary science teachers, the vast major-ity of them have enrolled in only a couple of university courses by choice. Their experiences which drove them to being a wonderful, caring, invested teacher of young children were sometimes accompanied by experiences which drove them away from science as it was presented to them as rational, non-engaging, in-humane, and unnecessarily abstract. Having my preservice teachers "unlearn" science in the same way that Ball (1988) described the need to "unlearn" math-ematics is key for novice teachers to hear and to see from these examples how science can be inclusive—even to children who have histories of science aca-demic failure. When my preservice teachers read Gallas, Ballenger, and others and explore video excerpts where children are thinking together, my novice teach-ers often regain their interest to teach science in K-6 contexts, which we all know is so desperately needed—and why we need so many more teachers like Ms. W to welcome our young teachers into their classrooms and to apprentice them. What we need to figure out from teachers like Ms. W are questions like:

- What is the requisite knowledge Ms. W is relying upon to orchestrate such learning?
- How is this requisite knowledge best passed on to others, and how agile is this knowledge in other contexts?

I will offer one additional insight regarding the use of these young children's dialogic representations with STEM professionals in higher education. I have been given the honor and the privilege of leading cohorts of practicing engineers through an educational research doctoral program where we explore what it means to prepare to teach inclusively or to capture vis-à-vis research methods the essence of what critical thinking looks like in context. With each group of my engineering faculty, we have read from Gallas' and Ballenger's work as we discuss what engineers would like to see in their own undergraduates' outcomes. They are consistently struck by the openness of students, the inquisitive questioning, and even the use of evidence which children bring to "science talks." My engineers ask questions like, "These students are talking like grown engineers. I would love to have these students eventually come to my undergraduate engineering course. What happens to students in k-12? Why can't we retain these kinds of dispositions in students?" My engineers were both right and wrong in their observations. As much research has demonstrated, expert scientists and engineers often engage in talk which reflects the norms of discourse that Gallas and others have demonstrated. Science and engineering are conducted in social spaces where ideas and evidence are co-constructed by participants and where norms of speaking, thinking, and acting are negotiated. By following the children's contributions, Ms. W is fostering an environment of public and collaborative inquiry as she must synthesize, compare, and evaluate the contributions toward a particular resolution, much in the same way a materials science engineer may begin with a problem and need to adapt their solutions iteratively depending upon external conditions (i.e., moisture, temperature, infestations). They are also right that children are often taught in school not to speak, think, and act in these kinds of ways.

What my engineers fail to grasp is that they speak in duplicitous ways as they welcome open dialogue but demand highly rigorous and narrow academic standards. Only when we explore further with engineers the kinds of exams, grades, math competencies, AP courses, placement tests, and other standards they demand for entrance to their programs do they begin to recognize their own culpability in supporting and steering a system which biases other kinds of discourse. It is no coincidence that the courses that engineering majors must take in their first three semesters are meant to weed out "lesser" students according to the narrow set of knowledge and skills represented by these entrance exams.

So, we are all faced with an interesting dilemma in our roles as P-16 STEM educators. We are faced with a tough choice that, if interpreted through a narrow lens, seems dichotomous. We can push for more of the same high-stakes tests,

AP Calculus scores, and other similar measures which privilege specific groups. The other choice is to embrace collaborative inquiry, culturally sustaining pedagogy, and different norms of discourse that we think better represents the thinkers we want to produce. I don't think our choices to train preservice STEM teachers should be based upon such dipolarity. I believe there is middle ground to be found—a place where teachers can both hold high standards and simultaneously tap into the wealth of assets diverse groups of children present teachers in schools across this country. But the typography of this teacher knowledge landscape falls in an under-studied realm of STEM education—the examination of the brilliance, insight, and agility of the Ms. W's in our world. I hope we learn to harness such knowledge and soon… for all our sakes.

References

Ball, D. L. (1988). Unlearning to teach mathematics. *For the Learning of Mathematics, 8*(1), 40–48.

Ballenger, C. (2004). Meaning and context: Studying words in motion. In R. K. Yerrick & W.-M. Roth (Eds.), *Establishing scientific classroom discourse communities: Multiple voices of teaching and learning research* (pp. 175–198). Lawrence Erlbaum.

Carlone, H. B., Mercier, A. K., & Metzger, S. R. (2021). The production of epistemic culture and agency during a first-grade engineering design unit in an urban emergent school. *Journal of Pre-College Engineering Education Research, 11*(1), 10. https://doi.org/10.7771/2157-9288.1295

Gallas, K. (1995). *Talking their way into science: Hearing children's questions and theories, responding with curricula.* Teachers College Press.

Hawkins, D. (1974). I, thou, and it. *The informed vision: Essays on learning and human nature* (pp. 48–62). Agathon Press.

Moje, E. B., Collazo, T., Carrillo, R., & Marx, R. W. (2001). "Maestro, What is 'Quality'?": Language, Literacy, and Discourse in Project-Based Science. *Journal of Research in Science Teaching, 38*(4), 469–498.

Randy Yerrick is Dean of the Kremen School for Education and Human Development at CSU Fresno. He has also served as Professor of Science Education at SUNY Buffalo where he served as Associate Dean and Science Education Professor for the Graduate School of Education. Dr. Yerrick maintains an active research agenda focusing on two central questions: (1) How do scientific norms of discourse get enacted in classrooms and (2) To what extent can historical barriers to STEM learning be traversed for underrepresented students through culturally sustaining practices?

Part V

Utilizing Technology in Science Learning Environments

Deborah J. Tippins

In today's world, technology speaks to all of us—teachers, students, parents, and citizens. While technological advances can sometimes seem remote from our everyday lives, recent developments and innovation in communications and other forms of technology can be very empowering for students. Almost with certainty we can say that children in today's classrooms will be unlikely to recall a time when there were no cellphones, computers, robots, tablets, social media, drones, digital cameras, data storage, nanotechnology, and so much more. Amidst the context of a cyberspace generation of youth, it becomes imperative for teachers and all of us, more than any other time in history, to consider the consequences of adopting different technologies in the classroom and bringing them into the hands of children. We should continuously ask ourselves who gains or who becomes further impoverished through the appropriate or inappropriate use of technology in the classroom. In other words, we must consider what is to be gained or what is traded when technology becomes a centerpiece of learning. Decisions about the use of technology in science teaching and learning is not something that can be solely in the hands of scientists, engineers, the business community, or the military. It is important to recognize that teachers can play a critical role in designing learning environments where even very young children can apply science to generating a solution or making a product, what we might refer to as the creative application of technology.

At its best, technology can provide the current generation of learners with new images of what it means to do science and authentic examples of who can do science. Whether learning about programming through robotics, using Lego Mindstorms to build robots, collecting and analyzing live weather data, viewing multimedia webcasts with high resolution animations, or applying systems thinking to local environmental issues, children have opportunities to develop as young

D. J. Tippins
Department of Mathematics and Science Education, Mary Frances Early College of Education, University of Georgia, Athens, GA, USA
e-mail: dtippins@uga.edu

scientists. In this sense, technology can provide children, as scientists, a means to become empowered as they learn and sophisticate skills of creative problem-posing and solving, teamwork and collaboration, communication, project planning, listening, and thinking.

However, the challenges that come with integrating technology with science teaching and learning in the early years cannot be overlooked. In the quest for a more technological science education, we cannot take learning through technology for granted, such that it becomes a placeholder for stagnate bits and bytes of information. Many scholars have pointed to the disappearing line between machines, humans, and other animals. Synthetic experiences with simulated animals, rocks, oceans, and rainforests that technology affords are not an adequate substitute for authentic interactions in outdoor environments and may contribute to a widening disconnect between children and nature, even potentially impacting their health and emotional well-being. Other scholars suggest that the inappropriate use of technology may in some ways diminish children's senses and feelings of awe and wonder that comes with first-hand encounters with nature. While technology is often viewed as a tool to foster children's problem-solving abilities, Bowers (2014) and other scholars emphasize the ways in which histories and cultural practices may be overlooked in the process. In addition, most educators are astutely aware of the issues of equity, access and loss of privacy that surround the use of technology. A concern for many educators is the rapid pace at which technology becomes obsolete, requiring continuous financial expenditures.

The four cases in this part highlight some of the challenges elementary teachers of science have encountered in their efforts to use technology as a tool for science learning. The cases portray both successes and failures and provide a starting point for rich conversations about the role that technology might play in elementary science teaching and learning. In *Fly Girls Face Failure*, Sally Creel provides an interesting example of student engagement with drones in an afterschool STEM program for girls. Her case captures the problems that can occur when young children have a view of technology as infallible and science as absolute. Tamieka Grizzle, in *Scratching and Rocking with Rocks*, recounts her experiences with kindergarten students as they attempted to code with Scratch Junior to learn about the physical attributes of rocks. Most recently, schools have been faced with numerous challenges that accompany the transition from face-to-face to online learning environments. In Ji Shen's case, *Transitioning to Online Learning: When Your A Student Gets an F*, an innovative second grade teacher struggles to help parents understand how to support their children in a synchronous learning environment including science lessons using Nearpod, Flipgrid, and similar platforms. Finally, Lautaro Cabrera, in the case *Sphero Struggle: Productive or Demoralizing?* describes a fourth-grade teacher's efforts to help students benefit from productive struggle during a robotics design lesson. Together, these cases illustrate some of the tensions that emerge in the technology propelled landscape of today's elementary classrooms. With the recognition that technology is kinetic, educators must be continually aware of the caveats and possibilities it affords for today's youth.

Reference

Bowers, C.A. (2014). The false promises of the digital revolution: How computers transform education, work, and international development in ways that are ecologically unstainable. Peter Lang.

Case: Fly Girls Face Failure

52

Sally Creel

Abstract

Fly Girls is a program that has chapters in classrooms across a Georgia school district. The program was developed as an enrichment program to support underrepresented populations in STEM, specifically females. Teachers and female students in grades 4–5 meet after school to participate in the Fly Girls program. To start, the group meets to learn about the purposes of drones and how to operate them. Then, during the first month of a cycle, the girls and their teachers meet to launch new missions related to a career that features a female scientist using drones. The subsequent month is the "landing event," where the girls showcase their career-related skills with drones. The program repeats every two months using this cycle. In this case, two girls were distraught when they unexpectedly encountered a problem with their drone.

The Fly Girls teacher sponsors met virtually after school to plan for the upcoming Fly Girls mission. The sponsors all work in different schools and departments across the school district. Dr. Creel is the district STEM Supervisor. Ms. Davis serves as the district science coach. Ms. Dean and Dr. Grizzle are innovation specialists at two different STEM-focused elementary schools. Since its inception, Fly Girls has been extremely popular with the 4th and 5th grade girls and their parents. Thus, the teacher sponsors were taken by surprise when they were faced with an unexpected challenge during a recent launching mission.

S. Creel (✉)
STEM and Innovation Supervisor for Cobb Schools, Georgia, United States
e-mail: sally.creel@cobbk12.org

Fly Girls Mission Planning Meeting:

Creel: Thanks so much for jumping on this call. We will be launching missions 5 and 6 at the end of next month. I'm excited that our career spotlight this time will be an ecologist using drones as an emerging technology to study wetland areas.

Dean: That's cool. How did we connect with her?

Davis: Our partner from the university sent out feelers to her personal network of STEM professionals. Caitlin responded and said she'd love to connect with our Fly Girls.

Grizzle: You said in your appointment for this meeting that she's a licensed drone pilot, right? And that she's been using drones to monitor the health of various wetland areas. Does this include all types of wetlands?

Creel: Yes. Caitlin sent me a slide deck of an overview she will provide the girls. Part of this overview describes in detail different types of wetland environments, including a pond, marsh, river, and a retention pond in a subdivision. She has images she captured with her drone of a drainage pipe that spills into a stream. This is an area she has been studying.

Dean: That should show how sediment accumulates from the drainage pipe, right? That would help teach weathering and erosion standards along with the ecology standards. Maybe that can be included as part of the mission. We can have the girls capture footage of a wetland area being impacted by weather and erosion. Perhaps they can also do a voice narration and add text over the video they capture to highlight the affected area.

Davis: We could model that in our intro video for this mission. I think it would be great if we also gave them the option of choosing to identify plants and animals that are found in the wetland area they explore. We can have them identify the many plant and animal species they catch on film.

Grizzle: That will work, and it gives them some choice and voice in the mission. I know a couple of girls have asked about free video editing software they can use. Any ideas for that?

Dean: I'll make the mission overview video using a free trial software. It will leave a water-mark on my video, but that will give them an example of what that free software looks like. Watermarks from the program are definitely ok.

Creel: Love it! Please send me the pieces you're developing for this mission within the next week, and I'll get them posted for the girls.

The Following Week:

Creel: The Fly Girls launch of Mission 5: Habitat Hunt went well. The ecologist did a great job of connecting with the girls and helping them understand her job. The girls asked great questions and seemed to be excited about this mission.

Fast Forward: Fly Girls Launch Habitat Hunt Landing Mission LIVE

Riley: Ms. Dean, that wet marshy area off the 3rd grade hallway would be considered a wetland area, right?

Michelle: I have a pond in my neighborhood. I could video that area one day after school.

Ms. Dean: Riley, the area off the 3rd grade hallway is a marshy area that is considered a wetland. Why don't we go take a look at it to see how we could use our drones in that environment? Yes, Michelle, the pond in your neighborhood would be a good option too.

The Fly Girls eventually select the marshy area off the 3rd grade hallway for their mission. The next day, they prepare to head out to the marsh with three different drones in hand. They check to make sure the drones have fully charged batteries, and, once outside, check for overhead hazards like powerlines and trees.

In small groups, the girls take turns flying the drones over parts of the marshy area and recording footage. They try to fly their drones low enough to see the boundary of the marshy area. One group finds a nest in a tree and maneuvers their drone above it to see if anything lived in it.

After all of the groups finish surveying the wetland with their drones, they head back inside to download their footage and see what they captured. Two of the groups are excited and start shrieking in amazement at the images on the computer from their drone. The third group keeps repeatedly unplugging and plugging their drone back into the computer. Ms. Dean goes over to investigate.

Maya: None of our video is on here. The card is blank.

Kaci: I know I pushed the record button. I remember the screen turning red to indicate that it was recording. Why isn't it on there?

Brooklin: It took me forever to get up to the right height to see inside that nest. It had eggs in it! Why didn't it work?

Rakshitha doesn't say anything at first, but she is looking at the other groups who are giggling and pointing out all the cool things on their videos. With a discouraged tone of voice, Rakshitha mumbles, "this was a big waste of time."

Maya: [nodding her head in agreement] I thought it might be cool to be a scientist... or an ecologist like Ms. Caitlin. But now I think it's just a stupid job. We worked so hard to get our drone to take a picture of the bird nest, but we still failed our mission.

Ms. Dean can see in their faces that they are upset and disappointed. It is almost time for the Fly Girls mission to end for the day. There is not time or enough battery for the girls to go back out outside and try again. Their parents will arrive soon to pick them up. Ms. Dean is concerned that the girls' drone experience might permanently diminish their excitement about science. She wonders how she might turn this negative experience into a positive outcome for this group of girls.

For Reflection and Discussion

1.1. Technology failures happen. How would you have supported the team whose technology failed? What would you have done if you were in Ms. Dean's position?

2.2. As a teacher, what could you have done to prevent the video failure from happening in the first place?

3.3. What other kind of images could the girls capture using their drones in the wetland ecosystems that might connect to science concepts that the team had not considered?

4.4. How might you support students who are feeling discouraged after facing a failure during a project that they consider important?

Sally Creel is a former elementary educator. She currently serves as the K-12 STEM & Innovation Supervisor for Cobb Schools. She has served on the board of the Georgia Science Teachers Association for over 20 years. Her educational interests include teacher leadership and cultivating community partnerships. Dr. Creel collaborates nationally to strengthen STEM ecosystems.

Commentary: Flying into Mistakes: Doing "Real" Science

53

Kimberly Haverkos

Abstract

This is a commentary to the case narrative, *"Fly Girls Face Failure"* written by Sally Creel.

We have spent the last forty years and more of science education research looking at the ways in which stereotypes around gender (and other identities) impact science performance in the classroom, in the workplace, in popular culture, and in research spaces. What we have not spent as much time exploring is the stereotypes wrapped up in the actual *act* of science—the ways in which we *expect* science to be performed. The Fly Girls case is a perfect example of this. The students' expectations around how science is supposed to work impacted their relationship with science (and thus their long-term engagement with science). We need to rethink how we teach, portray, and explain the act of science in order to both better engage students of different identities and to also set up realistic expectations for what science is and does.

As a former science teacher, I distinctly remember trying to teach the Nature of Science (NOS) during the first two weeks of the school year to my students. I knew it was important enough to pull out of the content and explicitly teach, but what I forgot to do was to wrap and weave those ideas of NOS into the rest of the school year, to demonstrate the ways in which science was tentative and inferential and how, with more observations, better inferences may arise. As I moved into teaching preservice teachers how to teach science, I recognized this mistake and now work to help preservice teachers weave NOS into all the activities that they do in a science classroom.

K. Haverkos (✉)
Thomas More University, Crestview Hills, Kentucky, USA
e-mail: haverkk@thomasmore.edu

One of the things that I find most interesting in tying the Fly Girls experiences to NOS, is that as I teach NOS to preservice teachers, a key tenet is "the kind of knowledge and expectations with which we approach any phenomenon may affect the way we interpret that phenomenon." As I read Fly Girls, I kept wondering, what are the expectations that we approach doing science with and how do they affect the ways in which we interpret our experiences with science? Context is key to understanding our observations and experiences. What is the context that these girls are bringing to their expectations of science and doing science? One of the ideas that came to mind is that there are no mistakes in science. The Fly Girls didn't know how to act when their camera didn't record because their expectations of doing science was that mistakes are not a part of the process of science.

Let's explore for a moment, the ways in which doing science is portrayed in popular culture. When and where are mistakes portrayed? One of my own children's favorite shows was Disney's *Phineas and Ferb*, which has some great science and engineering cultural expectations. However, in the doing of science and engineering the mistakes are often only made by the villain of the show. Who wants to be the villain? This is especially the case when the villain lives up to all the identity stereotypes of an evil scientist as well! And these expectations around the doing of science (and how identity expectations are linked to this) don't stop in the world of cartoons and popular culture.

Even if we explore the ways in which "real" science is done, mistakes are not centered, even if that is often how science gets accomplished. What gets published? Not the mistakes. What is success based upon in the world of science research? Not the mistakes. Our relationship with real science is based on the successes and the expectation that success is the only outcome of scientific endeavors. One of the real-life consequences of this can be seen in the reactions not only of the Fly Girls, but also of the general population in their response to the recent COVID-19 pandemic. Expectations were around scientific success without understanding the tentative nature of the process of science and that success in science is often built on numerous mistakes and failures first. These expectations affect how individuals interpret the doing of science which can lead to many alternate conceptions and misunderstandings that are detrimental to any type of scientific literacy we may be seeking to achieve in science education.

So, what am I asking you to consider as you read about the Fly Girls' experiences? I am asking you to throw the step-by-step process of the scientific method out the window or at least help everyone understand that science is not done in a stair-step process every time—particularly if the top step equates with "success." I am asking you to weave the NOS into every science activity. I am asking that we consider the damage that only sharing the successes in science and engineering has done to the expectations students, teachers, researchers, scientists, our communities all hold.

Specifically in a classroom, we need to ask ourselves the following questions:

- How am I creating space for my students to make mistakes and learn from them?
- How am I sharing my own mistake making with my students?
- How am I creating an environment in my classroom where failure is normalized as part of the learning process?
- How am I normalizing mistakes in science specifically? How can I share the failures of science that eventually lead to success so students expectations of doing science are aligned to the realities of doing science?
- How am I building the nature of science into all of my science education so that students can both experience the explicit and implicit ways that NOS is important in doing science?

Through this kind of reflective work, we can begin to create spaces that see mistakes as an expectation of science, failure as a possible interpretation that leads to more questions and more scientific activity. The Fly Girls' frustrations with their results are a great opportunity for all of us to rethink how our own expectations of doing science may be reinforcing stereotypes that move us away from the goals we have set for inclusive science education and a scientific literacy that is so much more than a stair step to success. We need to be more like our favorite science teacher of all times, Ms. Frizzle, and "Take chances, make mistakes, and get messy!"

Kimberly Haverkos is the dean of the College of Liberal Arts and Social Sciences at Thomas More University. A former middle and high school science teacher, she worked as an Associate Professor in the Education Department at Thomas More. She holds a Ph.D. in Educational Leadership from Miami University and researches the intersections of identity, science, and the environment.

Case: Scratching and Rocking with Rocks!

54

Tamieka M. Grizzle

Abstract

It is the end of September and kindergarten students in the Bronx County School District (BCSD) in New York City have learned how to describe objects according to their physical attributes. Specifically, the standard states: SKP1. Obtain, evaluate, and communicate information to describe objects in terms of the materials they are made of and their physical attributes. (a) Ask questions to compare and sort objects made of different materials. (Common materials include clay, cloth, plastic, wood, paper, and metal.). (b) Use senses and science tools to classify common objects, such as buttons or swatches of cloth, according to their physical attributes (color size, shape, weight, and texture). Students also know basic coding with ScratchJr, an iPad app. However, in this case students' use of the technology interests with their understanding of classification, creating some unique challenges for their teacher.

Dr. Grizzle is the K-5 STEM Lab teacher at Kingston Elementary School located in the inner city of the Bronx. Kingston Elementary school is a Title I school where 75% of the students are on free or reduced lunch. The student demographics include 62% Black, 23% Hispanic, 11% White, 3% two or more races, 1% Asian, <1% Native American, and <1% Native Hawaiian or other Pacific Islander. Dr. Grizzle is a 15-year veteran at this school and has taught 4th and 5th grade including inclusion classes and an all-boys class, which she looped with from 4th to 5th grade. She is also the sponsor of various K-5 after school coding and robotics clubs. She initiated the STEM Lab at Kingston Elementary with the permission of her principal, Dr. Simmons-Deveaux, who supported her vision of brining computer science to marginalized groups (students of color and women)

T. M. Grizzle (✉)
Kimberly Elementary School, Atlanta, USA
e-mail: tamieka.grizzle@atlanta.k12.ga.us

© The Author(s), under exclusive license to Springer Nature Switzerland AG 2023
S. Jeong et al. (eds.), *Navigating Elementary Science Teaching and Learning*,
Springer Texts in Education, https://doi.org/10.1007/978-3-031-33418-4_54

Fig. 54.54 Students working collaboratively to sort and classify rocks using ScratchJr to demonstrate their learning

who are underrepresented in this STEM field. She also believes that in order to have a better representation of marginalized and underrepresented groups in computer science, you have to start them young and introduce computer science as early as elementary school. Dr. Grizzle does not have a computer science background but has developed a passion for teaching elementary students how to code and has earned a Computer Science Endorsement. While other labs in the school district focus on STEM challenges using recycled materials, Dr. Grizzle chose to focus on computer science and creating coding projects based on the Science New York Standards of Excellence. There are no desks or chairs in the STEM lab (see Figure 54.54), which is intentional because Dr. Grizzle wants the students to have enough room for collaboration and group settings since the lab is simply a small classroom that was transformed for the purpose of STEM engagement. Dr. Grizzle has been the school's STEM Lab teacher for the past four years and sees each K-5 class on a weekly rotation for 45 minutes.

At this point in the school year, Dr. Grizzle's expectation is that the kindergarteners have learned how to describe objects in terms of the materials they are made of and their physical attributes according to the Science New York Standards of Excellence. Dr. Grizzle is a passionate teacher who sets high expectations for all of her students and scaffolds students' learning when they are struggling. She also expects teachers to do their part in making sure that they are teaching the science standards, so that those standards will be a review in the STEM Lab. Since she only sees her students once per week, she spent the first four weeks of school teaching her K-2 students how to use the ScratchJr interface and students created simple and fun coding projects. Students learned that the blue blocks are for movement (up, down, left, right, turn left, turn right, hop, and go home), the yellow blocks are for triggering the program script (sending messages, start the program on green flag or by tapping a character, etc.), the green blocks are for sound (recording their voices), the purple blocks are for looks (shrinking, growing, disappearing, and speech bubbles), the orange blocks are for control (wait, set speed, repeat, and stop), and the red blocks are for ending the program script (repeat forever, next scene, and end). Students also learned how to code introductory activities ranging from simple to hard. These activities can be found at https://www.scratchjr.org/teach/activities.

It is 8:00 am and 17 kindergartners from Mrs. Lakeita's class are entering Dr. Grizzle's STEM Lab.

> Dr. Grizzle: Good morning! Come on in and sit on your spots. I have something fun planned for you today!

The students take their seats on the carpet, quietly waiting for class to begin. By now, they know the rules and procedures of the STEM Lab.

> McKenzee: Are we going to do ScratchJr today? I like Scratch the cat.

> Dr. Grizzle: Yes, and we are going to apply what you have been learning in class about classification.

> Journee: Classification? What is that?

> Dr. Grizzle: You know, when you put things into different groups according to how they look.

> Journee: Oh, do you mean like sorting?

> Dr. Grizzle: Yes, Journee, exactly like sorting. Who can give me an example of classification or sorting?

> Ramon raised his hand.

> Ramon: You put things in order.

> Dr. Grizzle: Yes, good Ramon. Can you give me an example?

> Ramon: 1, 2, 3, 4……10.

At this point, Dr. Grizzle was wondering whether her kindergarteners may not know what it means to sort or classify. So, she continues asking probing questions to her students.

> Dr. Grizzle: Okay, nice job counting in order Ramon. Can someone else give me an example of sorting or classifying?

Gisela raises her hand.

Gisela: a, b, c, d….

Dr. Grizzle: Very good Gisela, great job saying your abc's. What about you, Journee? You said classification is like sorting, can you tell me what you mean?

Journee: Yes, it's like when we sort shapes in math. We put all the big triangles together and all the small triangles together.

Dr. Grizzle: Yes! Great explanation, Journee! Okay class, today we are going to work in groups to classify rocks according to their physical attributes. You will sort the rocks into three different groups. Once your group sorts the rocks, you will open up the ScratchJr app on the iPad and use the Paint Editor feature to take a picture of each group of sorted rocks, upload the pictures to the stage, code your character to move to each group of sorted rocks and create a recording for the character that describes the features of each group of sorted rocks. Don't forget to add a background!

Dr. Grizzle sees quizzical looks on the kindergarteners' faces, but she reassures them that they are going to do a great job. Camille raises her hand.

Camille: What do you mean physical attri….(cannot pronounce attributes)?

Dr. Grizzle: Like luster.

Camille: Huh?

Dr. Grizzle: Luster means something is shiny. Physical attributes mean how something looks.

Dr. Grizzle proceeds to put the students in groups of four and gives each group a box of rocks and an iPad to share. She tells the class to begin sorting the rocks and walks around the room to make sure the groups are on track. However, some students begin playing with the rocks, others are coding with ScratchJr, some rocks are finding their way into pants pockets, and Dr. Grizzle overhears a conversation of one student telling her classmate that she was late to class today because she had to take an Uber to school because her mother's car broke down and the car didn't have any gas. Dr. Grizzle continues to tend to each group and tries to get them back on track. She asks probing questions such as: What shapes do you see? Is your rock shiny or dull? What colors do you see? Is your rock smooth or rough? Are your rocks big or small? Finally, there is some semblance of restored order in the STEM lab as students are now focused on sorting their rocks into three groups. Until…

Journee: How do I put the picture of the rocks in ScratchJr?

Dr. Grizzle's eyes widen as she realizes that importing pictures into ScratchJr was not an activity that she had taught previously. So, she quickly conducts a short tutorial by projecting her iPad on the SMART Board and not surprisingly, at least one or two of these digital natives in each group catches on quickly! However, some students realize that they can take selfies and pictures of each other and begin doing just that. The chaos begins again. At this point, Dr. Grizzle has 30 minutes left until the end of class.

> Dr. Grizzle: One, two, three all eyes on me!

The students settle down once again and each group begins snapping blocks together to create programming scripts that will tell their characters what to do. With 15 minutes left, Dr. Grizzle decides to have groups share their ScratchJr programs. She is looking for the group's character moving from one group of sorted rocks to the next group of sorted rocks and listening to how the character describes each group of rocks. Ramon taps the green flag to trigger his program script. His group was able to get the character to move, but in the wrong direction. His group's character described the sorted rocks using the correct terminology mentioned earlier in class (luster, small, and rough) but the rocks did not match the terminology used.

> Dr. Grizzle: Good job coding your character to move and recording your voice, Ramon. Do you think your group described the rocks correctly?
>
> Ramon: Yes, in class we only used pictures of buttons and we only did it one time.

Next, Dr. Grizzle calls on McKenzee's group. McKenzee loves to code with ScratchJr and has become somewhat of an expert. Her group was able to use the correct amount of movement blocks to get the character to each group of sorted rocks in the correct sequence and direction. McKenzee tapped her character to begin her group's program script.

> Dr. Grizzle: Great job with sequencing your codes and getting your character to move all the way to the groups of sorted rocks. Why did your group sort all three groups of rocks with the same feature, round?
>
> McKenzee: Well, that's all we know. We don't know how to describe the rocks with other words.

Dr. Grizzle calls on her last group. Gisela's group did a good job demonstrating computational skills in programming their character to hop the correct amount of times to the rock sort and using the sound block to record their sort explanation.

> Dr. Grizzle: Gisela, I only see one group of sorted rocks and your character stated that all the rocks are small. Can your group tell me another way you could have sorted the rocks?
>
> Gisela: It was too hard. We didn't see another way to sort the rocks.

At this point, Dr. Grizzle is wondering how she could have made this lesson better so that students left her lab with a conceptual understanding of classifying rocks or classifying anything for that matter using ScratchJr. She also wonders how much knowledge the students had prior to her lesson. Her students really enjoy coding with ScratchJr and are able to snap the blocks together to create a program script. But when it comes to integrating their coding knowledge with the exploration of science concepts, the students find it a bit challenging. Dr. Grizzle also wants to know how integrating ScratchJr could better serve as a foundation for learning sorting and classification. In the end, Dr. Grizzle wants students to be able to identify the common property of a group of items without being reminded of the possibilities. That's when she knows that she has developed critical thinkers! How can Dr. Grizzle achieve this?

For Reflection and Discussion

1. What are some ways that Dr. Grizzle could have made sure her kindergarten students were ready for classification of objects?
2. Technology use is ubiquitous throughout STEM classrooms, and students should be producers and not just consumers of digital content that support mastery of standards. In what other ways can technology be used to build understanding of other content areas?
3. When working with primary students (K-2), what steps would you take to ensure students are prepared to integrate technology into their learning?
4. What would you have done to make sure that there was a seamless transition for students learning about classification to engaging in integrating technology to reinforce learning about classification?
5. Think about how Dr. Grizzle responded to students' difficulties in understanding what classification meant. Or consider the frustration students faced when not knowing how to import pictures into ScratchJr. How could you have handled these situations differently?
6. Dr. Grizzle faced some challenging moments during her instruction with her kindergarteners. Think about your strengths and/or weaknesses with using technology. What can you do to make sure you are integrating technology that is culturally relevant to your students?

Tamieka M. Grizzle is an instructional coach at Kimberly Elementary School, a Title I school in the Atlanta Public Schools District in Georgia. Prior to that, she was the STEM and Innovation Professional Learning Specialist at a Title I STEM and STEAM certified school. Dr. Grizzle also taught 4th and 5th grades, became intrigued with STEM in 2016, and initiated a K-5 STEM lab at a Title I elementary school, formerly known as Harmony Leland Elementary School. Addressing the need for representation of minoritized groups and greater diversity in STEM careers, she used her STEM lab instruction to integrate coding and robotics across diverse disciplines.

Commentary: Not Just "Scratching" the Surface

Holly Amerman and Caleb Amerman

Abstract

This is a commentary to the case narrative, *"Scratching and Rocking with Rocks!"* written by Tamieka M. Grizzle.

Over the last twenty years, a major shift in education has been in the effort to integrate technology into the classroom. What began as simply introducing video and overheads grew into an effort to make students not just "consumers" of technology, but "producers" of technology. Yet, most of us who go into education did not do so because we are natural computer engineers. We may have used technology to do our assignments in school or enjoyed the latest in apps and games as teens, but actually making this happen in a classroom, no matter the age of the student, can be quite a challenge. Oftentimes, it is the "added" requirements of things like "integrating technology" that can drive both veteran and novice teachers to their wits' end, but successful integration of technology into a lesson can help students move beyond the "knowing" of science and into the understanding and interacting with science, as well as enhancing their ability to use critical thinking skills, just as was Dr. Grizzle's aim.

Caleb is a precocious 11-year-old who has attended a state-certified STEM school in the southeast for the last six years. As he describes his first experience of learning with Scratch Jr. in Kindergarten, you can hear echoes of what was probably happening in the children's heads in Dr. Grizzle's lesson:

H. Amerman (✉)
Darlington School, Rome, Georgia
e-mail: holly.amerman@gmail.com

C. Amerman
University of Georgia, Athens, Georgia

S. Jeong et al. (eds.), *Navigating Elementary Science Teaching and Learning*,
Springer Texts in Education, https://doi.org/10.1007/978-3-031-33418-4_55

My kindergarten teacher put us in groups and handed out the iPads. He told us to open the app that looked like the one he has up on the board. My group clicked and clicked on it, but it didn't work. Finally, we got it open, and he showed us on the Smartboard about the blocks and actions and told us to make a story with the Scratch Jr. I really loved playing with it from the beginning, my group was really good, but we fought over whether to make the background blue or red. Then our teacher told us it was time for recess. We didn't use Scratch Jr. again until the following year.

Holly, too, has had her fair share of failures with using technology as a veteran teacher of high school science; the effort to integrate technology was a bit more natural, but no less rife with difficulty. Early on in her teaching career she ended a lesson by asking her 12th grade AP Biology students to save their Excel documents with their data from the lab, and send them to her as an attachment so that she could combine the data overnight for analysis. Every kid nodded and smiled as if they knew just what to do. After waiting up until midnight for the students' emails, only one of the 13 groups (over four classes) completed the task of sending their data as an attachment. Remember, this was not homework, it was just attaching an already completed and checked file from in class and sending it. Entering the room mad the next day, she learned that the students could not find the "attach button," and after just a moment of pulling up an email screen on the board and showing students the paperclip icon, suddenly 13 more emails flooded her inbox.

What is the lesson for integrating technology in all three of these cases–kids are great with technology, but they don't have a clue on how to use it for school, and should not be expected to, no matter whether they are kindergarteners, seniors, or even college students. While we will always have digital native students who just "get it" (and, by the way, most teachers under 40 are digital natives now too!), we must treat the learning tools we use the same way we treat any other content–figure out what our kids know in that area, build on that knowledge, and support them to mastery. Teachers can use the smartboard or other projection device, standard in almost all classrooms, by walking kids step by step through the technology, at least until they are certain the students are comfortable enough to use the technology. Rarely does this step take as much time in the real world as you expect it to, as the more computer knowledgeable kids "get it" faster, and the spread of understanding generally happens quickly. Any students not catching on quickly can be moved into groups with the more advanced users and learn collaboratively.

Another important factor to consider when integrating technology into a lesson is whether or not the technology adds or detracts from the aim of instruction. In the case of Dr. Grizzle, the clear answer is that the use of Scratch Jr. was not an enhancement of the lesson. In a classroom of students who clearly struggle with the idea of classification, do not understand the classification scheme they are being asked to use, and are too young to properly focus when other distractions abound, this was likely not the right time to introduce Scratch into the mix. The aim was good, but as often happens, things went a little off track. A better lesson design would have stretched the activity over two lessons, one on classification, and the second on learning about and applying the technology to the content. This seems like a simple change, but in an educational world that seems to pride itself in shoving far too many requirements into far too little time, this was an unlikely obvious choice for Dr. Grizzle.

Integrating technology into science instruction may seem like a natural fit since STEM, after all, does start with science! But the pitfalls to avoid are as important as the selection of and planning for any lesson. High expectations for students are important, but expectations of previous knowledge with new apps, such as Scratch and Scratch Jr. are not the same as believing students need to work hard and they are all capable of learning. In the world of attempting to elicit deeper student understanding, and not just "Scratching" the surface of learning, spending the time to teach the technology, just as you would with any science content, is just as important as any other instruction that will happen in your classroom.

Holly Amerman is a 15-year science education veteran with more than ten years in the classroom at the secondary level and five years as a science coordinator at the school and district level. She is currently a Ph.D. student in science education at the University of Georgia with a research focus on applying artificial intelligence and machine learning in K-12 science education.

Caleb Amerman is an 11-year-old sixth-grade student at Darlington School in Rome, GA. As the son of two science teacher parents, he loves science and spends much of his time building with Legos or building Minecraft worlds. He is a competitive swimmer and loves to travel, having already been to 26 countries and 36 states in his young life.

Case: Transitioning to Online Learning: When Your A Student Gets an F

56

Ji Shen

Abstract

This case points to the complexity of using innovative approaches for teaching science in the second-grade classroom during a transition from face-to-face to online instruction. The case illustrates the different expectations of and negotiation among the teacher, parents, and students. Mrs. Bell is an experienced and passionate teacher who has taught for more than 15 years. With a doctoral degree in education and as a veteran of educational innovation, Mrs. Bell is optimistic in continuing to bring high quality education to her students during the pandemic. Questions and issues start to pile up while students and parents experience difficulties in keeping up with the learning demands of her innovative activities while living in the pandemic.

Five weeks into the online instruction mode in Coral Bay elementary school in Marine City in the United States during a global pandemic, life wasn't easy for anyone, especially teachers. Late in the night, Mrs. Bell, a second-grade teacher, was processing her emails after spending a couple of hours grading her students' assignments. Hours of Zoom classes during the day made her more than ready to step away from the computer screen. But then, she saw the email from Lucas's dad:

Dear Mrs. Bell,

Hope you are well. We checked Lucas's Nearpod lessons and ensured that he completed the three lessons assigned last week: Tiger Shark, Plant Growth, and States of Matter. We heard from Lucas today that you reassigned two of these lessons for him to do. We are wondering

J. Shen (✉)
Department of Teaching and Learning, University of Miami, Coral Gables, USA
e-mail: j.shen@miami.edu

S. Jeong et al. (eds.), *Navigating Elementary Science Teaching and Learning*,
Springer Texts in Education, https://doi.org/10.1007/978-3-031-33418-4_56

if it is because you could not see his work in the system, or because he did his work poorly. Please let us know.

Also, when we first received your message on Monday about Lucas missing the Flipgrid assignment, we sat with him to make sure he completed all the steps. He also made the video and uploaded it. Afterwards, we took a screenshot and attached it on Microsoft Teams' assignment when he clicked the Turn-In button. We are not sure what went missing.

Thank you for letting us know that Lucas received a "C" and an "F" in two assignments. However, we do not believe that reflects fairly his effort and performance considering his record in the past. Can you further instruct us on what he needs to do to change the grades?

Finally, while we understand you want to bring the best educational experiences to your students, which we support wholeheartedly, we just want to bring it to your attention that we do not have enough time to monitor closely Lucas's work on the computer. We have a fulltime job and our younger one can no longer go to daycare due to the pandemic. The recent switch from Studioboard to Microsoft Teams has caused much confusion already. To be honest, we don't think Lucas can manage all the technical side of the assignments on his own. Hope you understand.

Thank you!

Madiella and Robert

A letter like this was not the first time Mrs. Bell heard complaints or concerns about her assignments from students' parents. But it was the first time she received a complaint from Lucas' parents. Lucas has been a top student. He consistently followed instructions, worked very hard, and always turned in the highest quality homework on time. At least before the pandemic. Lucas has not been doing so well recently since the pandemic. Mrs. Bell was aware of the fact that Lucas's parents cared very much about his education. Mrs. Bell has been communicating with Lucas's parents about the low-quality of his online work and missing assignments. Earlier in the week, Mrs. Bell had emailed Lucas's parents that Lucas could redo his assignments because he received two unexpectedly poor grades—a "C" on a science lab and an "F" for an assignment in which he copied paragraphs directly from the Internet. Apparently, there was something lost in translation (or instruction).

Both assignments required the use of computer technology. Mrs. Bell was well-known in the school for her innovative approaches to education. She had students engage in projects about preserving local environments that students displayed at the city's youth fair. She and her students conducted videochats with local scientists during class, and she invited parents to her class to talk about STEM in workplaces. Mrs. Bell was an advocate of using educational technology in instruction. For example, she assigned several lessons in Nearpod, a platform where teachers can upload lessons with activities that allow many interactive features, such as drawing diagrams, responding to questions, taking polls, watching videos, and manipulating computer models. Recently, she taught students how to conduct video presentation and discussions using Flipgrid, a platform whereby a teacher may create a "grid" with specific instructions, and students can post video responses in tiled grid display. To make sure students understand what is required in an online assignment, she posted the instructions with very detailed and specific guidance on both Studioboard and Microsoft Teams. She also promised to deliver physical copies of any assignment to students' homes if needed.

Even before the pandemic, Mrs. Bell had received complaints and resistance from some parents, mainly about the extra work needed to help their children keep up with all the different platforms and login information. For example, students had to pay close attention to small details such as using a specific web browser to make a particular program work. Some parents met with the principal to request that her instruction go back to "normal." One parent transferred her child to another classroom to avoid the extra work. On the positive side, Mrs. Bell received strong support from other parents who understood that these assignments and learning activities were designed to get their children to practice higher order thinking

In fact, Mrs. Bell's class was the very first in the entire school to engage in synchronous online instruction after schools were forced to deliver distance learning during the pandemic. At the same time, she had also been offering workshops to other teachers in the school on how to use digital tools. In addition to designing and delivering online instruction as well as mentoring her peers in using digital tools, Mrs. Bell had to field more questions and requests from parents and students about various issues related to her online assignments. She had been very patient in responding to everyone's needs and tried her best to resolve technical issues as soon as possible. She spent an appreciable amount of time on emails and text messages and sometimes she doubted her innovative instructional choices, especially during such a challenging time. Every time doubt arose, she reminded herself that her students were very much engaged in and excited about the science learning activities. Mrs. Bell also firmly believed that her students were learning 21st century skills through her online science instruction that would benefit them throughout their education.

But then Mrs. received the email from Lucas's parents. She did not expect what she perceived to be a push-back from Lucas and his parents. Of all of her students, she had been certain that Lucas would thrive in an online environment and enjoy using all of the new technology tools in the online activities. She was perplexed about why his grades were so low but chalked it up to the transition and gave him opportunities to revise and resubmit his work. But, with his parents' email, it was clear that there were other issues in play. She wondered, is a pandemic the wrong time to incorporate her innovative instruction? Other students who traditionally did not perform as well as Lucas were doing fine. Why are some students excelling and others not? Should Mrs. Bell continue her educational innovations during a time of challenge?

For Reflection and Discussion

1. Should Mrs. Bell continue with the innovative learning activities she had planned for her students during the pandemic? Why or why not?

2. Should teachers assume that online assignments will require students to get help from their parents or guardians? Why or why not?
3. How should technology be incorporated in science instruction for different types of activities in lower elementary school? What issues might arise and how could those issues be addressed?
4. Considering what happened to schooling (and life) during a challenging time such as the pandemic, how could science classrooms better prepare students for the unpredictable future?
5. Construct a letter that you would send in response to Lucas' parents.

Ji Shen is Professor of STEM education in the Department of Teaching and Learning at University of Miami. His scholarly work focuses on developing innovative STEM learning environments, alternative assessments, interdisciplinary approaches, and modeling-based instruction in science.

Commentary: Managing Expectations During Online Learning

Kimberly Dinsdale

Abstract

This is a commentary to the case narrative, *"Transitioning to Online Learning: When Your A Student Gets an F"* written by Ji Shen.

I clearly remember the day that schools closed due to the COVID-19 pandemic. Now having taught elementary students online for several months, I also realize how this experience will continue to impact the way students learn, how teachers plan lessons, and how families participate in their children's schooling. During this time, my own household went from two people to five as my daughter, my son, and his roommate moved in from their college dorms. Every room had people learning, teaching, and working online, all while trying to manage the uncertainty of the pandemic.

I also knew that my students' families were feeling the stress of the unknown and the increased workload of managing their children's schoolwork. In one of the first studies on the impact of the COVID-19 pandemic school closures, Garbe et al. (2020) concluded that balancing professional obligations and multiple levels of learners in a household led to feelings of stress and a lack of personal time for parents. Keeping children motivated to complete classwork while lacking an understanding of the pedagogy of the curriculum caused even more stress for parents during distance learning.

Distance learning also pushed teachers out of their comfort zones, as they had to create digital lessons, manage technology, and build relationships with their students online (Kim & Asbury, 2020). Teaching is much more than standing in front of a group of students and lecturing, as it involves planning, creating,

K. Dinsdale (✉)
Saratoga Union School District, Saratoga, CA, USA
e-mail: kdinsdale@saratogausd.org

and implementing lessons that are engaging and exciting; assessing and providing meaningful feedback; and communicating with students and parents. Most importantly, teaching involves having strong relationships with our students and families. Like many teachers, Mrs. Bell was pushed out of her comfort zone as she tried to continue delivering innovative online science lessons, manage questions from students and parents, and help colleagues quickly increase their technical skills in effective online learning practices.

Yet, whether online or in person, this case showcases why it is critical that parents and teachers communicate when schoolwork is not meeting grade-level expectations, or there appears to be a change in a child's attitude or work behaviors. The problem may be as simple as the student not understanding the directions, or the problem may be indicative of a larger issue. During the pandemic, some students were experiencing loneliness, isolation, and anxiety that affected their motivation to complete schoolwork. I had students in my own classroom who refused to turn on the video camera during class, complete online classwork, or allow their parents to help them. We know now that some of our students experienced mental health issues as a result of COVID-19 quarantine safety protocols and, whether in a pandemic or not, changes in attitude and work output must be immediately addressed. Belair (2012) found that parents who took an active role in helping students communicate with their teacher modeled effective communication, which translated into an increase in academic achievement and student motivation. It appears in this case, that both Mrs. Bell and Lucas' parents responded quickly and professionally to the change in Lucas' academic activities and tried to reach a solution that would benefit Lucas. It is evident that Mrs. Bell had high expectations of Lucas' ability to navigate the technology, so his inability to produce grade-level work gave her pause and made her reflect on the delivery of the science lessons or consider if there were other issues contributing to Lucas' struggles.

Additionally, Mrs. Bell was an educational leader in the development of curriculum with the infusion of technology. Innovative teaching techniques, coupled with technology, are essential to effective learning pedagogy and for meeting the academic and social needs of our students. Utilizing technology and multimedia tools is effective in differentiating curriculum and giving students voice and choice in their learning. Technology gives all students, regardless of income and geographic location, access to information that they may not typically have access to and creates opportunities for learners to acquire critical twenty-first-century learning skills. The use of varied multimedia also allows science teachers to engage and keep the attention of the Generation Z students. Though a learning curve in utilizing technology exists, flipped and blended classrooms can be beneficial in providing tasks to online learners and offer flexibility in learning choice (Merrill & Gosner, 2020). Mrs. Bell's use of online technology allowed her to quickly pivot to meet the needs of her students as they transitioned online from a face-to-face classroom.

Technology alone, however, can be very isolating and ineffective for some learners. Distance learning revealed these issues in my own classroom, even with students who previously enjoyed learning with technology. Some of these issues surfaced in part due to the repetition of online assignments, lack of students' digital skills, and the distraction of videos and gaming at the click of a button. These issues were compounded because parents were inexperienced in using online platforms, busy with their own work obligations, and now tasked with helping their children manage their online work. Additionally, there was the possible issue of students' lack of self-regulation and keyboarding skills needed to complete online assignments. Kumar & Rajasekhar (2020) concluded that distance learning curriculum needs to consider students' working memory and the cognitive load required to process visually and auditorily via Zoom or a Google classroom platform. I found that my students could not complete the same level of work that they would have been able to in a face-to-face classroom, and that I needed to modify the curriculum. Even the best laid out directions require interpretation, and without a teacher or peer to clarify instructions, some students seemed to be lost when starting an assignment. Many students, once they signed off from the online classroom, did not know how to complete assignments, and our elementary students often were not mature enough to know how to ask for help.

Like Mrs. Bell, I also found that some of my students thrived in a distance setting with online learning, while others did not. In reflecting on my practice, I realized that the real issue is that innovative learning techniques also require collaboration. The reason that twenty-first-century skills are incorporated in current pedagogy was the need for students to be able to critically analyze problems, come up with creative, out-of-the-box solutions, and, most importantly, be able to communicate their ideas into fruition. Learning theory informs us that giving students opportunities to socially collaborate provides peer scaffolding to increase learning for all students. Mrs. Bell's use of technology engaged her students and pushed them to be critical thinkers and innovative leaders. Yet, technology, without collaboration, takes away the students' audience and diminishes creative thought.

What I have learned from teaching at a distance is that teachers can build strong relationships and community, students can learn, and parent–school relationships can be strengthened. Distance learning, however, can be difficult for some children. Mrs. Bell was insightful and responsive in reflecting on her lessons and making modifications as needed to meet the needs of her students. Mrs. Bell showcased projects that provided students with an opportunity to create and tackle real-life problems and continue to engage in higher-level learning. Though some of the families in Mrs. Bell's classroom felt that the work was too difficult, making her doubt her innovative lessons, she at least tries to maintain high expectations for her online science instruction. As teachers we need to not be afraid to be pushed out of our comfort zone because it is what our students need from us. We have to model that it is ok to try something and fail, make changes, apologize for a lesson gone awry, and move forward. These are the critical skills that our students learn from us when we attempt new things whether those things are successful or not. On most days we get it right, and on some days we don't! Through the process of trying, we model humbleness, flexibility, and resiliency when faced with obstacles and give each other permission to make mistakes and start again.

References

Belair, M. (2012). The Investigation of Virtual School Communications. *TechTrends: Linking Research and Practice to Improve Learning*, *56*(4), 26–33. https://doi.org/10.1007/s11528-012-0584-2

Kim, L. E., & Asbury, K. (2020). 'Like a rug had been pulled from under you': The impact of COVID-19 on teachers in England during the first six weeks of the UK lockdown. *British Journal of Educational Psychology*. *90*(4), 1062–1083. https://doi.org/10.1111/bjep.12381

Kumar, V. D., & Rajasekhar, S. S. S. N. (2020). Too much but less effective: Managing the cognitive load while designing the distance learning instructional formats. *Journal of Advances in Medical Education & Professionalism*, *8*(2), 107–108. https://doi.org/10.30476/JAMP.2020.85990.1208

Merrill & Gosner (2020). *Teachers around the world tell us reopening is tough, but joyful.* https://www.edutopia.org/article/teachers-around-world-tell-us-reopening-tough-joyful

Kimberly Dinsdale is a principal at Saratoga Union School District in Northern California. She has been an educator for seventeen years working as an administrator and as an elementary and Gifted and Talented (GATE) teacher and served as an English Language Arts lead for curriculum development. She holds a Doctorate in Applied Learning Sciences from the University of Miami and enjoys sharing best practices with school communities to promote rigorous and supportive learning environments.

Case: Sphero Struggle: Productive or Demoralizing?

58

Lautaro Cabrera

Abstract

Bridget is in her second year of teaching 4th grade at Maple Elementary and has just started her Masters in STEM education after realizing that teaching science is her passion. During her last year of college and first year as a teacher, she participated in a course and professional development project to learn about computational thinking and computer science. These experiences, coupled with the supportive teaching staff and administration at Maple, have convinced her to integrate robots in her classroom. She's excited to give the kids the opportunity to develop persistence, learn about science, and practice their coding while they program their robots. In this case, Bridget shares the unexpected challenge she faced when two of her students encountered difficulties with "productive struggle" during a robotics design lesson.

Desks against the wall. Students standing, jumping, laying on the ground with rulers. The room covered with scraps of paper, sticks, and glue. The floor swarmed by robots emitting lights and sounds as they roll around. A cacophony of robotic chirps, student cheers, questions, and Eurekas! The sound of learning.

I created this lesson to foster my kiddo's learning about animal adaptation. My students had to pair up and program the robots so that they would mimic the behavior of an animal of their choice. I gave the kids paper cups, markers, glue, tape, and other crafting tools so that they could "dress" their Sphero robot. But the biggest challenge was to mimic movement—they had to use code, such as [roll 0° at 60 speed for 2 s], to make their robot walk, swerve, or run.

L. Cabrera (✉)
University of Maryland, College Park, MD, USA
e-mail: cabrera1@terpmail.umd.edu

After a short tutorial on connecting the Sphero to the iPad and placing blocks together to make a program, I gave students the green light to start. The room erupted.

"Let's make it go super-fast like a cheetah!" "I want to do mine first!" "Man, now I want to change my animal!" Doing their best to take turns, students began planning, testing, fiddling, measuring, experimenting, and loudly sharing every event in the process. As I walked around the classroom, trying to listen in to one conversation at a time, Alexis and Tim made me feel like a Spheros lesson was the right choice. Alexis had chosen a bunny and, while they realized that there was no way to make the ball-shaped robot hop, they could mimic the behavior by making it roll fast a specific distance, stop, and then roll again.

Alexis: Each time it should go the same because the jumps are all the same

Tim: Yeah, so make a hop like one floor tile long?

Alexis: Okay, so try roll 0° at 60 speed for 5 s

When the "bunny" rolled for way more than a floor tile, the pair tried again. This time, they changed the roll time to one second. It fell short by about half a tile. So they measured the distance, debated changing the speed or the time and eventually settled for the same code, but for 2 seconds. The Sphero was still short from a full tile "jump" by about an inch. "Argh!" Tim exclaimed, with Alexis making a complementary frustrated facial expression. But then, they quickly resolved to slightly increase the time of the roll and tried again, finally succeeding. "YES!" "We got a hop!!!"

I was pleasantly surprised. They had "failed" THREE times! But in what looked like a clearly frustrating moment, they immediately jumped back to problem-solving and persisted in their goal. I've seen these kids be challenged and defeated in my classes before. I could not imagine anyone trying the same exercise three times and not asking for my help! My plan to induce a productive struggle was working.

As I moved on to other groups, I felt like Alexis and Tim were not an exception. Everywhere I looked, students were sharing (yelling) new ideas to their partner, using rulers and fingers to measure distances, and using their own bodies to pretend like they were robots and gain insight into the kinds of movements they should program. I could feel the energy of real science learning—the beauty of organized chaos that results from tinkering, persisting, and being inventive.

But, when the room is loud and hectic, it's hard to notice the quiet and still.

As I continued my tour of the classroom, an image startled me. Nora and Lindsay were sitting, slumped on their chairs against the wall. The contrast with the frenzy in the classroom made me almost miss them in the motionless background. Lindsay was raising her left hand in the air, while using her right hand to support the other arm—as if she'd been waiting for hours. Their robot was stationary, with its blinking blue light indicating it was paired and ready to be programmed.

"What's going on?" I asked carefully as Lindsay dropped her hand in relief. "We don't know how to do it," she said. "I wanted to do a snake, but we can't make it turn," Nora added.

The pride I felt during my stroll through the classroom was quickly ceding to a deep concern. How long had these girls been waiting for help? Was my activity only good for those who already had the ability to persist? How could I motivate them to go on?

"Well, show me what you tried so far, and let's make sure your robot is paired," I said to buy me some time as they stood up and handed me their iPad. Their code had a block to make a sound and then a movement block that seemed random: roll 24° at 182 speed for 2 s.

"Okay, so what happened when you tried this code?" I asked to kickstart their thinking process. "It went far for a while. But then we looked, and there's no block to turn" Nora replied. "Yeah, and the snake is supposed to move one way, then turn and go the other. How do we do that?" Lindsay asked.

I knew exactly how to do that. I knew that they needed multiple blocks and a change in the first "heading" slot, where the degrees go. But I couldn't just tell them. I was committed to my method—if I give them the answer, then there's no productive struggle! I had learned in professional development that girls need more support in being brave and trying out new things—especially when doing computer science. Would I be encouraging bravery if I just handed them the blocks they needed?

Bridget: Well, let's see if we can use this block you already have to make it turn. Have you tried changing the numbers?

Nora: Yes, but it just goes more forward

Lindsay: And we don't know what the little zero is after the first number

Bridget: Have you tried changing it and see what happens?

Nora: Yes, but it doesn't turn. Can you show us how to make it turn?

"Argh!" I thought, channeling my inner Tim when he was considering the code for their bunny. What a great opportunity to learn about degrees! But, I stopped myself. Wouldn't a three-minute-long explanation about degrees defeat the purpose of the activity? I don't want to tell Nora and Lindsay that the first number of the block actually controls the direction of the robot—I want them to find out through trial and error. But, from the look on their faces, it's clear that Nora and Lindsay feel defeated; they seem to need a small win, not another challenge.

For Reflection and Discussion

1.1. How can Bridget maintain her commitment to "productive struggle" while ensuring that Nora and Lindsay succeed?

2.2. How might Bridget support students' development of persistence when they don't feel confident to try again?

3.3. How can Bridget balance providing adequate support for students with creating an environment where they can productively fail?

4.4. Considering that Bridget particularly wants girls to develop an interest in computer science, what additional considerations might Bridget take in her interactions with girls during the lesson?

Lautaro Cabrera or LC as he likes to be called, is an educational researcher who focuses on integrating concepts and applications of computing into elementary and middle school science education. His research aims to understand how students and teachers conceptualize computer science ideas and use them to advance disciplinary understanding. LC is originally from Argentina, has a Bachelor's degree in Psychology from Ohio Wesleyan University and a doctoral degree in Technology, Learning, and Leadership from the University of Maryland, College Park.

Commentary: Spheros Struggle: Where Failure and Success Intersect

59

Ellie Cowen

Abstract

This is a commentary to the case narrative, *"Sphero Struggle: Productive or Demoralizing?"* written by Lautaro Cabrera

I read one of my favorite quotes about productive struggle on a small sign on the wall of a preschool hallway, years ago. The sign read, "A child loves his play, not because it's easy, but because it's hard" (Benjamin Spock). After reading it twice, I walked back to the classroom where I was teaching third grade, pondering the simple and universal truth of this assertion: children at play are instinctively engaged in productive struggle. As with so many observations about young children, it holds true for adults as well. We don't water ski or host dinner parties or knit sweaters because these are easy things to do. Humans just love to grow, and struggling is an essential step in the learning process.

Bridget's story is loaded with significance, because in it we see a practitioner who both recognizes the power of the productive struggle and then encounters the unavoidable complexities of trying to harness that power in a real live classroom. Bridget has designed a lesson that checks all the boxes for powerful experiential learning: it prioritizes the student-centered exploration by placing it at the start of the lesson, it is open-ended and engaging, and it provides opportunities for observational assessment and feedback.

When Bridget approaches Nora and Lindsay, she notices that they have not yet met with the same success as their classmates. (This is already a win for Bridget, though she may not feel it in the moment: she is receiving valuable information

E. Cowen (✉)
Nashoba Regional School District, Bolton, USA
e-mail: ecowen@nrsd.net

about not only the girls' current task mastery, but also their perseverance in problem solving!) She asks several questions to get herself caught up on the journey that has brought them to this deadlock: (1) what have they tried so far, (2) what happened, (3) have they tried changing the numbers yet? Again, she is on the right track, committing to the practice of letting children fail in small doses, without giving in to the temptation to solve their problems for them.

But what happens next? There are a variety of options available to Bridget at this moment, each of which could prompt the girls forward without undermining the goal of letting them learn through their own trial, error, and re-trial.

One option is simply feeding them the next step. While Bridget is right that a mini-lecture on degrees probably wouldn't do much good in this scenario, a little scaffolding won't hinder the productive struggle process. Consider a lively game of hide-and-go-seek: when the seeker is stuck, the game is ruined if all the hiders come out of hiding and give themselves away. Luckily, some brilliant seeker once cried: "Make a noise!" and the rest is hide-and-seek history. Perhaps all that Nora and Lindsay need is a noise: a small, but straightforward, nudge in the right direction.

Another option is simply to change the goal posts. There are times when a teacher over–or under-estimates the readiness of her students for a challenge. A flexible, creative educator might decide to adjust the objective to fit the students. Bridget has designed an activity that is sufficiently open-ended to leave this option open. "Hmm," she might say. "A snake is tricky! Why don't you try a lizard instead? See if you can make it turn and then walk."

A third option is to sit back and allow this moment of failure–for now. As hard as it is to watch our students get stuck, sometimes the best practice is to let them not-know, for a little while. In her New York Times Magazine article "Why Do Americans Stink at Math?" Elizabeth Green proposed a lesson format that flies in the face of the popular "gradual release of responsibility" model that is sometimes referred to as "I Do-You Do-We Do." She argued that starting with "You Do," in which students work independently, then moving to a team or pair share (she calls this "Y'all Do"), and ending with "We Do," in which a teacher guides a whole group discussion, is better for building engagement and fostering perseverance (Green, 2014). In Bridget's lesson, the "You Do" portion of the activity is well underway. Some students are meeting with great success, while others are struggling. The beauty of the "You-Y'all-We" format is that in the next step, the "Y'all Do" step, they will teach and learn from one another. Any elementary educator knows that students are more interested in what their classmates have to say than what their teachers have to say. And having already spent several minutes trying (and yes, failing) to make their "animal" move, Nora and Lindsay are now perfectly primed to learn from the classmates who have accomplished what they have not.

Of course, a truly exceptional teacher is likely to combine methods. Perhaps Bridget will give the girls a prompt or change the goal, then walk away for a few minutes, giving them a little more time with the challenge, before bringing groups together to talk about their strategies. She will be careful to pair the children that struggled most with those that figured it out. And when they finally come together to discuss in whole group, Bridget can keep an eye on the students who had the most trouble. She might even encourage those students to share what they learned, so they, too, get a "win" in the end.

Ultimately, that's the beauty of a strong productive struggle lesson: failure is success, as long as you play your teaching cards right!

Reference

Green, E. (2014). Why do Americans stink at math? *New York Times Magazine.* https://www.nyt imes.com/2014/07/27/magazine/why-do-americans-stink-at-math.html

Ellie Cowen is a math specialist for the Nashoba Regional School District in Stow, MA. She has been a math teacher specialist, instructional coach, and consultant for fifteen years. Her work centers around designing powerful instruction and building student engagement in the learning process.

Part VI
Assessing Students' Learning

Deborah J. Tippins

Meaningful authentic assessment is an important part of the science teaching and learning process. In contrast to assessments premised on discrete definitions and categories, authentic assessment is grounded in a belief that young children need opportunities to demonstrate what they know rather than what they do not know. In an era of increased accountability, scripted curricular demands, and large-scale testing, we, as teachers, are called on to rethink assessment in ways that position children as empowered intellectuals. Implicit in this notion of the empowered intellectual is the recognition that children should have a voice in their own assessment. This kind of assessment environment happens when teachers spend time helping children represent their learning in many different ways. By encouraging divergent thinking and the asking of questions teachers can help children begin to assess their own abilities and aptitudes.

Assessment practices are influenced by teachers' beliefs about science teaching, learning, and curriculum. For beginning and experienced teachers alike, it may be difficult to let go of a notion of expertise in which the teacher is the sole authority of knowledge. While this is not always easy, it is important to recognize that teachers are also learners. How teachers view learning and how student and teacher roles can directly impact what they determine counts as evidence of learning. We see an example of this in Mandy Smith's case *Thinking outside of the (Check) Box: Honoring Creativity Within, and of, Young Children's Science Learning*. In this case, two preschool-kindergarten teachers find that portfolio assessment can be an authentic way of representing what children are learning over time. This form of assessment provides opportunities for the student and teacher to work together in selecting student work artifacts for the portfolio which can best represent science learning. However, the teachers in this case wonder how to explain their assessment philosophy to a visiting Headstart administrator who believes that student learning can be reduced to a mere number. Jennifer Stark, in the case *What Happened to the Puddle?*, relates the story of Elliot, a fourth-grade teacher who also struggles with the question of what counts as evidence of science learning. As part of student exploration of evaporation in a puddle, Elliot uses knowledge of key vocabulary

D. J. Tippins
Department of Mathematics and Science Education, Mary Frances Early College of Education, University of Georgia, Athens, GA, USA
e-mail: dtippins@uga.edu

words as the main criteria for assessing students' understanding of states of matter. In the process, he comes to understand that rote mastery of vocabulary does not guarantee learning.

Assessment can be complicated, especially for the early childhood educator who may believe personal and social learning are equally important as academic science learning. And often, the goal of helping students learn about themselves as learners cannot be easily separated from academic goals. Thus, it is important to recognize that assessment can have many different purposes. At the core of the issue is the distinction between assessment as learning and assessment of learning. What is often referred to as formative assessment is an approach to enhancing learning by recognizing and responding to students in ways which provide both the teacher and student with feedback. Summative assessment is typically used when a teacher wants to sum up or make a judgment about student learning after a given period of time.

There are a wide range of assessment practices which are integrally linked to instruction and the activities that accompany it. It is not uncommon to see teachers using, "thumbs-up," "traffic lights," and similar strategies. However, assessment must go deeper to truly develop an understanding of children's prior knowledge and ideas about science phenomena. Concept maps, K-W-L charts, student science journals, and predict-observe-explain are just some of the strategies that teachers might use to probe students' sense-making in deeper ways. Valarie Bogan and Selcen Guzey in their case *Sunscreen Design for Assessing Students' Engineering Practices and Science Learning* highlight the struggles of Mr. Jensen, a fifth-grade teacher who wonders about students' level of understanding during an integrated STEM unit in which students are asked to design a natural sunscreen. He is dismayed when students' designs and presentations of their sunscreen projects do not seem to reflect any application of science content. Likewise, Michelle Petersen also uses a culminating project to assess her fifth-grade students' abilities to demonstrate their knowledge of models and modeling in the case *"Arts and Crafts" with a Side of Science*. While Michelle strives to let students have ownership in their assessment, she is nevertheless perplexed by the superficial level of understanding demonstrated after a year-long effort to focus on models and modeling. Similarly, Elizabeth A. French in the case *Where the River Flows* illustrates the complexities of using projects as a form of assessment.

Across all these cases, it is important to keep in mind that assessment is situational. Assessment must always consider the context in which sense-making takes place. With the recognition of the increasing diversity of today's classrooms, we must also keep in mind the presence of diverse sets of worldview and cultures that contribute to the uniqueness of each assessment context.

Case: Thinking Outside of the (Check) Box: Honoring Creativity Within, and of, Young Children's Science Learning

60

Mandy McCormick Smith and Sophia Jeong

Abstract

We live in an era of accountability characterized by an increasing emphasis on value added measures of teacher effectiveness and large-scale testing. Amidst a context of accountability and achievement discourses, Delia and Adrian, two preschool-kindergarten teachers, take on the challenge of using portfolio assessment to support diverse children's science learning. As they begin to experience success in using this assessment strategy, they are dismayed when a visitor from the state Head Start agency expresses dissatisfaction and uncertainty with the teachers' use of portfolio assessment.

Delia is an early childhood teacher with over 15 years of experience, employed as a classroom lead teacher in a preschool-kindergarten classroom within a large, urban school district. Delia's preschool-kindergarten classroom centers around the engagement of children in a year-long STEM project steeped in the Reggio Emilia traditions of emergent, child-centered curricula. Adrian is a classroom aide in Delia's room and is engaged in his own professional learning through his involvement in Delia's STEM project. Delia and Adrian developed the STEM project together with a focus on an integrated and creative approach to lesson planning where each week's activities were designed in ways that allowed children to explore materials in their own self-directed ways. Both Delia and Adrian believed that whatever the students created should have personal meaning to them and the learning process. As such, they rallied against *crafts* or any prescribed outcome.

M. M. Smith
The PAST Foundation, Columbus, OH, USA
e-mail: msmith@pastfoundation.org

S. Jeong (✉)
Department of Teaching and Learning, The Ohio State University, Columbus, OH, USA
e-mail: jeong.387@osu.edu

The STEM project sought to emphasize process over product through all class-room activities as well as through the assessment of children's understandings. Adrian, a self-identified artist, was instantly comfortable with a process over product approach to science teaching and learning. However, both teachers struggled with how they might assess students' inquiries.

> Adrian: To properly see growth we need to understand the process it takes for each child to get there. By doing so, we can honor each child's unique path.

> Delia: This feels very "loosey-goosey" or a bit "new-age" to me. How should we assess what the kids are learning? How can we connect the eventual assessment to the content learning standards?

> Adrian: Inquiry can not be fairly assessed in a checkbox form, can it?

As Delia and Adrian pondered their assessment dilemma, they sought out some professional readings from the *Science and Children* journal to learn more about portfolios. Delia had been introduced to the idea of portfolio assessment in her teacher preparation program, but wanted to refresh her memory and share with Adrian what she had learned about portfolios.

> Delia: Portfolios are used to show an individual's understanding at given points in time, like snapshots, and assembles them in ways to share the child's growth over time, similar to movies. The portfolio is a type of journey that takes both the teacher and student on the longer path of learning. Along that path children may make leaps and bounds one day, and struggle to concentrate on anything the next day. The student's path is authentic, sometimes moving forward and sometimes moving back, but all of it very real. Portfolios take into account the idea that learning is the compilation of a student's understanding over time.

> Adrian: So, do we decide what kind of evidence to put in each child's portfolio? As educators, isn't it up to us to connect children's efforts to the outside-world's standards?

> Delia: A portfolio is more than just a compilation of understandings. It is the learners' personal journeys and holds their independently voiced explanations. The information within the portfolio goes far beyond what can be obtained with a standardized testing instrument. The portfolio demonstrates a student's all-around growth over time in such areas as language, reasoning and inquiry skills. It shows how students' curiosities change and interests develop over time. Portfolios provide an opportunity for all voices to be heard, not just those that are the loudest. They allow us, as teachers, to step into the shoes of students and see the learners' perspectives. If we truly value child-centered, inclusive education the individual learner should be at the center of assessment and should have a say in selecting the evidence that will go into the portfolio.

> Adrian: There are amazing things happening in our classroom every day. We could not stop it if we tried—it is the curious nature of children. They tinker, play, laugh and try again. What if we sat back and observed the wonder of children's worlds as the gold standard and tried to fit our adult worlds and standards into their checkboxes, instead of asking children to limit their curiosities to a checkbox? I really like this idea of portfolio assessment and think we should give it a try.

Delia and Adrian were excited as they began to introduce portfolio assessment into their preschool-kindergarten classroom. Yet they quickly realized that portfolios can be messy. They wondered: How should we store students' creations over time? How should we document a student's observations when placing white carnations into different colors of water? Eventually Delia and Adrian reached a compromise solution, although they were not sure if it would be sustainable over time, especially when Adrian moved on to a classroom of his own and Delia did not have the extra set of hands. They decided to write students' ideas directly onto their projects or onto index cards. These dated index cards were attached to each activity and Delia and Adrian had discussions with each child to select which artifacts would ultimately end up in the portfolio. In this way, each child's portfolio began to build over time.

The children, their parents/guardians, colleagues and building administrators all started to take note of the incredible growth they observed in students across time. But more importantly, Delia and Adrian observed firsthand the changes in their students with the creation of a space for the young children to take ownership in their own learning and assessment. Adrian pointed to Miguel, an older student in the classroom, as an example of the growth they could see in many children.

Adrian: Miguel often struggles to express his ideas with peers and teachers. His emergent language abilities sometimes overshadow or hide his innovative ideas. Sometimes because Miguel lacks the proper "adult" word, the other children run out of patience with him. The weekly explorations and self-directed inquiries give Miguel multiple mediums to bring his ideas to life. I've observed how he is now more involved and engaged in projects. Our STEAM project, together with the use of portfolios, has given Miguel a new avenue for communication.

Delia: Yes, I've seen Miguel's enthusiasm and have observed how fast his language skills are developing, particularly as he tries to describe scientific phenomena. Since we are emphasizing process over product it makes sense to also recognize the sense-making process that students like Miguel experience. The product may not come neatly packaged in a clean score, but who benefits from a score, and is it the data we need? We do need to connect learning experiences to content standards, but if we only document student understanding using traditional assessments, the picture is incomplete and removed from our brilliant sense-making students themselves.

Delia and Adrian finished the week of teaching with a sense of elation and freedom. They were confident that their approach to considering students' science work over time was intricately connected to the knowledge and experiences children bring to the classroom, their growth, and the connections they make with peers and adults in the classroom. Delia and Adrian were not naïve, recognizing that some educators may not initially see the value in emergent curriculum and portfolio assessments. However, the following week they were taken aback when the state agency Head Start representative visiting their classroom dismissively stated: Show me the test scores!

For Reflection and Discussion

1. The emergent, child-centered philosophy of approaches to teaching such as Reggio Emilia seem to run counter to current reforms calling for increased accountability through test scores. How would you balance these seemingly competing tensions if called on to explain your teaching philosophy to parents, other teachers, administrators or representatives from state agencies?
2. How would you introduce portfolio assessment into an early childhood science learning environment? What are some of the factors you would take into account in planning for the use of portfolio-based assessment?
3. What are some of the benefits and constraints that might accompany portfolio-based assessment?
4. What are some other ways that you might assess students' STEM learning in a preschool-kindergarten environment similar to the one in this case?

Mandy McCormick Smith is a former K-12 classroom teacher, STEM elementary teacher educator, and is currently the Director of Research at the PAST Foundation, a non-profit organization focused on STEM Education in Columbus, Ohio. Smith has 10+ years of experience in research and evaluation with purposeful skills that intersect in numerous areas, including research design, evaluation, statistical analysis, testing and assessment protocols.

Sophia Jeong is an Assistant Professor of Science Education in the Department of Teaching and Learning at The Ohio State University. Her work draws on theories of new materialisms to explore ontological complexities of subjectivities by examining socio-material relations in the science classrooms. Her research interests focus on equity issues through the lens of rhizomatic analysis of K-16 science classrooms. She is passionate about fostering creativity, encouraging inquisitive minds, and developing socio-political consciousness through science education.

Commentary: Can You Score Uniqueness, Creativity, and Imagination?

61

Sevil Akaygun

Abstract

This is a commentary to the case narrative, *"Thinking Outside Of the (Check) Box: Honoring Creativity Within, and Of, Young Children's Science Learning"* written by Mandy McCormick Smith and Sophia Jeong.

The letters x, y, z, and α do not only refer to symbols used in the alphabet, but also to the names of generations born after 1965, 1981, 1997, and 2010, respectively. These generations have unique characteristics, but also unique needs, as the world and the society they live in differ. Since the beginning of the twenty-first-century, the world has been rapidly changing to become more globalized; citizens need to have particular skills, also called twenty-first-century skills, including creativity, critical thinking, communication, and collaboration. Accordingly, the education system has evolved in a new direction where curricula place more emphasis on these skills. In addition, Science, Technology, Engineering, and Mathematics (STEM) education has started to be implemented from kindergarten through high school. Because the students in the kindergarten today will be the citizens of tomorrow, they will need to gain twenty-first-century skills as part of their formal education. Therefore, the three pillars of instruction—*aims, instructional methods,* and *assessment*—should be aligned with these twenty-first-century skills. Assessment of learning generally is not easy; assessing twenty-first-century skills is not easier. While some educators prefer to use portfolios, and some insist on tests, both have merits and drawbacks.

Alternative methods of assessment, including portfolios, have been suggested to assess twenty-first-century skills as they showcase the process instead of just

S. Akaygun (✉)
Department of Mathematics and Science Education, Bogazici University, Istanbul, Turkey
e-mail: sevil.akaygun@boun.edu.tr

evaluating the product. In portfolios, students can demonstrate and communicate their critical thinking, creativity, and imagination over a period of time by filing work, including drawings, essays, or photographs, that convey their uniqueness. For instance, in Turkey, when students apply to a Fine Arts High School, they submit a portfolio of their work created over months or even years. The referees or educators who evaluate the portfolios not only assess students' performance and abilities, but also how they improved and developed their identity over the time of collection. Portfolios can convey rich information about the students. The preparation of portfolios in kindergarten, particularly at a school where the Reggio Emilia approach has been adopted, is also appropriate. The Reggio Emilia approach follows an emergent and child-centered curriculum focusing on students' strong potential and how they grow in relations with others. While the students work on their STEM units, they can prepare their portfolios by collecting, documenting, and presenting their artifacts over the course of the units. At the end of each unit, the portfolios can give a complete picture of each student.

However, some educators strongly argue that assessment should be objective and generalizable so that comparisons and decisions can be made based on the analysis of test results. These educators actually have a point. Many countries find it very important to compare their students to students in other countries using worldwide tests, such as Program for International Student Assessment (PISA) and Trends in International Mathematics and Science Study (TIMSS). This comparison allows them to make inferences for their science and mathematics education, as well as students' reading and comprehension skills. For instance, educators might like to know more about the science and mathematics education in the countries like China, Korea, or Finland, where students recently received the top scores in these exams; this kind of information might help other educators change their systems accordingly. The questions in these tests do not measure factual knowledge or rote memorization, but instead measure higher-order thinking skills, such as critical thinking, problem-solving, or designing experiments.

While both methods of assessment have merit, they also have limitations. Portfolios may bring too much extra work for teachers. In general, teachers are usually busy with preparation, planning, teaching, and grading; with portfolios, they will need to think about organization and logistics as well, including where to store the binders and how to involve students in selecting their work. In addition, evaluating complete portfolios is not easy: assessing an engineering design drawing of a tower made up of pasta noodles and marshmallows, or a story about a bridge, is a challenge without having very clear rubrics for evaluation.

Standardized tests also have drawbacks, even though the assessment of them is easy. The tests are given to everyone on the same day, which may not be the best day for every student. The physical or physiological needs of the students may affect the way they perceive and answer the questions: some students may be sick; some can be sleepless; some can be hungry; some may find the room where they are taking the test too noisy or too cold. Therefore, the physical and emotional environment may affect their performance. In addition, some students may have text anxiety: no matter what is asked in the test, they may be nervous to answer any question, even their names. Last but not least, although questions measuring higher-order thinking skills can be asked in tests such as PISA and TIMSS, it is still not possible to measure uniqueness, creativity, and imagination.

As portfolios and tests both have merits and limitations, some suggestions can be made to improve their value and usability. Regarding the portfolios, professional development programs can be provided to teachers to improve the effective use of portfolios in their classes. Communities of practice can be helpful, where teachers of the same discipline or level collaborate to prepare guidelines, rubrics, and instructions for students, as well as implement and evaluate portfolios. Third, school administrators may help design environments suitable for implementing portfolios, such as adding a cabinet to the classroom and providing each student with a binder and supplies. Fourth, parents can be involved in the preparation and evaluation of the process. Students can present their work to parents at an end of the school year event where school administrators and others can be invited. When my nephew, Can, was in second grade, he prepared a portfolio in his English class and presented it to his parents at the end of the semester with enthusiasm and joy, and his parent listened to all his work with pride. Student presentations and parent involvement add another layer to the value of portfolios since students are actively involved, and their voices are heard during their presentations. Another layer can be added by including parents' and peers' feedback to the portfolios. Finally, the portfolios can also be prepared electronically by using learning management systems or applications where students can upload their pictures, drawings, videos, voice recordings, and more. This approach may increase the type of artifacts included in the portfolio and can make the portfolio accessible anytime and anywhere. Tests could also be added to the portfolio, as they can convey the level or score attained by students at a certain time. In other words, they can still be kept, just not as the main source of data—as an artifact providing another type of information about the student.

Sevil Akaygun is an associate professor of chemistry education at Bogazici University, and a former science and chemistry teacher. Her research interests include visualizations in chemistry education, STEM education and nanotechnology education. Dr. Akaygun has served on the national chemistry curriculum development programs organized by the Turkish Ministry of Education.

Case: What Happened to the Puddle?

62

Jennifer C. Stark

Abstract

This case focuses on the experiences of Elliot, a student teacher in a fourth-grade classroom at a high-poverty, urban school, who sought to address barriers to students' science comprehension. After teaching an inquiry-based lesson on changes in the states of matter, he realized that his students still struggled with using the key vocabulary to explain a familiar phenomenon, the evaporation of a puddle in the playground. He worked with his university supervisor to identify additional strategies to enhance student comprehension and expand their discussions about states of matter.

While reading his students' journal entries at the end of the day, Elliot felt a wave of disappointment followed by confusion. His fourth-grade students seemed so engaged in the day's 5E lesson on changes in the states of matter, which was gratifying since his university supervisor was present that day. He thought they would be able to fully answer the journal prompt at the end of the lesson, "What happened to the puddle we saw at the playground last week?" He found instead that students wrote incomplete explanations such as "The sun dried it up" and "The water went back up into the sky." Few students used any of the key vocabulary from the lesson, such as "evaporation," "water vapor," or "heat" in their explanations. These responses suggested to Elliot that students struggled to apply the science concepts and vocabularies from the lesson. He wondered what went so wrong.

When designing his lesson, Elliot took care to research and select different representations of the changes in the states of matter due to heating and cooling. He knew that young students needed firsthand experiences with concrete examples

J. C. Stark (✉)
School of Education, University of West Florida, Pensacola, FL, USA
e-mail: jstark@uwf.edu

© The Author(s), under exclusive license to Springer Nature Switzerland AG 2023   327
S. Jeong et al. (eds.), *Navigating Elementary Science Teaching and Learning*,
Springer Texts in Education, https://doi.org/10.1007/978-3-031-33418-4_62

of the abstract concepts in the lesson to support their conceptual understanding. He also knew that one opportunity was likely insufficient to help students solidify their understanding of the concepts, so he identified multiple opportunities for concrete experiences both before and during the lesson. Elliot hoped that the experiences prior to the lesson would also supply relevant background knowledge for his class.

About a week before his lesson, Elliot pointed out a puddle on the playground to his students, which he knew would be evaporating over the next few days. Each day when they went to recess, he commented, "Let's check out our puddle." Toward the end of the week, students made observations such as, "It's so much smaller now. It looks more like mud than water." The day before his lesson, he gave students popsicles to eat and placed several popsicles on the hot sidewalk to melt during recess. Elliot opened his lesson with a discussion of the puddle and the popsicles, asking students to describe and explain the changes they saw. Next, he facilitated a structured inquiry investigation focused on the question, "What happens to butter when it is heated and then cooled?" Students observed the melting and freezing of butter when they placed it in cups of hot water and then ice water. They recorded their observations in their journals using drawings and written descriptions. When the butter was placed in the hot water, Ally wrote that "the butter got gooey" while Sophie wrote, "the butter first turned mushy and then melted completely." When the melted butter was placed in the ice water, a Khyla wrote, "it got hard again and turned bright yellow."

In a class discussion, Elliot debriefed students on their observations and emphasized that heating a substance makes the particles move faster, and cooling a substance makes the particles move slower. He used an online simulation illustrating these changes to support his explanation. Additionally, students completed guided notes as Elliot discussed each change in state (i.e., melting, evaporation, freezing, condensation). Elliot then demonstrated the melting and evaporation of water when he heated ice cubes in a beaker on a hot plate for several minutes. He held a cool mirror above the beaker after the water began to evaporate to show condensation. After the demonstration, Elliot encouraged students to work together to elaborate on their notes by adding examples of changes in state of matter due to heating and cooling from their own lives. Michael wrote down "heating chocolate causes it to melt," Ana noted "after swimming, my hair dries because of evaporation," and Hunter jotted down "the grass is wet in the morning because of condensation." Finally, students responded individually to the journal prompt about the puddle.

Elliot was noticeably distressed when he met with his university supervisor, Jane, for a post-observation conference. Jane started the conference by attempting to laud how all of the students participated in the investigation and recorded thorough observations in their journals. Jane noted,

> The students put a lot of effort into making accurate and detailed sketches with labels and explanations. Some even went back at the end and added more details without any prompting. It's clear that they know how to make and record good observations.

Elliot brushed aside her positive comment and pulled out copies of student responses to the final prompt. He exclaimed, "They did the activity ok, but they couldn't answer this question at the end of the lesson. I don't think they made the connections they need to make to do well on the unit test. The test is heavy on vocabulary. They need to know the words!"

For a moment, Jane was taken aback by Elliot's reaction to her affirmations of the lesson. Jane then inquired, "Tell me more about what type of responses you had hoped to see." Here, Elliot struggled. First, he stated, "I wanted them to provide more scientific explanations of the phenomenon of the puddle disappearing and use the vocabulary from the lesson." When she pushed him to give an example of the kind of response he desired, he said,

> I think that the puddle at the playground evaporated because the heat from the sun made the water particles move faster and change state. The liquid water in the puddle turned into gas, little by little, until all of the liquid water evaporated. The water that was in the puddle is now in the atmosphere in the form of water vapor.

Jane asked Elliot to list the vocabulary words that he used, and he came up with the following list: liquid, gas, heat, particles, evaporate, and water vapor. She commented, "Sometimes students have the conceptual understanding but are still in the process of learning the formal language of science. Have you ever studied a language in high school or college?" When he nodded, she continued, "Think about how difficult it was for you to learn another language and how much practice it took to master even basic vocabulary." She asked Elliot, "What are some ways to help students use vocabulary in their explanations in your next lesson?" Elliot thought for a moment and said,

> I think I need to give students more opportunities to review and practice using the vocabulary. I could give them a list of terms to use in a concept map or word web... I could have students talk about what they are thinking before they start writing, and maybe they could refer back to the concept map... I could even teach them to use the rubric for their journal entries for self-assessment or peer assessment. The rubric specifically mentions 'accurate use of science vocabulary.'

Elliot left the meeting with Jane still feeling discouraged, but at least he had an idea of a way he could help his students tomorrow. He was eager to talk with his fellow student teachers to see which strategies they were using to foster student comprehension and integration of science vocabulary.

For Reflection and Discussion

1. How did Elliot attempt to support student comprehension and integration of vocabulary in his lesson?
2. Besides those identified in his post-conference with Jane, what are some additional strategies that Elliot could use to support student comprehension and integration of vocabulary in science lessons?

3. Besides those identified in his post-conference with Jane, what are some additional strategies that Elliot could use to support students in expressing or demonstrating what they know?
4. What assumptions does Elliot have about assessment and evidence of student learning?

Jennifer C. Stark is an associate professor of science education in the School of Education at the University of West Florida in Pensacola, Florida. She teaches coursework for undergraduate and graduate students and supervises teacher candidates in their culminating clinical placements. Since 2015, she has been part of a team of general and special education faculty working to embed the principles and guidelines of the Universal Design for Learning framework in undergraduate teacher education programs at her institution.

Commentary: Being Your Own Worst Critic

63

Stacey Britton

Abstract

This is a commentary to the case narrative, *"What Happened to the Puddle?"* written by Jennifer C. Stark.

Elliot's story clearly demonstrates that he cared about the students learning both content and processes. It was also evident that he understood the need to show concepts in a variety of ways. What is also apparent is that, as a new teacher, Elliot held expectations that were different from what his students were willing to, and possibly capable of sharing; he believed that all students can understand scientific concepts and are capable of sharing their knowledge. In this case, Elliot was working with children in a high poverty, urban school, and he may not have been aware of the impact home-life and surroundings have on students' willingness to talk outside of the norm. There were many instances where the students expressed their understanding and were actively involved in the instructional activities he planned. However, the student responses did not fit a "script" which would signal to Elliot that he was successful.

First of all, Elliot was much harder on himself than was warranted. A very important thing he forgot was that his fourth-grade students were engaged in the learning process and actively understood what was happening to the states of matter. Young children, and fourth graders are still young, are more apt to share their understandings in ways that are familiar to them such as using terminology that is known to them. If Elliot could more clearly delineate the goals of the lesson, whether it be vocabulary mastery, conceptual understanding, or rote learning, this could dramatically change his perspective of what students actually learned.

S. Britton (✉)
Department of Early Childhood Through Secondary Education, University of West Georgia, Carrolton, GA, USA
e-mail: sbritton@westga.edu

Through the engaging activities, students were able to successfully explain the states of matter using drawing and narration; the only items not mentioned by the students were the movement of molecules and how the proximity and movement caused a change in state.

As teacher, it is important to ask yourself about the expectations you set for lessons. Knowledge of student behaviors, their prior knowledge and their level of interest are essential—and small victories, such as their visible excitement in seeing science happen, cannot be diminished. In this case, a pre/post-assessment could have provided beneficial feedback on the students' knowledge base and aid Elliot in help them build the capacity for using new vocabulary. As a teacher, we don't always remember that we have to talk like a scientist. In this case, the children were acting like scientists—as evidenced by the commentary provided by Elliot's university instructor—but they may not have used language appropriately. The larger, more important idea here is that students could conceptualize where water went during evaporation and the processes that changed substances from solid to liquid.

Additionally, these fourth-grade students DID science. For many students, this experience may have been the first time the teacher had talked with them about why puddles "disappear" or where the water comes from and goes. They took nature walks and observed how the water behaved over time. They were able to bring in experiences from outside of school that helped them connect to the content. The excitement and responses of the students indicate that learning happened, but as is evidenced in conceptual change literature, this does not alter your students' ability to explain their understanding overnight. The abstract nature of particle behavior is not something that is common for a fourth grader, or any elementary student, to understand or even consider. Knowing the cognitive levels, along with prior knowledge and changes in understanding, would have been a more effective way for Elliot to gauge learning. We have to meet students where they are and celebrate the sense-making that takes place, no matter how "elementary" we think it may be.

Stacey Britton is an associate professor at the University of West Georgia. She currently works with preservice and in-service elementary teachers sharing her love of science and helping them make connections between the everyday world and the classroom. Her research focuses on eco/social justice issues in the classroom and with local schools in experiential environmental education.

Case: Sunscreen Design for Assessing Students' Engineering Practices and Science Learning

64

Valarie Bogan and S. Selcen Guzey

Abstract

This open case presents the dilemma of assessing students' engineering design practices and science learning. Mr. Jensen, a beginning fifth-grade science teacher, hopes that the integrated STEM unit in which students design a natural sunscreen, will keep them engaged, help them learn about engineering design practices, and enhance their understanding of cell cycles and skin cancer. Mr. Jensen decides to use the final project, a "Shark -Tank" style class presentation of students' engineering design products, as an assessment tool. However, he struggles with effectively assessing students' engineering practices and science learning . Mr. Jensen desperately needs more effective strategies to help him evaluate students' understanding of both engineering practices and life science content.

Mr. Jensen was so excited when he finished his teacher preparation program and landed a job at a local elementary school. He would be teaching fifth grade in a small school that was surrounded by cornfields. He loved the rural setting and was excited to start teaching the students. Soon after he was hired, the school informed him that they were implementing a new integrated STEM program into the science curriculum . To prepare for this new program, he was expected to participate in a three-week professional development (PD) workshop during the summer to explore ways to intentionally and explicitly integrate engineering into his science instruction. A key focus of the first two weeks of the PD was on strategies for

V. Bogan (✉)
Curriculum Development, National Radio Astronomy Observatory, Charlottesville, VA, USA
e-mail: vbogan@nrao.edu

S. S. Guzey
Department of Curriculum and Instruction and Department of Biological Sciences, Purdue University, West Lafayette, IN, USA
e-mail: sguzey@purdue.edu

integrating engineering design into science instruction. The third week of the PD focused a model integrated STEM unit in which teachers engaged in. Teachers were asked to implement the unit in their classrooms in the subsequent school year. The unit was contextualized around skin cancer and used an engineering design process to develop natural sunscreen. Mr. Jensen felt confident during the first two weeks of the PD; however, he became a little apprehensive during the last week of the PD because he would be a first-year teacher in the fall and did not have experience with integrating STEM disciplines. The other teachers had been in the classroom for years and certainly knew more than he did about the best ways to teach students.

The integrated STEM unit was an interesting mix of cell biology and engineering. The first two days of instruction introduced students to the context of the overall unit and the engineering design challenge. Students learned about cells and cell cycles and applied what they learned to designing and optimizing a natural sunscreen to prevent skin cancer. Students also learned about the engineering design process—specifically, the iterative nature of the design process and its components: understanding the problem and background, planning and implementing solutions, and testing and evaluating solutions. The next part of the unit focused on cell biology concepts, which Mr. Jensen was prepared to teach. However, unlike a "regular" cell biology lesson, he would need to remember to help students link the science to engineering by continually asking them to think about and apply the scientific knowledge to their engineering design challenge. Also, the science activities required students to go beyond memorizing scientific facts; students engaged in several inquiry -based activities to explore cell division and skin cancer. In the unit's final lesson, students designed, tested, evaluated, and redesigned their natural sunscreen. Mr. Jensen was a bit worried that there were just so many things the students had to do and so much he had to keep straight, but he was up for a challenge, and this unit seemed it could be engaging for students and directly relevant to their everyday lives.

Mr. Jensen started the unit midway into fall semester. In his eyes, the unit was progressing smoothly, and the fifth graders were soon ready for the last part of the unit—the sunscreen design challenge. Mr. Jensen launched the lesson by discussing with students what they needed to accomplish in class. He and the students read the "client memo" from a local company, Derma Worldwide, which introduced the challenge of designing a safe, effective, and affordable sunscreen formula. The memo read:

Dear Engineers,

While you can find many different types of sunscreen products on the market, some of them are full of harmful chemicals and some are not as effective as others. Over the last few years, in partnership with scientists at a nearby university, engineers at our skin care research and development company have developed and tested new, once-a-day, natural sunscreen products to meet the demands of many consumers. We are at capacity in our ability to keep up with all the work in the area of sun protection products. I want to enlist your help to provide us with a safe and effective sunscreen formula.

Pamela M. Lie, M.D., Ph.D.

Chief Executive Officer

Derma Worldwide

After a brief discussion of the memo, each student began to research differ-ent ingredients found in sunscreen formulations (e.g., shea butter, coconut oil, beeswax, zinc oxide). They identified the main ingredients that they thought were most important for their sunscreen formulation and wrote notes in which they explained the reasons for their decision (e.g., how a specific ingredient protects against the sun; why that ingredient is suitable for skincare). Students also deter-mined the cost of their sunscreen formulas using the cost sheet that Mr. Jensen shared with the students. Next, the students formed small groups and shared their ideas and cost worksheets with each other to negotiate group formulation and cost. When each group was finally ready to test the effectiveness of their sunscreen formula, they conducted UV testing with color-changing UV beads. These white beads change color when exposed to UV light, so if the sunscreen is effective, the beads will not change color. After students performed their tests , they used the data to modify their formulation and continued the testing/redesign cycle until they were pleased with their final formulation.

Finally, it was time for students to plan, prepare, and practice their presentations for the "Shark Tank" event. The presentation to the class and guests needed to be at least five minutes long and required every group member to speak. Students were asked to share: (a) their product formula, (b) why they chose their formula, (c) how their product protects against skin cancer and helps maintain normal cell function, (d) total product cost, and (e) how the cost of their formulation compares to other products. Mr. Jensen shared with the students the rubric he would use for evaluating presentations and for providing feedback.

He thought that he could evaluate student science learning and engineering practices based on the information they would provide on why they chose their materials in terms of effectiveness and cost and what they know about the cell cycle and skin cancer. The rubric for the shark tank presentation included these elements of the project; thus, Mr. Jensen used it as a summative assessment tool.

Mr. Jensen invited several colleagues and the school principal for the student presentations. He was so excited, and the students were very motivated. Mr. Jensen reminded students that "Good presentations share the important information while keeping the audience engaged. Also, your presentations must be persuasive." The first two presentations were engaging and interesting; however, they all mostly focused on telling the audience about the ingredients of their sunscreen formula and their formula's cost. Before the third group started their presentation, Mr. Jensen reminded students about the need to also present connections to the sci-ence they learned—cell cycle and cancer. Mr. Jensen was optimistic that the next presentation would be better with that reminder.

The next group began their presentation, and it was just like the preceding ones; they talked about the ingredients, cost and compared their product to the others on the market. At the end of the presentation, Mr. Jensen started to question the students when one group member blurted out, "Oh yeah. This is sunscreen, so you put it on your cells. Cells are really cool because they go through that cycle that lets them clone themselves. You know if that cloning goes wrong, then you get skin cancer." Mr. Jensen sighed but circled the line between a three and a five on the rubric for the row about connections to the cell cycle. While the group did not do a great job explaining, they did share about the cell cycle and skin cancer. They even attempted to connect the cell cycle to skin cancer, which earned them a four on the rubric. However, Mr. Jensen was dissatisfied with that score because the group did not demonstrate an understanding of how their sunscreen protected cells. The last three groups were similar. They spent most of the time talking about the ingredients they chose, the cost of the formula, and how it compared to commercial sunscreens. At some point, one member blurted out something about cells and skin cancer, which got them at least a three on the rubric, but no group demonstrated understanding of how the concepts were connected.

Colleagues and the principal expressed that students did a great job with their presentations. All the presentations were interesting and visually appealing. However, Mr. Jensen was disappointed with the lack of science content in student presentations. It seemed that students did not apply science content to their design; they simply mixed the cheapest ingredients to convince the audience and researchers that their product was the best. However, students successfully completed the science activities, and they thought hard about the ingredients for their sunscreen formula. Students also did not demonstrate that they engaged in engineering practices to solve the challenge. Clearly, the presentations did not reflect any involvement in iterative problem-solving. Mr. Jensen used the rubric to score the presentations as he planned, but he quickly realized that the rubric he constructed was not a great tool to measure student learning of science and engineering practices . He now noticed that his rubric did not emphasize students' science and engineering learning and had a stronger focus on cost than he remembered.

At the end of the unit, Mr. Jensen spent significant time reflecting on what he needed to do to better assess student learning and science and engineering skills. While he implemented an integrated STEM unit that integrate science and engineering content and practices, he recognized that he needed to work on developing corresponding ways for students to provide evidence of their learning content and practices when engineering and science are integrated. He was so excited at the beginning of the class presentations. He had hoped that students would clearly show their understanding of cell division and skin cancer and demonstrate their problem-solving skills within their presentations. And looking at the scores produced from the rubric, the numbers suggested that students successfully completed their final project—the primary assessment of the unit. Yet, Mr. Jensen knew that this was not the case, nor did he know how better to assess students' engineering practices and science learning (Fig. 64.1).

Aspect		1	3	5
Presentation	Hook	Little to no apparent hook to gain audience's attention. Students do not go beyond introducing themselves or stating their names or group name. It is not necessarily clear what the presentation will be about.	Adequate to good attempt at grabbing audience's attention with small introduction to project. Introduction is appropriate. It is clear what the presentation will be about.	Excellent example of a hook to gain audience's attention, introduction is appropriate and entertaining. Audience members are engaged. It is clear what the presentation will be about and how it will proceed.
	Sunscreen Formula	Students provide little to no details of their formula, they do not elaborate on choices about formula.	Students provide the details of the formula; however, it is not clear the amounts they used or why they chose those materials.	Student provide all the details necessary for their formula and a clear rationale for why they chose all of the materials.
	Sunscreen Cost	Students do not share the cost of producing the sunscreen. There are not attempts to relate its cost to other products	Students share the cost of producing the sunscreen and there are some attempts to compare it to other sunscreen products.	Students share the cost of their product and compare and contrast it to more than one other sunscreen product.
	Connections to the Cell Cycle and Cancer	Students make little to no attempt to connect sunscreen product to the cell cycle.	Students make connections to the cell cycle but some of these connections are incorrect or inadequate.	Students make excellent and accurate connections to the cell cycle, explaining how their product will affect cells in the cell cycle.
	Use of Persuasion	Group makes little to no attempt to persuade audience of product.	Group makes good attempt to persuade audience of product but did not succeed, or did not seem fully invested in product themselves. Persuasive techniques are either logical or emotional.	Group succeeded in persuading audience that their product is the best and they are invested in their product. Groups use of persuasion is both logical and emotional.
Determining Cost Worksheet		Worksheet is somewhat complete, but missing answers to follow up questions, or answers are highly inadequate.	Worksheet is complete but minimally done, answers reflect the minimum amount of work to answer the questions	Worksheet is well done and complete. Answers are of high quality demonstrating critical thinking skills.
Group Behavior		Few group members were an active part of the presentation, few group members were well behaved during the presentation, group members did not give feedback to other presentation groups.	Most group members were an active part of the presentation, most group members were well behaved during the presentation. Group members provide some feedback to other groups during presentations but inconsistently.	All group members were an active part of the presentation, all group members were well behaved during the presentation and provided feedback to other groups on student feedback pages.
			Team Score	/35

Fig. 64.1 Rubric for presentations

For Reflection and Discussion

1. What ideas for assessing integrated science and engineering content and practices would you suggest to Mr. Jensen? How would these assessments better reflect learning compared to the presentations and grading rubric that Mr. Jensen used?
2. What are some ways in which Mr. Jensen could revise his rubric to better reflect the learning outcomes for this unit?
3. What aspects of engineering design should be assessed in a fifth-grade integrated STEM unit? What should fifth graders learn about engineering design and engineering practices —in other words, what is developmentally appropriate?

Valarie Bogan is a curriculum developer at the National Radio Astronomy Observatory. She holds a M.S. in Biology and a B.S. in Marine Science. She received her Ph.D. in Science Education from Purdue University in 2021. Before attending Purdue, She spent 10 years teaching science to middle and high school students.

S. Selcen Guzey is an Associate Professor of Science Education who holds a joint appointment in the Departments of Curriculum and Instruction and Biological Sciences at Purdue University. She holds a Ph.D. in Science education, an M.S. in Science Education, and a B.S. in Biology. Her research and teaching focus on integrated STEM education.

Commentary: The Integration and Assessment of Science and Engineering Practices

65

Kristina M. Tank

Abstract

This is a commentary to the case narrative, *"Sunscreen Design for Assessing Students' Engineering Practices and Science Learning"* written by Valarie Bogan and S. Selcen Guzey.

Engineering is intentionally presented alongside science within national reform documents that inform science teaching and learning, such as the *Framework for K-12 Science Education* (NRC, 2012) or the *Next Generation Science Standards* (NGSS Lead States, 2013). The NGSS "represent a commitment to integrate engineering design into the structure of science education by raising engineering design to the same level as scientific inquiry when teaching science disciplines at all levels, from kindergarten to grade 12" (NGSS Lead States, 2013, Appendix I). This means that elementary school teachers are expected to plan, teach, and assess engineering as part of their classroom science instruction.

The inclusion of engineering alongside science provides all students with key knowledge and skills that they will need to engage as workers, consumers, and citizens in our twenty-first-century society. Additionally, the integration of science and engineering provides other important benefits for elementary students, including engaging students in more realistic problem-solving; increasing student achievement, engagement, and motivation; and providing opportunities for interdisciplinary learning. However, this type of integrated instruction can be more difficult than just planning, teaching, and assessing a science-only unit. It is common to see both beginning and veteran teachers struggle with the integration and assessment of engineering into their elementary science instruction.

K. M. Tank (✉)
School of Education, Iowa State University, Ames, IA, USA
e-mail: kmtank@iastate.edu

S. Jeong et al. (eds.), *Navigating Elementary Science Teaching and Learning*,
Springer Texts in Education, https://doi.org/10.1007/978-3-031-33418-4_65

In this case, Mr. Jensen was provided the opportunity to attend a 3-week summer professional development. While in-service education opportunities in science and engineering are becoming more available, they are still not as common as we would hope, with fewer than 60% of elementary teachers participating in a science-focused professional development (Banilower et al., 2018). Professional development, like the one that Mr. Jensen participated in, has been shown to help teachers integrate engineering into their science instruction, but not all teachers have access to these opportunities. Other barriers elementary teachers face with integrating engineering include limited background and experience with engineering, limited time to teach science (with or without engineering), and limited access and exposure to quality curricular materials and resources.

Before we dive into some of the challenges that Mr. Jensen faced during his unit, let's not overlook many of the good things that Mr. Jensen did when implementing and assessing this integrated science and engineering unit. When including engineering, it is important to have a meaningful, motivating, and engaging context that will be relevant to students. Mr. Jensen did this by having the students read a "client memo" from the fictional company Derma Worldwide to introduce the challenge of designing a safe, effective, and affordable sunscreen formula. After introducing the problem, Mr. Jensen had his students participate in an engineering design task that tied to the context introduced in the memo. As part of their participation, Mr. Jensen intentionally engaged his students in the different phases of an entire engineering design process, from introducing the problem and background research, to developing group plans, to testing and iterating on their designs. This design process helps students develop engineering understanding, use practices, and problem-solving skills that engineers use, but it also allows students to learn from failure and have the opportunity to redesign.

Another positive aspect of Mr. Jensen's unit was that the unit included instruction in the related science content prior to students' engagement in the engineering design task. Mr. Jensen taught with student-centered pedagogies as he "required students to go beyond memorizing scientific facts; students engaged in several inquiry-based activities to explore cell division and skin cancer." Additionally, Mr. Jensen designed the unit in a way that promoted communication skills and teamwork through their cooperative work on the design task, but also as they worked to "plan, prepare, and practice their presentations for the 'Shark Tank' event."

Even though integrating science and engineering content and practices was an area of struggle for Mr. Jensen, there were positive aspects to highlight. When designing the unit, Mr. Jensen was intentional about the science content that he wanted students to learn and the engineering design task that would be connected to that science content. He created a performance-based summative assessment (the "Shark Tank" presentations) with a specific scoring rubric that helped to communicate expectations to students. Performance tasks require higher-order thinking and can provide students with a wider range of options and opportunities to demonstrate their understanding, which can be a great method of assessment following inquiry-based instruction. Additionally, his students were able to create and deliver "interesting and visually appealing" presentations that were based on the design task (e.g., they "talked about the ingredients, cost and compared their product to the others on the market"). However, as Mr. Jensen's case demonstrates, evaluating students' understanding of both engineering practices and life science content is challenging.

Some of the challenges exemplified by Mr. Jensen in this include integrating engineering into elementary science instruction and the assessing science learning and engineering practices. These are common struggles that often go hand in hand. A very common challenge of planning and implementing integrated science and engineering units involves the engineering task: It may not be closely connected to the scientific ideas, or it may not require students to apply scientific ideas to solve the task. The challenging nature of developing design activities that promote science learning requires teachers to deeply understand both science and engineering content, but also possess the pedagogical skills to draw students' attention to how the science ideas are relevant to a particular engineering context. In fact, it has been suggested that elementary teachers need multiple years of experience implementing integrated science and engineering instruction before they can effectively connect design activities to science content learning.

This leads into a second challenge faced by Mr. Jensen: How do you accurately assess student learning of science content and engineering practices? When designing an assessment tool for open-ended assessments, such as performance assessments, you need to clearly articulate the focus and desired outcome of the task. However, if the science learning is not clearly connected, or if students can complete the design task without the science content, then assessment of those loosely connected or non-existent science concepts will be even more difficult. How do you accurately assess student learning in something that you did not ask students to demonstrate? As was seen with Mr. Jensen's case, "it seemed that students did not apply science content to their design; they simply mixed the cheapest ingredients to convince the audience and researchers that their product was the best."

So how do we overcome some of the challenges in integration and assessment of science and engineering lessons? Teachers play a critical role in developing or modifying design activities to emphasize the relevant science ideas, while also helping students recognize and make explicit connections between the science and engineering during instruction. The use of published curricular materials or those developed collaboratively with other teachers and researchers (as seen with Mr. Jensen's unit) have been found to be more successful, especially during early years of implementing this type of instruction. It also helps to recognize that some content areas within science, such as physical science, can be more readily connected to engineering than others. Finally, a key to improving lessons includes realizing that instruction does not often go as planned, engaging in critical reflection, and focusing on ways for students to provide evidence of their learning.

References

Banilower, E. R., Smith, P. S., Malzahn, K. A., Plumley, C. L., Gordon, E. M., & Hayes, M. L. (2018). *Report of the 2018 NSSME+*. Horizon Research, Inc., Chapel Hill.

National Research Council. (2012). *A framework for K-12 science education: Practices, crosscutting concepts, and core ideas*. The National Academies Press, Washington.

NGSS Lead States. (2013). *Next generation science standards: For states, by states*. The National Academies Press, Washington.

Kristina M. Tank is an associate professor in the School of Education at Iowa State University in elementary science and engineering education. Her research is centered around how to better support and prepare elementary educators to integrate STEM in a way that supports teaching and learning across multiple disciplines.

Case: "Arts and Crafts" with a Side of Science

Michelle J. Petersen

Abstract

This case discusses an attempt to assess scientific modeling skills within the science curriculum. As a veteran science teacher, I spent much of the school year providing my fifth-grade students with an assortment of model-based lessons aimed at developing their understanding of scientific models. As student groups began to work on their culminating project, a years' worth of planning seemed for naught as most students slipped back into the idea that models are smaller versions of the real thing. Frustrations and questions arose in my mind. How do I fix this? What could I have done differently so that students do not revert to their naïve understanding of models? How could I foster their understanding of models while still providing them with developmental and creative freedom?

Ask a middle schooler to describe a model, and the majority of the time, they will mention toys that they or their sibling played with, such as cars and planes. Ask those same students to create a model, and they will discuss Styrofoam, paint, and labels. My goal was to change their thinking about what could be considered a model and, in doing so, grow as a teacher.

Modeling has moved to the forefront of science education with specific skills identified in the Next Generation Science Standards. Modeling skills range in difficulty from using models to model development. For example, fifth-grade students should be able to "develop a model to describe the movement of matter among plants, animals, and decomposers, and the environment" (NGSS, 2013, para 1). To aid students' modeling abilities this school year, I decided to devote additional time to lessons that would expose them to various types of models and provide opportunities to utilize them. For example, lessons included simulation models that

M. J. Petersen (✉)
Georgia, USA
e-mail: someday766@gmail.com

© The Author(s), under exclusive license to Springer Nature Switzerland AG 2023
S. Jeong et al. (eds.), *Navigating Elementary Science Teaching and Learning*,
Springer Texts in Education, https://doi.org/10.1007/978-3-031-33418-4_66

were available online related to plants, natural selection, and circulation. To practice analogical thinking, students compared organelle functions to the functions of items in their daily lives. Students even explored the connection between genes, DNA, and traits by making a graphic organizer, a type of visual model. Based on warm-ups and discussions at the beginning of the year, I knew that students typically viewed models as a different size than the real thing. Thus, I made a concerted effort to avoid using physical models within the classroom. So, throughout the entire year, my overarching goal was to help students achieve understanding that there are various types of models and they can be created and used for various tasks.

As a science teacher for 16 years in a growing suburban town in the southeastern USA, I have experienced many shifts in science education. The recent shift in the standards to more content application skills versus strictly knowledge acquisition has been an exciting challenge. As the standards began to incorporate skills alongside the content standards, I worked diligently to create lessons that focused on the application of content versus rote memorization of the information. Modeling was one of the skills that, for me, was the most difficult to incorporate into lessons because I, just like my students, had preconceived notions of what constitutes a scientific model. It was not until I took some graduate-level education classes that I began to learn more about the various types of models. Using that knowledge, I wanted to make sure my upcoming lessons incorporated exploring and using multiple types of models.

During this school year, my lesson planning took on a life of its own because I wanted each major unit to include activities that utilized various models, culminating with one comprehensive modeling assessment. In my classroom, assessments vary depending on their purpose. Formative assessment could be simply daily activities, vocabulary quizzes, content quizzes, and other activities that provided me with information about how well students were progressing in their understanding of a concept. Summative assessments required students to use the knowledge that they had gained during the unit. These assessments included activities such as laboratories, concept maps, hands-on manipulatives, and construction. For this particular assessment, I asked small heterogeneously grouped students to "develop an interactive model that shows how a community of organisms change over time as resources and populations fluctuate." Students were also tasked with determining how to collect data from their model that could provide evidence of their changes in populations. As I began to introduce the activity, murmurs of excited discussion began to erupt throughout various parts of the room. My excitement grew as I heard bits and pieces of their discussions. But my excitement soon waned as I realized students were defaulting to the apparently ingrained thought that models look like smaller versions of the real objects.

As the groups worked, David, Wendy, Corey, and Isabella excitedly summoned me to their table to describe their model. "We're making the animals. We should make them out of pipe cleaners," Wendy stated as other group members blurted out ideas for using stuffed or plastic animals in their model. Corey finalized their idea with the suggestion of a cardboard box containing all the parts of their ecosystem in miniature form. I subtly reminded them that they need to be able to collect mathematical data that shows how populations can change over time. I wandered away to help other groups, confidently thinking that they would draw on a year's worth of activities to aid their model development. A few minutes later, I was energetically summoned again as Wendy described how they could make the box with movable parts, like making a jaguar that could eat a sloth. Confidently Isabella stated that it would be "like a pop-up book," and David, always ready to work, asked, "So wait, whose gonna start with the arts and crafts?"

As Corey began to draw her idea out on paper, Isabella asked, "Are you making like a 3-D cube?" Corey replied, "Yes, it's supposed to look like the inside of a cardboard box. After their group decided to use a shoebox, the creation of animals was next. Wendy suggested making the jaguar "out of, like, play dough" while Corey wanted to make the animals out of pipe cleaners because you can "form the shape and then paint it black or whatever." As Wendy, Corey, and Isabella continued their discussion about how their model would look, David quietly provided his opinion, saying, "I don't think that's what we have to do." David appeared to be the only one who understood that models do not always appear as smaller versions, but his comment was quickly forgotten when I asked the group about their model. Isabella quickly described their model by saying that they "are making the animals out of, like, pipe cleaners" and creating their box to be "like a pop-up kids' book."

Since my goal was to discover what models the students were able to create, I stayed out of their decision-making processes as much as possible by simply stating "You are free to choose" and "it's up to you" when they asked what their model should look like. As I listened to their conversations evolve into discussions of appearance instead of functionality, I inwardly groaned and fought off the urge to bury my face in my hands. As I circulated around the room and listened to conversations, I wondered if there was a way to salvage this final assessment from turning into what David had astutely labeled "arts and crafts."

For Reflection and Discussion

1. How could Michelle salvage the final assessment?
2. What could Michelle have done earlier in the year to better ensure that her fifth-grade students understood various types of models?
3. Instead of staying "you are free to choose" and "it's up to you," what could Michelle have said to her students to help them change their modeling development direction without removing their autonomy?

Reference

NGSS Lead States. (2013). *Next generation science standards.* Retrieved from https://www.nextge nscience.org/pe/5-ls2-1-ecosystems-interactions-energy-and-dynamics

Michelle J. Petersen is an educator with experience ranging from special education to gifted/advanced. She strives to make science interesting and accessible to all students. She draws students into science by utilizing a variety of instructional strategies such as scientific modeling, technology, hands-on lessons, real-world and open-ended explorations, storytelling, and literacy activities.

Commentary: The Good, the Bad, and the Misunderstood: Developing and Using Models

67

Ayça K. Fackler

Abstract

This is a commentary to the case narrative, *""Arts and Crafts" with a Side of Science"* written by Michelle J. Petersen.

It is noteworthy that Michelle has made every effort to interest her students in developing and using models. Michelle addressed an important naïve conception about what models are: "Models are smaller versions of the real thing." The idea that models are copies of reality should come as no surprise because of insufficient guidance by the national K-12 science standards and teacher education programs on how to implement model-based lessons. This case provides several important insights into how elementary school students think about models and how teachers can go about using models or modeling in their teaching. It also offers an opportunity for future teachers of science, to reflect on their model-based instruction and the ways we address students' naïve perceptions of developing a model in science classrooms. Consider the following points Michelle highlights in her case.

First, it is important to explicitly teach students what models are and what they are for. At the beginning of the school year, Michelle took some time to discuss how different types of models can be used for different purposes. According to the case, the students were explicitly asked to "develop an interactive model that shows how a community of organisms change over time as resources and populations fluctuate." Then, the students were "subtly" reminded that they also needed to collect data from their model. These instructions could have possibly confused students. For instance, were they expected to make a model? Or were they asked

A. K. Fackler (✉)

Department of Learning, Teaching, & Curriculum, University of Missouri, Columbia, MO, USA

e-mail: ayca.karasahinoglu@gmail.com

© The Author(s), under exclusive license to Springer Nature Switzerland AG 2023   347

S. Jeong et al. (eds.), *Navigating Elementary Science Teaching and Learning*,

Springer Texts in Education, https://doi.org/10.1007/978-3-031-33418-4_67

to engage in the process of modeling by producing empirical data with their models? If it was both, then, perhaps the instructions should have been more explicit. Additionally, what does an interactive model mean to Michelle's fifth-grade students? Should we expect students to use a software program to make a model? Or does interactive refer to the idea of the change in a community of organisms over time?

Second, Michelle had her students use their model to collect data, which is a great way to teach how models can be used in scientific processes and to open up the lesson for further elaborations on the topic. As mentioned, Michelle asked her students to collect mathematical data from their models. The instruction for this task might not be easy to understand for fifth graders. Perhaps Michelle could encourage students to define some variables that represent something measurable or predictable and name these variables. Instead, students were asked to make a model that shows how a community of organisms changes over time as resources and populations fluctuate. In this case, what was it about a community of organisms that students might want to measure or collect data on? Rather than having a variable called *resources*, students might be able to change the variable's name to the availability of food, water, shelter, or competition for resources, or even more specifically, the number of predators. Giving students the freedom to solve problems independently is a good strategy, and sometimes less is more. However, in this case, telling the students "You are free to choose" and "It's up to you" may have led to instructions that were not entirely clear to them. What student models should look like depends on the purpose of their model (explanation, prediction, or data collection). There is a fine line between giving the students freedom in deciding how to do a task and giving implicit instruction that students only half-grasp.

Third, David's perception of models, as "arts and crafts," shows that he thinks models represent *things* (in this case animals) rather than an event, process, or system. Arts and crafts are good materials for creating a poster of things, but may not work well in developing a model of phenomena. Even though Michelle extensively discussed different types of models (e.g., simulation models or analogical models) with the class and avoided using physical models, the students ended up having "models that look like smaller versions of the real object." It seems that David and his classmates perceived the task as making a representational model. When students consider developing models as a representational task, they are more likely to create replicas by focusing on explicit features (material aspects) instead of conceptual features (structural aspects) of the phenomenon. This distinction between material aspects and structural aspects (form vs. function) situates models in the context of their use. Models do not merely represent the structural aspects of a phenomenon. Rather, models should explain how a system works (its function) as well as the form (a specific shape, size, or character) itself. Consider this familiar example, *Oreo Cookie Moon Phases*. Teachers often have their students make a model that shows the phases of the Moon by using Oreo cookies. Students' models with Oreos may help them learn how to match a moon phase name with a moon phase appearance. However, students will probably not be able to explain what causes the phases of the Moon, how/if lunar phases are related to lunar eclipses, or why there are no eclipses at every full and new moon by using their Oreo model.

Finally, by introducing one of the science practices, developing and using models, to her fifth graders, Michelle challenged herself to make sure that students had a chance to learn how to think with models in science. She also reflected on her experience teaching about models in this case. Michelle took her students' ideas about models seriously and tried to address them throughout the school year rather than telling her students what she wanted them to know. Elementary science teachers tend to use models as an instructional tool to focus on specific characteristics of an event, object, or process. However, Michelle, in this case, used models and modeling to facilitate student learning about a complex system, namely changes in a community of organisms over time as resources and populations fluctuate. Even though David's focus on "arts and crafts" has diverted his attention away from the actual modeling task and its purpose, Michelle set up a learning environment that invited students to express their existing ideas and use their inquiry skills.

Ayça K. Fackler is an Assistant Professor of Science Education in the Department of Learning, Teaching, and Curriculum at the University of Missouri. Her research interests include scientific modeling practices, bi/multilingual learners in science education, qualitative research methods, educational assessment, and science denialism.

Case: Where the River Flows

68

Elizabeth A. French

Abstract

In this case, Liz, a third-grade teacher, struggles to assess what her students learn about water pollution through explorations of a local stream. The stream was in fact closer than most students were aware–it was piped from the woods behind the school right under the playground. As part of her water pollution unit, she incorporated many formative assessments embedded in activities and mini projects. Liz planned to have students develop service-learning projects as the final unit assessment, envisioning that they could easily implement them around the school. She believed it would be a win–win–win situation for her teaching practice, student learning, and the stream itself. But when the students started discussing their summative assessment project ideas, Liz got something very different than she expected.

The usual flurry of after-lunch activity was well underway in Liz's third-grade classroom. Students returned lunch boxes to their backpacks, sharpened pencils, and told jokes to friends while slowly making their way back to their seats for their science lesson. Liz waited until every student was seated before excitedly announcing that they would be heading outside to start our lesson today. After the cheers died down, Liz discussed a few outdoor safety expectations with her students who grabbed clipboards on the way out the classroom door (Fig. 68.1).

Liz gathered the students at the edge of the school property next to a chain-link fence along the sidewalk. A few students bragged that they walk this section of the sidewalk every day to and from school. Few students had taken the time to stop and look through the fence down into the ditch and saw that something was

E. A. French (✉)
Jefferson City Schools, Jefferson, GA, USA
e-mail: liz.french@jeffcityschools.org

© The Author(s), under exclusive license to Springer Nature Switzerland AG 2023   351
S. Jeong et al. (eds.), *Navigating Elementary Science Teaching and Learning*,
Springer Texts in Education, https://doi.org/10.1007/978-3-031-33418-4_68

Fig. 68.1 Students enjoying the outdoor lesson

moving at the bottom of the ditch under the overgrown kudzu that held up food wrappers and other miscellaneous trash.

"There's water down there!" Jayden yelled. The rest of the class pressed against the fence, some standing on tiptoe, to get a better view of the small stream flowing right under their feet. A few students scrunched up their noses at the little stream. Yahir and Caleb were the first to comment out loud what many of them were thinking, "That looks gross," and "It smells!"

"Why is the stream like that? What happened?" multiple students queried.

The class followed Liz through the open fields used for field day and kickball back toward the school building, stopping along the way to peer down grates that were placed in the ground along the edge of the playground. About six feet down, students observed running water and noticed that a straight line could be drawn from the grates to the visible stream seen from the chain-link fence.

Liz had first noticed the grates herself during the previous school year while on recess duty. She traced the stream through the elementary school campus until hitting the fenced wooded area above the bus entrance behind the school. Returning one evening after school, Liz managed to walk around the fencing and found that the wooded land behind the school funneled stormwater runoff downhill to meet a very small trickle of water down a steep bank. This stream started on the school grounds. She realized this stream would offer the perfect opportunity to reach out to students for their unit on the effects of pollution on the environment. Right here on the schoolyard was proof that pollution caused by humans could negatively impact a natural part of the environment (Fig. 68.2).

Fig. 68.2 Student diagram of school property with stream

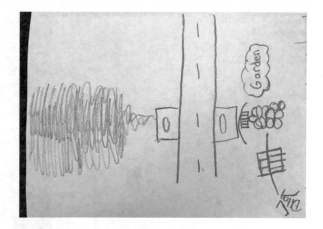

Later, back in the classroom, students had many questions about the small stream behind the school. What lived in the stream? Did it have a name? Why was there trash in it? Why did it look so gross? The students listed their questions in their science notebooks and created posters with their most important questions using chart paper.

Over the next few weeks, Liz referred to these questions and planned activities with the class to discuss watersheds and water pollution. Using their student laptops, the students worked in groups to find pictures of streams they thought looked "healthy" and "not healthy." Liz shared some of these pictures with the projector so the whole class could make observations and discuss differences between the pictures and the school's stream. Daejah's group shared that they thought the stream in the schoolyard looked unhealthy because it was brown and mostly covered in one plant that looked like a big blanket. Christian and Quentin also pointed out that they didn't see any animals near the stream, like dragonflies or frogs, which was a major bummer.

The following day Liz led the students out into the wooded area of the school where the stream started to build on their initial observations. In small groups, the students collected water samples and clumps of leaves called leaf packs that were submerged in the water. The water samples were tested and unfortunately yielded negative results, the oxygen in the water was low. "Who knew stuff in the water needed oxygen?" remarked Caleb. There were almost no insects found in the leaf packs either.

Back in the classroom, the students used models and spray bottles filled with water to visualize how stormwater can move over land and through a community. The water from the spray bottles picked up trash and other pretend pollutants for the experiment and pushed it all into the bodies of water on the models (Figs. 68.3 and 68.4).

To view the topography around the school, the students used maps and again walked around the fence with Liz to study the watershed of the stream on their school grounds. The class noticed that much of the trash was moving downhill

Fig. 68.3 Examples of models and diagrams students used to discuss stormwater pollution

Fig. 68.4 Examples of models and diagrams students used to discuss stormwater pollution

into the school property from a grocery store that had a large parking lot without any barrier between the edge of the concrete and the slope. A four-lane highway was in front of the grocery store.

After a rainy morning bus ride and indoor recess, Yahir and Christian pointed out that there were many areas on the playground and near the school's entrance where muddy puddles formed during a storm because there wasn't any grass. The muddy water also flowed right to the stream. The class hadn't yet covered the standards discussing erosion, but the students were already noticing this phenomenon at their school.

This unit was going exceptionally well, Liz thought to herself. Her students were basically experts on water quality at this point! The students had kept science journals where they shared their ideas and questions throughout the unit. As a summative assessment, Liz thought a service-learning project would be a good way to help students learn that they can make an actual difference in their local ecological community. Liz planned to have student groups design their own solutions to the water quality issues they discovered in their school's stream. Students would be assessed based on their abilities to clearly describe the pollution issues and how their projects could address these problems.

Liz was excited to split the class into small groups and lead a productive brainstorming session to discuss practices for controlling stormwater pollution. Easel pads were laid out for each group, and she sprung for the good chisel-tip scented markers for students to use while they designed their pollution solutions.

The students were also excited and ran into the classroom more focused than usual after lunch with their big news. About half of the class—with Christian and Lakea leading the effort—had decided that they already had the perfect solution to their stream's sorry state. They decided that to get clean water, all they needed to do was to pour store-bought gallons of water into the stream. No fish in the stream? No problem. They could buy some at the pet store and just pop them in the water outside. There, solved.

At first, Liz was amused by these ideas. She prepared to explain how the water bought at the grocery store wasn't exactly right for the environment, and that a goldfish would be an invasive species, when Christian pulled up the "flyer" on his laptop. Unbeknownst to Liz, Christian and a few friends had decided to create a flyer asking students in the whole school to bring in gallons of water and "aquatic wildlife" for the stream. The class that brought in the most stuff to dump in the stream outside would win a pizza party. And he had sent the flyer out to teachers!

Liz's earlier amusement faded to distress. Had the students been holding the misconception that store-bought "clean" water for human consumption was the same as clean water for a waterway the whole time? What other issues were students confused about that were not yet discovered? Would the students design projects that were well meaning but ultimately not actually feasible, wrecking her plan to showcase service learning in a project-based assessment?

When the last student had left for the day, Liz sat at her desk and pondered her plans for assessing student learning. It was apparent that students were struggling to make the leap from observation and experimentation to thinking and analyzing critically to generate possible courses of action. It seemed like the science journals, watershed pollution models, and classroom stormwater discussions were somehow falling short as examples of what their service-learning projects might include. Did the students realize that the formative activities were designed to help them complete their service-learning projects? If the other student projects revealed similar alternate conceptions, Liz wondered how she could get a clear picture of what students really understood about their local watershed.

For Reflection and Discussion

1. How can teachers who may not have access to natural resources on their school grounds connect to local pollution issues?
2. How can teachers scaffold student learning in ways that move them beyond observation and experimentation?
3. What are some ways Liz included formative assessment in the unit to ascertain students' understanding of stormwater runoff and pollutants? What other formative assessments might Liz have included and why?
4. How might Liz anticipate alternate conceptions and redirect students' thinking without squelching enthusiasm?
5. How might Liz have framed the summative assessment in a way that students more directly connected the formative experiences with the project-based learning project?

Elizabeth A. French is the BioSTEAM coordinator for Jefferson City Schools and works to plan and implement biology and environmental science programs. Liz has a background working in non-formal environmental education settings.

Commentary: Stream Studies Are Engaging

69

Alec Bodzin and Kate Popejoy

Abstract

This is a commentary to the case narrative, *"Where the River Flows"* written by Elizabeth A. French

This case exemplifies a placed-based learning project that is highly engaging with children. Place-based learning activities connect learners to their immediate environment. Engaging children in a local stream study near their school is an authentic learning activity. The close proximity of the stream to the school makes Liz's learning activity personally meaningful and relevant to her third-grade students' daily lives. We see that Liz's students have taken ownership of an authentic environmental issue–they perceive the stream as not providing a healthy habitat for animals and wish to address this problem by coming up with a way to make the stream healthy. When children take ownership of an environmental issue to solve a problem, the learning context holds much potential to promote a pro-environmental ethos with students. This became apparent in this case. Liz's goal is to work with her students to come up with a service-learning project to address the health of the stream, and the ideas that students discussed are intended to have a positive impact on the environment.

A. Bodzin (✉)
Department Education and Human Services, Lehigh University, Bethlehem, PA, USA
e-mail: amb4@lehigh.edu

K. Popejoy
Popejoy STEM LLC, Whitehall, PA, USA

The learning activities in this case involved authentic data collection. Students gathered data by making visual observations and by using chemical testing kits or probeware to measure the amount of dissolved oxygen in the stream. Dissolved oxygen is just one type of indicator to indicate the health of the stream. The students concluded that the school's stream is not healthy based on observing murky water, expansive coverage of one type of plant in the stream, and low dissolved oxygen readings. Liz also showed students pictures of a healthy stream. The students' conclusions were valid based on what they observed and the content background knowledge that they had. In reality, it's possible that the stream is healthy. Dissolved oxygen levels fluctuate over the course of the day, and to our knowledge, only one data collection reading was made by the class. Typically, in stream studies, more than one measure is made to validate the accuracy of the reading. In addition, murky water may also occur in healthy streams. For example, after a rainfall event, sediments get stirred up in streams and make it appear murky. In this case, we are not provided with weather conditions that occurred prior to the data collection.

During the case, it became evident that the children may have some alternate conceptions about water quality and perhaps the local watershed. A local watershed includes the area of land where precipitation collects and drains into a common outlet such as a river, lake, bay, or other body of water. The school stream is part of a larger watershed system, and student ideas about the health of the stream were generated by focusing on their immediate environment while not considering the larger watershed and how their local stream is situated in that larger watershed. In addition, one major alternate conception that arose was that the children thought they would address a stream clean-up by pouring gallons of clean water into the stream. Liz is concerned that this conception of the children might derail her service-learning project. However, we see that having the students learn about where water comes from became an opportunity to address their alternate conceptions. First, a basic introduction to the water cycle (see Fig. 69.1) can be presented to let children know that water is continuously cycling in our environment. Second, Liz can discuss with her students the environmental issues associated with pouring gallons of water from plastic jugs into the stream. Third, Liz uses this opportunity to expand her use of maps for teaching and learning. The students developed physical maps of their immediate area. Liz now has the opportunity to use direct instruction with Google Maps or Google Earth to show her students how the local stream next to the school fits in the larger watershed. Liz can use explicit modeling while projecting an image of the stream and the surrounding area on a Smartboard or a large monitor. Liz can trace the local stream on the display with her students using guiding questions to observe what is near the stream, and can also ask what might contribute to the health of the stream and its watershed.

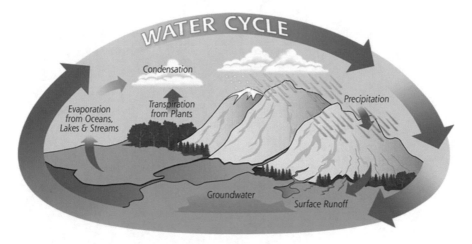

Fig. 69.1 Water cycle. Image retrieved from NASA, https://gpm.nasa.gov/education/water-cycle

Alec Bodzin is a professor in the Teaching, Learning, and Technology program and is a core faculty member of the Lehigh Environmental Initiative. His research interests involve the design of immersive virtual reality learning environments, engagement, and learning, learning with spatial thinking tools including GIS, learning design, design and implementation of inquiry-based science and environmental curriculum, learning technologies, game-based learning, and environmental literacy and education for sustainable development. Dr. Bodzin teaches science methods, environmental education, and curriculum and instructional design courses.

Kate Popejoy is a science education researcher with substantial experience in elementary science education, technology integration and geospatial thinking and reasoning. She has been the recipient of three National Science Foundation grants related to secondary science teacher preparation (Noyce), and to socio-environmental science investigations using GIS tools in secondary classrooms (ITEST). She also currently serves as the Executive Director of the Association for Science Teacher Education (ASTE).

Part VII
Addressing Socioscientific Issues

Chelsea M. Sexton

Whether discussing climate change, a global pandemic, or the best way to arrange a school cafeteria, we find that science cannot solve problems in a silo. The further we move into the twenty-first century, the more we see entanglement of science, technology, and society. With the prevalence of social media, students at a young age are growing in awareness of the challenges of society. Explanations of phenomena are only a quick search away, which can be a plague or a panacea based on the source of the information. Sometimes the search yields credible science; other times it returns the guy down the street who hosts a YouTube channel. As educators, we need to be preparing our students with science as they see it in the real world. While science textbooks offer classic examples that seem uncomplicated and easy to teach/learn, they lack the element of reality. After all, for the most part, our students will not be viewing the world solely through the lens of a test tube or petri dish.

As illustrated in a few of the cases in this part, integrating socioscientific issues in the elementary classroom is not an easy feat. By their very nature, socioscientific issues are complex and often laden with controversy. Sometimes students do not see the connection from the textbook to the real-life situation reflected in socioscientific issues. Other times, students will show interest in an issue that a teacher is unprepared for and unwilling to entertain. Sometimes students worry that decisions made during theoretical class projects will have real effects for generations to come in their neighborhood. Other times, students put their blood, sweat, and tears into a project, only to not be taken seriously enough. Integrating societal problems into a science classroom can be trying for new and veteran teachers or even the students themselves.

Even though the process can be involved, socioscientific issue instruction can provide a more cohesive and robust learning experience for students. As the adults and decision-makers of tomorrow, children in today's classrooms will be called on to approach complex problems, whether they ultimately join a scientific profession or not. As citizens of the twenty-first century, they will need to take a stand on the environmental health and economic and social justice issues that impact their

C. M. Sexton
Department of Mathematics and Science Education, Mary Frances Early College of Education, University of Georgia, Athens, GA, USA
e-mail: cmsexton@uga.edu

daily lives. Building skills from a young age about science, technology, and ethics can create a more scientifically literate population.

Instruction with socioscientific issues often includes the practice of argumentation. Argumentation is a process of articulating, supporting, and discussing a data-based claim about a phenomenon. While many may call this a debate, argumentation is built to be a structured practice in which students communicate, compare, and (scientifically) challenge claims built by each other. As socioscientific issues do not necessarily have one correct answer, argumentation can allow students to understand pros and cons of different claims and solutions related to the issue. Argumentation also encourages the idea of revising claims more than winning and losing a debate. As socioscientific instruction will not have a strict answer or simple conclusion, argumentation allows students to take ownership of the lesson and learning rather than the teacher controlling the narrative.

In this part, we meet three educators implementing socioscientific issue instruction in elementary classrooms and two fifth-grade students who are faced with problems identified in science that do not seem to have answers. Each case presents an opportunity for the exploration of socioscientific teaching in diverse contexts and reflects complicated and more simple issues.

In *Don't Drink the Water*, Lynn and Sabrina are two young women of color who want to discuss further social issues that stem from scientific problems in a community like their own. They are fifth graders in the New York City public school system learning about water pollution. Between their class discussions and private dialogue at lunch, they become very interested in exploring the challenges created by environmental racism. On the other hand, in *But Why Do We Need to Study This? The Case of Arsenic in the School's Soil*, Mr. Joseph Pavez wants his students to see the importance of an authentic science issue in their community. He plans a culminating project for the matter unit to include soil chemistry tests after a nearby elementary school discovers toxic chemicals on their school grounds. In the process of instruction, he struggles to help his students connect the science from his unit and the social issues at the neighboring school.

In *School-Cafeteria Make-Over Real-World style*, Ms. Lisa Weatherbee faces challenges when her students endeavor to improve their school cafeteria. While her students are very engaged and fully buy into the project, Lisa wonders if anyone will take the students' research findings and recommendations seriously. In *"Are We Getting Rid of a Soccer Field? But I Have Nowhere Else to Play Soccer!": Implementing an SSI-BASED Ecosystem Unit in the Third Grade*, Mrs. Laura Zangori begins a lesson about the importance of pollinators and ways to support their well-being. She has the opposite problem from Lisa as her students are worried that they will be making a final decision regarding the future of a beloved recreation area at their school that will have long-term consequences.

These cases exemplify a variety of attitudes and reactions that teachers and students can have when facing the challenge of using science to inform decision-making for a societal problem. Exploring these tensions within the case and commentary, the authors provide novel approaches for socioscientific instruction which can spark discussions among teachers who currently implement or are considering making authentic societal issues central to their science curriculum.

Case: Don't Drink the Water

70

Lisa Mekia McDonald and Felicia Moore Mensah

Abstract

Lynne and Sabrina are two girls in the same grade five classroom. They attend a local public elementary school in New York City (NYC) located in the Washington Heights neighborhood. This open case addresses how children of color may benefit from having racial affinity groups to address the discomfort of engaging in difficult topics such as race within their school setting and science curriculum. An affinity group is a group of people with common interests, backgrounds, and experiences who come together to support each other and to discuss ideas and situations relevant to them as a group. Lynne identifies as biracial, and Sabrina identifies as Latinx. Conversations about difficult topics such as race typically are not discussed in the science block of time in their classroom. However, questions of racial identity development, affinity spaces, and racial discourse within the classroom come into focus and play a role in science teaching and learning.

Lynne and Sabrina have been best friends since they were in the first grade, and this is the third time they have the same teacher. Lynne racially identifies as biracial, with a Black mother and White father, and Sabrina identifies as Latinx. Racial affinity groups are not part of their school's curriculum or most public schools within New York City. Conversations about difficult topics such as race are typically only addressed through the civil rights curriculum during the social studies

L. M. McDonald · F. M. Mensah (✉)
Teachers College, Columbia University, Science Education, New York, NY, USA
e-mail: fm2140@tc.columbia.edu

L. M. McDonald
e-mail: lmm2238@tc.columbia.edu

time block on Tuesdays and Thursdays. However, students often talk about these topics at lunch among themselves.

Lynne and Sabrina are eager to go to lunch so they can have some time to themselves, connect with their other friends from the neighborhood, and talk about their favorite thing to do after school—writing poetry. They attend an after-school writing program with other girls of color from surrounding neighborhoods. The girls like to spend time with other Black and Latinx girls who are not in their class or school. They enter the cafeteria with their classmates, and both girls look over to see what is for lunch. They sit next to each other and wait for their table to be called to pick up their lunches. After getting their lunches, they sit and talk with their friends and, of course, the book of poems they are currently writing. Across the lunch table, someone yells, "This is stupid!" Another student yells, "You're stupid like the president!" The other student responds, "I can't be stupid like him because I am not White or Mexican." Both girls look at each other and sigh. Lynne looks at her friend and asks, "Sabrina, are you ok?" Sabrina shakes her head and hunches her shoulders.

Another conversation pops up at the table regarding race and the president. Sabrina exclaims, "I hate talking about race and politics. It's sad." Lynne continues, "Well, that conversation over there was stupid because not all White people are stupid. My dad isn't." Sabrina adds, "Mexicans are not stupid either. Ugh, I am tired of hearing about politics in class too."

Lynne adds, "It makes me sad, too, that we are learning that Black people were slaves and White people got everything. I also feel good about it because we get to learn more about Black people, but what about what Black people did that was good? And do you know what else is annoying? When someone says something like, are you Black or White, or what color is your skin?" Sabrina sits quietly, taking it all in. "You know, they ask me that too." Both girls continue to eat their lunch in silence. Their teacher, Miss Abby, comes into the cafeteria—"Everyone, clean your tables and line up so we can return to the classroom for our science lesson."

Ms. Abby is a veteran teacher with 15 years of teaching experience in public schools. She has been at her current school for seven years and teaches fifth grade. She enjoys having Lynne and Sabrina as her students because they are quiet girls and do not cause much trouble. Back in the classroom, Ms. Abby settles in the class and asks the students to get out their science notebooks for the lesson and to meet her on the carpet.

Ms. Abby starts the science lesson by putting some review notes on the Smartboard. "I want everyone to copy the notes into your science notebook. You can use these review notes for the science test in three days." The science lesson focuses on the properties of water. Lynne looks at the Smartboard and begins taking notes. She starts doodling, but Sabrina nudges her to focus. Lynne and Sabrina continue to take notes, and Lynne raises her hand and asks, "Ms. Abby, how does water impact us in NYC? My mom said a report went home to families about the lead in pipes in New York and New Jersey, and now we shouldn't drink the water from the fountains in the school." Ms. Abby replies, "Lynne, that is a great question, but we are not focusing on that in this unit." Lynne takes a deep breath, puts her head down, and stops writing notes. Sabrina nudges her again, but Lynne looks away.

Although Ms. Abby is teaching a lesson about water and its properties, her instructional style dismisses the social justice aspect of science within the lesson and the question Lynne poses. Ms. Abby's overall teaching style does not enact culturally relevant pedagogy across all disciplines, but mainly in science, and students have few opportunities to work together. Miss Abby continues explaining, "Water is found everywhere on earth. Living in NYC, we are part of an urban watershed, and this affects how we get our water." Ms. Abby then shows a map of an urban watershed from the Midwest and highlights runoff into a river. She adds, "The runoff can cause contamination." Lynne writes a question in her notebook, "What about Newark, New Jersey, and their water? My cousins live in Newark, New Jersey!" Sabrina sees Lynne's notebook, and she raises her hand to ask a question. "Ms. Abby, in the urban watershed, does this mean that Newark and New York City have the same water? We don't live in the place you are showing on that map. And can the water be so bad in New York and New Jersey that we shouldn't drink the water?" "Well, Lynne," Ms. Abby says, "this is a good question. We can look at this tomorrow."

Once the science lesson is complete, Ms. Abby gives the class free choice-time. Lynne chooses writing time and asks Sabrina to join her at the writing station so that they can continue to work on their book of poems. The class is buzzing with conversations about after-school activities, games, and the outbursts from lunch. As they are writing their poems, Lynne states, "This is a bad day. I can't wait for it to end." Sabrina agrees and says, "I know. Me too. And they are still talking about what happened at lunch. This is a horrible day." The girls do not like what they are hearing from their classmates about Black and Mexican people taking jobs from real Americans and how dark people are not as smart as light people. They overhear a classmate nearby say, "Black people made the water dark and dirty. That's why we're learning about water in science."

Ms. Abby comes over to check-in with the girls. She asks them, "How are you doing?" They both reply in unison, "Fine." After a few seconds, Ms. Abby says, "Hey, Lynne, our next science activity will focus on pH of our water. It should be fun, right?" Lynne shrugged her shoulders as if to communicate, "Maybe." Before turning away, Ms. Abby shares with Lynne and Sabrina that in social studies tomorrow the lesson is going to be about change-makers in the Civil Rights Movement. Lynne and Sabrina widen their eyes with mediocre excitement and look at each other, but Lynn says under her breath, as Ms. Abby walks off, "I know about MLK and Rosa Parks already." Ms. Abby makes her rounds to check-in with other groups who are working on different activities. One group of four males continue to talk about race and the incident from lunch. Ms. Abby ignores their conversation and moves on to visit another group while making her rounds. Lynne and Sabrina are listening but continue writing their poems.

For Reflection and Discussion

1. In what ways did the conversation in the lunchroom carry over to the science classroom for Lynne and Sabrina? In what ways may socio-geopolitical issues affect elementary students' science learning, both inside and outside the classroom?
2. Should discussions of race have a place in the science classroom? If so, how can teachers foster a classroom community and curriculum that support children in addressing difficult topics such as race? If not, why?
3. How can elementary classroom teachers incorporate culturally relevant teaching in science around socioscientific issues, environmental justice, or environmental racism?
4. What could a truly culturally relevant science and social studies unit look like for Ms. Abby's racially diverse classroom?

Lisa Mekia McDonald is a science teacher at Avenues the World School in New York City. Her research focuses on intersectionality of race and identity development, critical race theory, science identity development with young children and their families, and culturally relevant teaching practices along with diversity, equity, and identity within science teacher preparation.

Felicia Moore Mensah (@docmensah) is the department chair of Mathematics, Science and Technology and a professor of science education at Teachers College, Columbia University, New York City. Her research addresses issues of diversity, equity, and identity in science teacher preparation and teacher professional development, with culturally relevant teaching, multiculturalism, and critical race theory guiding her teaching and research. Her most recent work focuses on the preparation of teacher educators for racial literacy.

Commentary: Murky Water or Clouded Judgment?

71

Rashida Robinson

Abstract

This is a commentary to the case narrative, *"Don't Drink the Water"* written by Lisa Mekia McDonald and Felicia Moore Mensah.

As our nation continues to reckon with its long history of racial injustice and inequality, kids are bombarded with images and stories that revolve around topics of race and racism, and they have questions. This case represents the daily missed opportunities for teachers and students to engage in social justice and racial equity conversations in the context of science instruction in many elementary school classrooms. Miss Abby missed several chances to have an open and honest conversation with her students about race and stereotyping and engage her students more deeply in the science content that she was teaching.

While we can agree that children should not be exposed to inappropriate or excessively graphic information, parents and teachers should discuss race and racism with children openly, and these conversations should start from an early age. By the age of six, children can be conscious of social stereotypes and begin to naturally attribute traits, both positive and negative, to their ethnic or racial groups (Taylor, 2020). Like the girls in this case, young students of all racial backgrounds are constantly impacted by what they hear and what they see in the news, at home, in their neighborhoods, and at school. Without proper context and relevant information from adults, white children can develop bias about children who are different from them, and children from diverse racial and ethnic groups can develop feelings of inferiority from exposure to popularly touted stereotypes and disparities about themselves as well as other racial and ethnic groups.

R. Robinson (✉)
Teachers College, Columbia University, New York, NY, USA
e-mail: rmr2179@tc.columbia.edu

Since we now know that children's awareness of racial differences and the impact of racism begins at a very early age (as young as age 3 or 4), it follows that children are taking note of racial differences and may begin evaluating, comparing, and excluding their peers of different races (Cole & Verwayne, 2018). Given the current political and social climate, it is more important than ever to create safe spaces (like classrooms) for children to explore, question, and talk about their feelings regarding these topics. As such, teachers are obligated to not only teach and learn with children about these critical issues but to engage parents in this endeavor as well. Schools, in collaboration with families and the local community, have an important role to play in helping children learn what unjust social structures are, and how racial hierarchy is normalized, transmitted, and internalized by children. We want education to foster young children's positive racial identities of themselves and children that are different from them.

This case also speaks to an all too familiar scenario in which the students are left to make sense of complex topics like race and inequity on their own, and as a result, children feel frustrated and hurt. For example, Miss Abby, like many teachers, rejects the idea that a science classroom is a place for discussions about race, racism, and equity. As educators, we must look for and embrace opportunities to engage our students in critical conversations about race, racism, and equity and the role that these things play in all of our lives, even in the science classroom (Mensah, 2011).

Lynn and Sabrina are bright, curious girls who want to understand why deficit views are perpetuated on students of color like themselves, as these views conflict with their life experiences. Miss Abby, like all educators, is in a unique position because she has the opportunity to help her students unpack implicit racial bias and engage in culturally relevant pedagogy to help her students understand and take pride in their differences. However, she passes on every opportunity to do so, even choosing to use examples the students cannot identify or relate to within her lesson; at the same time, she dismisses their questions which have relevancy to the lesson.

Although early childhood educators may want to make space for learning about race and racism in their classrooms, they might feel unprepared to approach these complex issues, especially since parents and teachers may hold differing views, shaped by their experiences with issues of race and racism. Also, many adults may see race conversations as inappropriate for the elementary classroom and/or the science classroom (Husband, 2010). Elementary teachers may also be unsure of how to offer learning experiences that allow children to observe and celebrate their unique identities across subject areas. For instance, mathematics and science may seem less connected to these topics compared to social studies.

However, any subject, including science, should engage in culturally based pedagogies that identify and leverage the knowledge, practices, and resources of students and their communities. Making sense of the world continues to be universal across diverse cultures, and people bring diverse experiences and sense-making tools to science learning. All these characteristics can be leveraged for the benefit of the entire class. Science education is about teaching students the methods they will use to analyze and understand the world around them, which most certainly includes societal structures and values.

Instead of treating science and social justice as unrelated entities, science education should involve using scientific knowledge and skills to make the world a fairer and more just place for everyone, regardless of the age of the students. Science teachers can empower young students by making their science classroom a place where students can build skills and use knowledge in ways that speak to what the world asks of them. Teachers can show their students how to navigate socioscientific issues and challenges—such as whether or not it is safe to drink the water at school—and critically examine related social factors, even at a young age. Openly discussing these issues benefits science learning by providing students with a shared desire to make positive changes as a class, which activates their ownership of the content and motivates them to reach related science learning goals. For example, in this case, Lynne and Sabrina, along with the rest of their class, are residents of New York City. They can use what they are learning about water and its properties to figure out if the water in their school is safe to drink by doing water-testing. They can bring water samples from their homes and test the water for purity. Designing filtering systems, testing them out, and reporting the results to the school community are culturally relevant activities for science learning. Students become much more interested in learning when they have a say in what they learn while making meaningful connections to the content at the same time.

It is also important for students traditionally marginalized from learning science to feel as though the teacher is invested in their well-being not only as a learner but also as a whole person (Ladson-Billings, 1995). When teachers take an active interest in their students as people with concerns, wants, and needs, the students are more likely to thrive and excel in the subject matter, as this creates a nurturing learning environment that makes all students feel equally acknowledged and validated.

Elementary science teachers can offer their students the support to think about what is going on in the world around them socially and politically while fostering academic growth that will allow them to use science to take action for social justice in the present and the future. Science should be used to empower students by providing students with the tools to confront and solve real problems that affect them, the people they love, and the communities in which they live. To offer the proper support for their students to thrive, elementary science teachers need to become comfortable not only with their ability to adequately teach science concepts and skills but also to facilitate difficult discussions.

References

Cole, K., & Verwayne, D. (2018). Becoming upended: Teaching and learning about race and racism with young children and their families. *Young Children, 73*(2), 1–16.

Husband, T. (2010). He's too young to learn about that stuff: Anti-racist pedagogy and early childhood social studies. *Social Studies Research and Practice, 5*(2), 61–75.

Ladson-Billings, G. (1995). Toward a theory of culturally relevant pedagogy. *American Educational Research Journal, 32*(3), 465–491.

Mensah, F. M. (2011). A case for culturally relevant teaching in science education and lessons learned for teacher education. *The Journal of Negro Education, 80*(3), 296–309.

Taylor, K. R. (2020). *Crucial conversations: It's never too early to talk about race.* www.slj.com/?detailStory=crucial-conversations-never-too-early-to-talk-about-race-antiracism-racism-libraries-learning-protests

Rashida Robinson (@DocRahee) is an assistant principal at Eleanor J. Toll Middle School in Glendale, California and a graduate of the doctoral program in science education at Teachers College, Columbia University. Her research focuses on issues involving diversity and equity, as well as curriculum and teaching methodologies utilized in young girls' school and extracurricular science education experiences that foster the creation of a science identity.

Case: But Why Do We Need to Study This? The Case of Arsenic in the School's Soil

José M. Pavez

Abstract

Joseph Pavez, a recent biology education graduate, was in his first-year teaching elementary school. He was in the midst of teaching a unit on physical science to a group of fifth graders at a medium size urban elementary school. His students were learning about mixtures and pure substances, and they were experimenting with different techniques to separate mixtures into their components. Joseph decided to have a special hands-on activity to conclude the unit in an engaging way. What Joseph didn't know was that he was about to question his assumptions about relevant science teaching after hearing some surprising comments from his students during the activity.

Three weeks before finishing a unit on matter, I decided to have a more hands-on activity to wrap up some of the topics we had been learning about each day. At that time, a terrible story was on TV news, printed in the local newspaper, and reported on social media: health authorities had found arsenic and lead in the soil at San Sebastian Elementary School, a small school located in a rural area adjacent to our town. The presence of arsenic was due to the activities of smelting industries located nearby. These industries apply heat to ore in order to extract a copper and other metals. Further investigations also found different concentrations of these metals in the students' blood. Arsenic is an extremely toxic substance, and it has been associated with fetal malformations, lung damage, and cancer, among other complications.

All this information was being discussed on TV and rapidly changing as additional facts were becoming known. I thought that maybe my students would be interested in studying the composition of the soil surrounding our school during

J. M. Pavez (✉)
School of Education, Western Illinois University, Moline, IL, USA
e-mail: josemanuelpavez@gmail.com

science class. Since soils are a mixture of particles of different sizes, we could easily separate them in the classroom. I decided to use this current event as a way of bringing together everything the students had been learning about matter.

Two weeks before finishing the unit, I asked my students to bring in newspaper clippings with information about what was going on at San Sebastian Elementary School and other nearby schools having similar issues as a preamble to our hands-on activity. Some of the students were confused when I explained that we would be doing an activity to test the health of our soil:

"Did we finish the unit on matter already?"

"What does this activity with soil have to do with matter?"

I didn't explain much at that moment, setting aside their questions for the time being. During our next science period, students shared the information they had found with all class members, and we discussed implications for students' health and for the school community. These animated discussions generated different reactions in my students:

"I think what happened in San Sebastian school is really sad, and no child should go through this."

"We could write a letter to our local authorities and ask them to help people at this school, and to take action to prevent this situation from happening again."

"I think we should raise some money to help these students to relocate in a safe place."

"We should pray for these kids and their families."

It was very interesting to me, as a novice teacher, to see how my students got involved at an emotional level with this socioscientific issue. They not only brought in information about the issue, but also pictures of kids at that school that they found in newspapers. These pictures seemed to make the stories my students shared a little more vivid.

Silvia shared, "These are some of the students that were affected in that school, and their families are devastated. As of today, they haven't been relocated to a new school, and they are uncertain about the health issues they may have in the future."

When Silvia finished sharing, Marco replied:

"Guys, imagine if this would happen in our school!!"

Students' facial expressions seem to reflect a bit of fear for a moment until Sergio said, "We don't have industries nearby, so we shouldn't be worried."

My students were very empathetic with the situation at San Sebastian School. But because we rapidly went from discussing the topic of mixtures one week to discussing the topic of chemical contamination of soil and social implications the next week, many of the students thought that we were done with the unit on mixtures. At the end of class that day, I reminded students that next week we were going to finish up the unit with a hands-on activity: We were going to analyze the composition of the soil at our school, using one of the techniques we had discussed in previous weeks. Many of the students expressed excitement about the activity. I had a gut feeling that today's discussion was a great preamble for next week's hands-on activity. Or at least, I thought it was.

The following week, the students entered the classroom continuing to talk about the situation at San Sebastian school. I explained to the students how we would be working in groups, and each group needed to collect soil samples from a different area in the school and bring them back to the classroom. The students were very excited, probably because we were going to be outside of the classroom and doing "real" science.

"We are finally doing something fun, Mister Pavez," Marco noted with excitement in his voice.

"Yes, we're finally doing science!!" added Silvia.

I gave them handouts with the different activities for separating and studying the soil components. We were going to use three different techniques for this purpose: sieving, decantation, and filtering. I passed out the necessary equipment for collecting the soil samples. Then we went outside and collected soil from different areas around the school. We brought the samples back to the classroom, and I gave students the necessary equipment for conducting the different separation techniques. I planned for the students to use sieves, coffee filters, and plastic graduated cylinders for the separation techniques. I was most excited to see the soil columns in the cylinders that students made because it was the most complex technique I had planned.

Everything seemed to be going well until I decided to check on one of the groups, "Hey, everything going good here?"

"Yeah, Mister Pavez, but we are just wondering why we have to study soil," questioned Chelsea.

"Yeah, we don't see the point in doing this activity," Alfredo replied.

I was taken by surprise and wasn't prepared for these comments at that moment in time. I thought the discussion we had about the soil contamination at San Sebastian school was enough to help students make sense of this activity and see it as relevant to their everyday lives. I thought that my plan for connecting ideas about matter and mixtures with the science behind a real, local situation would be obvious to students. But apparently, it wasn't.

This particular group of students was performing the activity according to my instructions; they were following the steps perfectly and completing the other activities on the handout. Nevertheless, they seemed to be looking for something else. Maybe they needed something to motivate them or more fully capture their interest? This situation also made me question myself, "why did I decide to do this activity?" "What makes an activity relevant from a student's perspective?" In the moment I replied, "it is important for you to learn this because it is part of the curriculum," adding "it is going to be on our unit test." But deep in my heart, I knew that these answers were not what my students were looking for.

For Reflection and Discussion

1. How might you respond to the group of students who did not see relevancy in the soil testing activity?

2. How might you plan this activity differently if you were facing a similar type of situation?
3. What other strategies could Joseph have implemented to help students make more sense of the final activity and value the importance of studying soil composition?
4. How can we make science activities more meaningful and relevant for students?

José M. Pavez is a former middle school science teacher and secondary biology teacher. Originally from Chile, he has over ten years of teaching experience from 5th grade to graduate level. Currently, he is an assistant professor of elementary science education at Western Illinois University. His main research interests focus on science teacher education.

Commentary: Whose Science is of the Most Worth? Making a Case for Problem Posing Instead of Problem Solving

73

Rouhollah Aghasaleh

Abstract

This is a commentary to the case narrative, *"But Why Do We Need to Study This? The Case of Arsenic in the School's Soil"* written by Jose M. Pavez.

There should be a poster in every classroom with a quote from Elliot Eisner saying, "standardized teaching, from an educational perspective, is an oxymoron" (2002, p. 7)". How could we standardize this complex human relationship that we call teaching and learning? Is there a one-size-fit-all curriculum for all classrooms? How could teachers follow the curricular requirements mandated by state, district, and school and meet students' needs? What happens when a teacher tries to make a meaningful, culturally relevant lesson that is tailored to an individual classroom, but the students are still disengaged?

In Fall 2016, I was in charge of a science club at an after-school program that hosted urban LatinX children in a southeastern state. My team and I could sense the anxiety among the kids as we got closer to the presidential election just like everywhere else in the country. The heated election season and the polarized sociopolitical atmosphere did not allow people to dismiss the sentiments from both sides. In our after-school program, the rise of right populism and anti-immigrant sentiments were scary to many of the children. Of course, we never asked but it seemed safe to assume that several students in that group were undocumented immigrants or had undocumented immigrant parents, siblings, or relatives in their households. Hate crimes had increased since the election. Discussing the result of the presidential election was important since many students felt fear. Two teachers were removed from a nearby high school (the high school many of the after-school

R. Aghasaleh (✉)
School of Education, Humboldt State University, Arcata, CA, USA
e-mail: aghasaleh@humboldt.edu

students will attend one day) for making anti-immigration comments following the election. Emergent bilinguals, such as the children in our science program, often internalize the xenophobic discourse that surrounds non-native English speakers. The young children in our program were very savvy when it came to their knowledge about those stories in the media. They were constantly exposed to social media, which often presented the most extreme, negative, and even false societal views. This xenophobic discourse was amplified during the 2016's election season. It was clear that many of these young students, even at an early age, feared deportation, discrimination, and other forms of violence. Many students experienced their friends and families facing deportation due to the increased Immigration and Customs Enforcement raids that occurred in 2016. The after-school science program was a safe space in which we could provide a constructive and creative environment for students to comfortably explore these topics. For the children in our after-school program, like the girls in Jose's case, these tensions were a reality that no one could ignore. From mainstream news channels to social media everywhere was some sort of discussion around the election and its winner. Amidst the uncertainties facing emergent bilingual children daily, we recognized that our after-school science program needed to be grounded in an instructional approach that would provide students with both a voice and a space where they could safely explore science. After many discussions, we collectively settled on problem-based learning as an approach that seemed appropriate for the young children in the program and relevant to their everyday lives.

Many progressive science teachers who use inquiry-based teaching practices in their lessons are familiar with problem-based teaching. It forms the lesson around a problem that the teacher (or textbook) has identified as significant. The lesson includes a series of hands-on learning activities and ideally an experiment or two to compliment it. Many teachers have also adopted practices to make these problems culturally and linguistically relevant based on the extensive literature from critical educators who coined concepts such as critical pedagogy (Freire, 1968/1970), multicultural education (Banks & Banks, 1989; Nieto, 1992); culturally relevant pedagogy (Ladson-Billings, 1995), culturally responsive instruction (Gay, 2000); culturally and linguistically responsiveness (Hollie, 2015); culturally sustaining pedagogy (Paris & Alim, 2018). For instance, teachers may include a soccer video clip to discuss force and motion to attract the supposedly soccer-loving LatinX kids or they may include some sewable electronics kits to make science relevant to girls. While these superficial efforts to make science teaching relevant are engaging, these approaches mostly focus on "how" we teach and less often on "what" we teach. Furthermore, they rarely question "where" this content knowledge comes from.

So, what if we start by having students identify problems instead of bringing a problem identified by the teacher to the class to be solved? This could be a real and authentic problem that has occupied students: An issue that *they* think is worth of inquiry, investigation, and learning. Problem posing involves the students in finding a problem or a question that they attempt to solve throughout the lesson. During problem posing, the students have a framework they can use to relate all of the concepts they cover throughout the lesson/unit to their original problem. It may initially seem like a radical suggestion that the formulation of the problem is more essential than the solution, but this argument has a long history. If students are to be good at solving problems, they must also be good at finding problems.

Curriculum scholars have been asking this question, "what knowledge is of most worth?" (Spencer, 1860). Obviously, school curriculum cannot include all types and perspectives of scientific knowledge. One approach to making the scientific knowledge more inclusive, relevant, engaging, and authentic is to put the students first so that students can bring in their lived experiences, thoughts, feelings, and wonders to the science classroom. To this end, in the after-school program, we had students brainstorm their own problems. The students got together in groups and bounced ideas off one another about different problems they had/knew in the world. These problems ranged from personal issues to global ones. Discussing with peers about their problems allowed the students to reflect on their home culture (and the culture of their peers), their lived experiences, as well as their interests and aspirations. What surprised many of my colleagues the most was the extent to which the students took national and global issues and made them more personal (Aghasaleh et al., 2018). Simultaneously, when asking students why they picked their problem-posing topics, the reasons given were often personal and emotional. For instance, their projects on presidential elections, sexual violence, police brutality, and Black Lives Matter became as scientific as their projects on global warming, robotics, and engineering.

The following is a reflection from a teacher candidate who experienced the problem-posing approach in our after-school science program.

> Another assumption I mistakenly (potentially prejudicially) made was that most of the students at the afterschool program would choose topics related to their shared Latin American culture. An example of this preconception came from the teachers who ran the afterschool program, most of whom were Latinx. Upon hearing how our program would use problem posing for the students' projects, they thought that many of them would pick soccer-related issues. However, when the students were brainstorming their topics soccer rarely (if ever) came up, and none of the students selected it for their topic. Upon reflection, I recognized that despite their shared Latin American culture being the students' reason for being at the after-school program, the students were also a part of many different cultures, some of which they may feel are more important to them personally than their Latin American heritage. After seeing the students brainstorm and select their topics, I was even more grateful that we had them select their own topics instead of having them focus on instructor-selected problems that would potentially have little to no relevance to their day to day lives. (Aghasaleh et al., 2019, p. 37)

In this case, Joseph had thoughtfully designed a lesson on matter and mixtures around a real issue of soil contamination. He assumed that the San Sebastian school problem would be relevant, engaging, and authentic as it was happening at a nearby elementary school. To his surprise, he realized that Chelsea and Alfredo's group were not engaged as they questioned the importance of the lesson and demanded explanation for why they needed to learn about contaminated soil. Although Joseph was doing his best to identify a relevant problem, he struggled to account for the students' perspectives in identifying the problem focus of the lesson. Although our professional preparation as teachers and overwhelming accountability culture have suppressed both teachers' and students' creativity, there is still a *little* room to teach against the grain. And my hope is that we can do a lot with the *little*.

References

Aghasaleh, R., Enderle, P., Puvirajah, A., Rickard, J., Boehnlein, A., Bornstein, J., & Hendrix, R. (2018). Teaching computing to urban Latinx youth: Make science teaching great again. *Curriculum and Teaching Dialogue Journal, 20*(1/2), 143–147.

Aghasaleh, R., Enderle, P., & Puvirajah, A. (2019). From computational thinking to political resistance: Reciprocal lessons from urban Latinx middle school students. *Journal of Activist Science and Technology Education, Special Issue on Topologies of Activism, 10*(1), 29–44.

Banks, J. A., & Banks, C. A. M. (Eds.). (2019). *Multicultural education: Issues and perspectives.* Wiley.

Eisner, E. W. (2002). *The educational imagination.* Merrill.

Freire, P. (1968/1970). *Pedagogy of the oppressed.* Continuum.

Gay, G. (2000). *Culturally responsive teaching: Theory, research, and practice.* Teachers College Press.

Hollie, S. (2015). *Strategies for culturally and linguistically responsive teaching and learning.* Teacher Created Materials.

Ladson-Billings, G. (1995). Toward a theory of culturally relevant pedagogy. *American Educational Research Journal, 32*(3), 465–491.

Nieto, S. (1992). *Affirming diversity: The sociopolitical context of multicultural education.* Longman.

Paris, D., & Alim, H. S. (2018). *Culturally sustaining pedagogies: Teaching and learning for justice in a changing world.* Teachers College Press.

Spencer, H. (1860). What knowledge is of most worth? In H. Spencer (Ed.), *Education: Intellectual, moral, and physical* (pp. 21–96). D Appleton & Company. https://doi.org/10.1037/12158-001

Rouhollah Aghasaleh is an assistant professor in the School of Education at California State Polytechnic University, Humboldt. Their scholarship lays rests at the intersection of critical middle grades education, cultural studies of curriculum, and new materialist feminism that addresses issues of equity and its impact on the education system. They are the editor of an award-winning Brill volume, Children and Mother Nature: Storytelling for a Glocalized Environmental Pedagogy.

Case: School-Cafeteria Make-Over Real-World Style

74

Cassie F. Quigley, Danielle Herro,
and Lisa M. Kreklow Weatherbee

Abstract

Students in a fourth-grade classroom engage in a STEAM problem scenario about how to redesign the school cafeteria in this case. Ms. Weatherbee, the teacher for this group of fourth-graders, feels that the project will provide opportunities for connecting ideas through multiple disciplines. These ideas include engineering design in terms of the new layout for the cafeteria, noise reduction, food science with healthy choices, and math using budgeting and percentage. Ms. Weatherbee lets her students select a topic to investigate (food, space, use, logistics, etc.) as well as a way to disseminate their results. At the end of the project, the students present their ideas to the school board, principal, and the school administration, only to realize that their presentation was perceived as more for "show" instead of enacting real change. Ms. Weatherbee struggles with supporting the students during this time; however, the students were persistent and came up with new ideas that she had not thought of for making the change themselves.

The K-5 elementary school is located in a growing community in Southeastern Wisconsin. The area has a large refugee population (increased from 8 to

---

C. F. Quigley (✉)
School of Education, University of Pittsburgh, Pittsburgh, PA, USA
e-mail: cquigley@pitt.edu

D. Herro
College of Education, Clemson University, Clemson, SC, USA
e-mail: dherro@clemson.edu

L. M. K. Weatherbee
Westside Elementary School, Sun Prairie School District, Sun Prairie, USA

S. Jeong et al. (eds.), *Navigating Elementary Science Teaching and Learning*,
Springer Texts in Education, https://doi.org/10.1007/978-3-031-33418-4_74

18% in 2019 and is currently 21%), and the district administrators are interested in curricular innovations that engage students in deeper learning and foster a sense of community. To that end, the school administration and teachers focus on problem-based learning, discipline integration, and student engagement. The driving question at the heart of the students' STEAM project is: How can we improve the cafeteria experience at Northwood Elementary school?

Amar, Winston, and Wei are huddled together in the back of the science classroom. Each holds a clipboard and a pencil for recording ideas. Strewn across the floor are their drafts that they have discarded as they discuss options for the cafeteria redesign project. The class is tackling the problem of how to improve the cafeteria in four ways: noise reduction, aesthetics, food choices, and use of space. The boys' chosen topic is noise reduction in the cafeteria. Northwood's cafeteria doubles as the gymnasium for the school and so it is a large room with tall ceilings. During this brainstorming session, the students discuss how loud the room is, making it difficult for them to enjoy their lunch period (Fig. 74.1).

Amar leans over Wei's drawing to get a better view and asks, "How does putting the table there help us hear better?" Amar responds, "If we move the tables so they are facing one another, then our voices project forward." Wei gives him a quizzical look as Amar continues his explanation. "Watch this. I am talking to you now. Can you hear me well?" Wei nods. Then Amar steps behind him, "What about now? Which way can you hear me better? When I am front of you or behind?" Wei responds, "In front." "Seeee!" Amar exclaims and Winston giggles. "So that is why we should face the tables toward each other." Wei agrees, but adds, "Yes, that makes sense to me but by doing that, we are going to lose two seats when we squish them together."

Now, it is Amar's turn to be confused, but Winston jumps in, showing him his design. "Look, Amar. If the tables are all alone, six of us can sit here and six of us can sit here," he explains while pointing to the two tables separated. "But here" Wei adds, "I don't think you can get 12 around it when the two tables are smushed together." Winston starts to count, "1, 2, 3, 4" "Wait, that's right, only 8. So, we need to add more tables. Can we do that?" Winston simultaneously raises his hand as he asks the question of Ms. Weatherbee (Fig. 74.2).

Ms. Weatherbee, upon seeing Winston's hand, walks over to the group. The boys excitedly talk all at once about their idea. Explaining how they can hear each other better in the new formation because of the way the sound travels, they express their need for more tables. "Can we add more tables?" they ask. She responds, "I think you have a good rationale for why you would need more tables. What are some things you would need to figure out before putting it in your plan? Talk about those and then you can report out your plan to the class, ok?" The boys nod their heads and get to work.

This line of questioning and furious scribbling on pads of paper began when Ms. Weatherbee made a class announcement *Have you noticed that the Westside Elementary School Cafeteria could use a make-over? From the food to the waste to the physical space, there are many ways we could improve this room. Wouldn't it be wonderful if the cafeteria was an inviting space for us all? Well, good news, the*

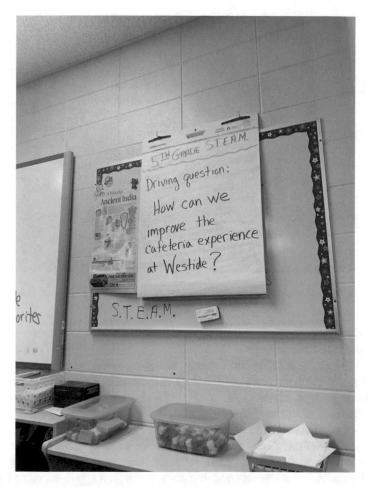

Fig. 74.1 Driving question for students' cafeteria inquiry

principal has asked that the fifth grade be part of the team that will design a plan to improve our cafeteria.

This STEAM problem scenario is designed to promote inquiry involving authentic tasks around redesigning the school cafeteria. Students engage in a variety of investigations related to their line of inquiry with the recognition that they can solve the problem in different ways. While Amar, Winston, and Wei are studying noise reduction, another group is examining how breakfast options could be healthier and sustain the students until lunchtime. Still another group is looking into ways that the school could compost food waste and recycle. And a fourth group is exploring the ways that the cafeteria could feel like a lunchroom despite its alternate use as a gym. The students brainstorm these issues using a shared Google Doc, and then choose the topic that is of most interest to them. In this way,

Fig. 74.2 Students share and discuss their plan for improving seating in the school cafeteria

Ms. Weatherbee feels she is laying the groundwork for enacting inquiry around an issue relevant to their lives (Fig. 74.3).

At the end of the science class, Amar, Winston and Wei present their thoughts to the class for feedback and describe their suggestion of more tables. Their classmates are eager to offer feedback, noting, "I like the idea of sitting at a big table with my friends. It would make me feel like I am at home at our dinner table." Syrah worries, "Those would be really big tables, could we still hear each other?" At the end of the discussion, Syrah and her partner, Chloe, decide they need to test out their ideas with the tables in the cafeteria. They begin to make a plan to do that. This plan includes testing over several days using an app to measure the decibels in the cafeteria as well as a student survey to understand their schoolmates' opinions.

The students present their findings in a variety of formats using technology to a panel of people including the school administration. When it is Amar, Winston, and Wei's turn to present, they offer data about their findings around table placement and noise reduction. Wei explains, "by placing the tables this way, students are able to hear one another better. See the results of our survey. This change would require six more tables in the cafeteria." The panel provides positive feedback and the boys are pleased saying "I think we are going to be able to get the tables and make the changes." Ms. Weatherbee overhears them and offers words of encouragement, "this is really very exciting. Clearly you all understood the problem, collected and analyzed the data, and presented it very well." She catches herself, and does not share the rest of her thought, "but I don't actually know if the district will do anything about it." When she posed the question to the administration, they said

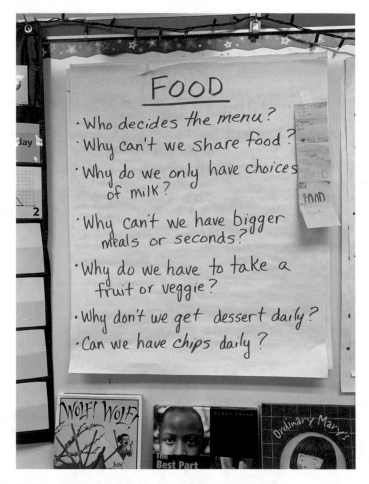

Fig. 74.3 Brainstorm about improving the food in the school cafeteria

they would look into purchasing the additional tables. However, weeks went by, and she did not hear back from them.

When students are activists, they often face barriers. But during these stalls, leadership can provide pathways to persist. Mitra and Serriere (2012) explored student voice in an elementary school, highlighting the case of "Salad Girls," three young students who invoked changes in the food choices at their school cafeteria. By conducting an inquiry that involved gathering data school-wide and communicating the results to the decision-makers at the school, the girls were ultimately able to galvanize changes in the school menu. Similar to Ms. Weatherbee's students, the Salad Girls faced standoffs. Likewise, the Salad Girls had the support of the principal, who encouraged them to meet with district-wide officials.

Not to be discouraged, Amar, Winston, and Wei continued with their quest, following up with the cafeteria manager, and discussing this issue long after the project was completed. However, the administration did not respond to the students' data collection or their findings. Ms. Weatherbee is now having doubts about the use of real-world problem-solving in her classroom in the future.

For Reflection and Discussion

1. How could Ms. Weatherbee support real-world problems in her classroom even if students' ideas will not result in change?
2. Should Ms. Weatherbee frame the scenario at the beginning of the STEAM project to help the students understand the limits of their inquiries? If so, describe what this might look like. If not, describe why you would not recommend it.
3. What role could youth activism play in a science classroom?
4. How can we create contexts for our youth to engage in real-world scientific issues?

Reference

Mitra, D. L., & Serriere, S. C. (2012). Student voice in elementary school reform examining youth development in fifth graders. *American Educational Research Journal, 49*(4), 743–774.

Cassie F. Quigley is a professor of science education in the Department of Instruction of Learning at the School of Education at the University of Pittsburgh. She received her doctorate in curriculum & instruction at Indiana University in 2010. During her time as a high school biology and physics teacher, she often witnessed students who did not connect with school science. Because of this experience, her research focuses on transforming science education to promote multiple knowledge types and ways of knowing.

Danielle Herro is a professor of digital media and learning in the Learning Sciences program at Clemson University. She earned her Ph.D. at the University of Wisconsin, Madison. Dani teaches learning sciences seminar courses, STEAM education courses, and courses focused on theories and learning with social media, games and emerging technologies.

Lisa M. Kreklow Weatherbee graduated with a Bachelor of Science in education from the University of Wisconsin-Madison in 1994. Her teaching career includes diverse experiences including Milwaukee Public Schools, Pardeeville Public Schools and Sun Prairie School District outside of Madison, Wisconsin. Teaching through strong relationships, relevant problem solving, and engaging manners is a passion which she strives to attain each day.

Commentary: Struggling to Make a Difference Through Students' Research Informed Activism

Deborah J. Tippins

Abstract

This is a commentary to the case narrative, *"School-Cafeteria Make-Over Real-World Style"* written by Cassie F. Quigley, Dani Herro, and Lisa Weatherbee.

Ms. Weatherbee teaches in a school, which on the surface, values STEAM problem-solving, interdisciplinary learning, and the building of classroom community. She encourages and enables her 4th-grade students to engage in real-world problem-based learning centered around improving the quality of the school cafeteria experience. She invites her students to express their ideas and opinions about factors that might impact issues such as quality of the food, arrangement of space, and noise level in the cafeteria. Students, working in groups, are provided with the opportunity to investigate and conduct research on an aspect of the cafeteria environment. Ms. Weatherbee hopes that these student-led investigatory projects will serve as a motivation to suggest that students might make a difference by bringing about actual improvements to their school cafeteria. Clearly, there are many important goals associated with this type of STEAM problem-based learning approach to instruction—fostering socioscientific reasoning, learning science concepts, creating student ownership of science learning, and experiencing the nature of science, to name a few. More importantly, Ms. Weatherbee successfully creates a space where students, with support and freedom of choice, can deeply explore an issue that is important to them. In contrast to artificially contrived contexts for problem-based learning, Ms. Weatherbee purposefully allows students to focus on an issue that can help them feel a genuine connection to their school community.

D. J. Tippins (✉)
Department of Mathematics, Science, and Social Studies Education, Mary Frances College of Education, University of Georgia, Athens, GA, USA
e-mail: dtippins@uga.edu

© The Author(s), under exclusive license to Springer Nature Switzerland AG 2023   385
S. Jeong et al. (eds.), *Navigating Elementary Science Teaching and Learning*,
Springer Texts in Education, https://doi.org/10.1007/978-3-031-33418-4_75

In this case, we can see how students are deeply invested in their projects, encouraged, and excited with the promise of bringing about real change to the cafeteria environment.

In an attempt to honor the voices of students and perhaps enhance their public speaking skills, Ms. Weatherbee organizes a culminating activity whereby each student group has the opportunity to share the results of their research and make recommendations for improving the quality of the cafeteria experience. Similar to other groups, Amir, Winston, and Wei confidently present their findings and recommendations concerning the relationship between the placement and arrangement of cafeteria tables and noise reduction to a panel comprised of school leaders and administrators. As the days pass after the panel presentation, the students eagerly await the response of the administrators, hoping to soon see their recommendations taken seriously through changes to the cafeteria. Meanwhile, Ms. Weatherbee increasingly begins to doubt the efficacy of her real-world problem-solving approach as the lack of response from administration becomes evident.

This case points to the need for school leaders and administrators to support teachers in their attempts to carry out instructional approaches which bring authentic problem-solving to a new level. In this case, both the students and their teacher share ideas which should be treated with the respect they are due. Perhaps unconsciously, by ignoring the ideas and solutions shared by the group of fourth graders, the administrators leave students feeling a sense of powerlessness to speak up in the future. Students need to know that others will listen to and value their ideas. They need to feel secure in expressing their ideas without the fear of being judged, knowing that their proposals will be considered seriously. Even the small changes to the cafeteria students propose, when taken seriously by adults, can serve as preparation for future activism. Real-world problem-solving is complex, particularly in light of teachers' attempts to move students' ideas to action. Direct experiences with an issue meaningful to students such as the quality of their lunch experience in the school cafeteria may ultimately lessen the divide between what students are taught in the classroom and what they learn in the real world.

Deborah J. Tippins is a professor emeritus and a distinguished research scholar in the Department of Mathematics and Science Education at the University of Georgia. Her scholarly work focuses on encouraging meaningful discourse around environmental justice in science education. She draws on ecojustice philosophy and the anthropology of science education to investigate questions related to citizen science, sustainability, culturally relevant science, and science teacher preparation.

Case: "Are We Getting Rid of a Soccer Field? But I Have Nowhere Else to Play Soccer!": Implementing an SSI-Based Ecosystem Unit in the Third Grade

76

Laura Zangori

Abstract

Laura teaches in a third-grade classroom in the city surrounding the university where she works. She planned a socioscientific issue (SSI) unit on ecosystem interrelationships. Specifically, her SSI unit focused on the reduction in Monarchs migrating through the state due to a decline in prairies. She asked students to consider whether or not a way to work on this problem would be to turn one of the two soccer fields on their playground into a prairie to support Monarch migration through the state. This open case explores a situation in which Laura discovers that her students took their position on this issue very seriously and assumed the outcome of the soccer field was in their hands.

"Ms. Zangori, Wait a minute! Are we getting rid of one of our soccer fields? But I have NO WHERE ELSE TO PLAY SOCCER!" Some of my third-grade students looked stunned and worried, while others were smiling at the thought of removing one of our two soccer fields and replacing it with a natural prairie.

My class had just completed a discussion within a socioscientific issue (SSI) unit about prairie conservation. Due to prairie reduction in our state, monarch migration throughout the state had declined substantially. In the class discussion, each student presented their position on whether or not we should convert one of the two soccer fields on the school playground into a prairie to support increases in Monarch migration numbers. At the end of the discussion, the students voted on whether they agreed or disagreed with the removal of a soccer field to add a natural prairie. As the discussion came to a close and the votes were tallied, I discovered that my students thought their class decision on this issue was final

L. Zangori (✉)
Department of Learning, Teaching and Curriculum, University of Missouri, Columbia, MO, USA
e-mail: zangoril@missouri.edu

and would actually occur at the school. How was I going to tell them that their decision was only hypothetical?

The idea to teach an SSI unit in my classroom came from a new initiative within the school district to teach science content through place-based education. There was also a heightened focus within the elementary school to teach argumentative writing in third grade. The science content focus for third-grade was ecosystems. I thought combining place-based education and argumentation into my ecosystem unit would meet these overall initiatives and make the unit more relevant to my students. To do this, I attended a professional development on design and implementation of SSI units within a local context. Within this type of SSI unit, students take a position on a local issue and use their scientific knowledge to justify their position, which aligned perfectly with what I would be teaching in argumentative writing! And connecting the issue to the school to make it place-based would provide a concrete issue for my students that was personal to them and they would have an opinion about. Yet the school district did not have an SSI-focused science unit for elementary students. The more I explored this idea, the more I discovered how little information there was about how to design and implement an SSI lesson in third grade. Despite the lack of information, the third-grade teaching team that I was a part of decided to try this approach with our students. Since there were no elementary SSI units for us to use, we chose to write the lessons ourselves.

The teaching team first discussed different issues we might use for the lesson focus. We came to a consensus to center the unit on the state reduction in Monarch butterfly migration numbers due to the decreasing number of prairies both within the state and in the city where we lived. In addition, this issue fit with the district-wide science project initiative to remove invasive plants harmful to Monarch migration. The SSI would also build on the science knowledge the third-grade students had learned in second grade about the relationship between a plant and a butterfly.

Next, the teaching team worked to frame the issue as place-based. We situated the issue within the context of the school playground asking students if one of the two soccer fields on the playground should be turned into a prairie to support Monarch migration. The soccer field issue then guided the design of the unit to focus the science content on ecological relationships, using the relationship between the Monarch butterfly and milkweed to talk about the different kinds of mutual relationships that occur within an ecosystem. According to our scope and sequence, this unit would be the first science unit we would teach during the school year. We chose to have students write their positions and justifications twice on the issue—once at the beginning of the unit and once at the end of the unit.

In the daily classroom schedule, we had 30-min for the science block. I used my first 30-min science block to introduce the issue. I played a short segment of the IMAX movie Flight of the Butterflies (Slee et al., 2012) which showed Monarch migration from Mexico to Canada. I then led students in a discussion about where and why Monarchs migrate. Our discussion included the relationship between Monarch butterflies and milkweed that grows in prairies, drawing on students' prior knowledge from second grade about this relationship. Lastly, we

Fig. 76.1 Pictograph of monarch migration decline in Missouri

worked as a class to construct a pictograph of the number of Monarchs migrating through the state using data supplied by the Missouri Department of Conservation (see Fig. 76.1). I chose to use our 30-min morning writing time for students to write and justify their position on this issue. I asked students to write a letter to the school principal. In the letter, they were to take a position on whether one of the two soccer fields on the playground should be turned into a prairie to attract Monarch butterflies. Students were asked to use their current science knowledge to justify their position on the issue.

When I read through their initial writings, I found that one-third of the students discussed why a prairie might be necessary, talking about how pretty butterflies were and how butterflies and other small animals might live in the prairie.

The remaining two-thirds of the students discussed the importance of the second soccer field because they liked to play soccer, and they needed the second soccer field since the fourth-grade students also played soccer during their shared recess time.

After this initial writing, we worked through the science lessons the teaching team had written. Throughout the lessons, I regularly brought up the Monarch migration issue to remind students about our focal issue and to talk about relationships between plants and animals within a prairie. Throughout the lessons, I assumed the students knew the question about the soccer field was hypothetical and the letters they had written and would write again at the end of the unit about whether to turn a soccer field into a prairie for Monarch migration, would not actually affect the soccer field.

After we completed the last lesson of the unit, I used the next 30-min science block for a whole class discussion about the soccer field issue. This was the first time since the beginning of the unit that students had an opportunity to voice and justify their position on whether or not to convert one of the two soccer fields on the playground into a prairie. For this discussion, I was most interested in providing the students with a chance to share their ideas about the soccer field with their classmates. I also saw this as an opportunity for the students to use their new knowledge about interrelationships in ecosystems to justify their decision regarding the soccer field. Finally, I hoped this discussion would prepare them for their final argumentative writing that would take place the next morning.

I called the students to sit in a circle on the carpet in the middle of the classroom where I joined them. I reminded students that we were going to discuss why so few Monarch butterflies were migrating through Missouri and if we should take action on this issue by turning one of our two soccer fields into a prairie. I pointed out our pictograph that we had completed in our first lesson

Lila was first to speak and called out "It's been over 20 years! And the butterfly numbers keep going down!" I was excited to see how well she interpreted our graph and was also excited at the great discussion I was sure to follow. I reminded students that we had discussed the important relationships that Monarch butterflies had with milkweed because milkweed is where butterflies lay their eggs and is also food for Monarch butterflies. I then asked students to share their position and justification on the SSI issue—should we turn one of our two soccer fields into a prairie? My students erupted in yells of "YES" and "NO"! It took students a little bit to quiet down as the students were passionate about their positions.

Lila shared first, as she was adamantly shouting "YES!" Lila said, "If we don't have more prairies, one day the Monarch species may be completely extinct, which may destroy some other species." I was so excited to hear her science justification about her position. Other students gave similar answers, talking about the ecosystem content we had discussed during our unit such as what pollination does and how food chains are an important part of pollination.

Then I called on Mike, who was sitting with a group of students that had passionately said "NO!" Mike said he knew about a relationship between the Monarch and milkweed, but he could not play soccer at home because he did not have a yard. It was important to Mike to keep both soccer fields at school so that he could still play soccer with his friends during recess. Other students also shared their concerns about the loss of a soccer field, which included whether community groups that used both soccer fields on weekends and after school would still be able to play soccer if one of the soccer fields was turned into a prairie. After all of the students shared, I asked them to take a vote and give me a thumbs up if the soccer field should be turned into a prairie. I saw over half of my students give me a thumbs up, and I announced "It looks like we think we should turn the soccer field into a prairie! Tomorrow morning during writing time we will write our final position letters to Mrs. Geinhardt" (the school principal). Some of the children erupted with "Hooray"!

I was in the middle of silently congratulating myself on how well my third graders did with the SSI unit and used their scientific justifications on their position, until I noticed Mike. He looked stunned, and the boy he was sitting next to started to cry. Mike said, "This is the only place I have to play soccer!" Another student came up to me and asked me how much longer they would have the soccer field before the principal destroyed it to make it a prairie. And another girl said

"But the fourth graders won't let us play with them on the other field!" I suddenly began to understand that my students thought that we were really going to implement this decision and remove a soccer field! What was I going to do now? Should I have even framed my SSI lesson this way with my students?

For Reflection and Discussion

1. One way a teacher could have handled the unit was to tell children up front that they would not be implementing their solutions in real life—i.e., they would not really transform a soccer field into a prairie. However, in doing so, how might telling them up front that any solution is only hypothetical change students' perceptions or participation in the lesson? What are some other ways to frame this unit that could have an implementable solution and not a hypothetical solution?
2. What are some other challenges that a teacher could pose for this SSI unit that would engage students in using their knowledge about interrelationships in ecosystems to take a position and build an argument?
3. Should environmental issues, such as decreases in species or effects of climate change be place-based and in local contexts? Why or why not?
4. How could we have discussions in elementary classrooms about important issues, while honoring how those issues affect each student in different ways?

Reference

Slee, M., Barker, J., & Cuervo, R. (2012). *Flight of the butterflies* [Motion picture]. SK Films.

Laura Zangori is an Associate Professor of Science Education at the University of Missouri. She teaches and works with students ranging from elementary through undergraduate classrooms. Her research focuses on how to support teachers and their students in using systems models and modeling practices to understand biological systems and apply their scientific understanding to grapple with complex issues with environmental, societal, and political implications. Laura completed her Ph.D. in Science Education at the University of Nebraska-Lincoln in 2015.

Commentary: Teaching with Socioscientific Issues

77

Troy D. Sadler

Abstract

This is a commentary to the case narrative, "*Are We Getting Rid of a Soccer Field? But I Have Nowhere Else to Play Soccer!*': *Implementing an SSI-Based Ecosystem Unit in the Third Grade*" written by Laura Zangori.

I started thinking about teaching with socioscientific issues (and implementing issues-based learning experiences) when I was a high school biology teacher. When I transitioned from a classroom teacher to a teacher educator and science education researcher, I maintained my interest in socioscientific issues-based teaching and learning. In the past 20 years since that transition, I have worked on numerous curriculum development projects and research studies focused on teaching science through student explorations of complex issues that matter beyond the walls of science classrooms. Issues like the loss of biodiversity (including Monarch butterflies), climate change, global food security, and the emergence of new diseases like COVID-19 represent the kinds of societal challenges that make learner development of scientific literacy so critical. We know that it is difficult for individuals of all ages to make sense of these challenging issues, and there is ample reason to think that as a society, we have a long way to go toward making more informed decisions about these issues. Therefore, I have consistently maintained that science classrooms should be places in which learners have opportunities to consider these complex issues and work toward understanding how they might use their science knowledge and experiences to inform their decisions and actions.

T. D. Sadler (✉)
School of Education, University of North Carolina at Chapel Hill, Chapel Hill, NC, USA
e-mail: tsadler@unc.edu

© The Author(s), under exclusive license to Springer Nature Switzerland AG 2023
S. Jeong et al. (eds.), *Navigating Elementary Science Teaching and Learning*,
Springer Texts in Education, https://doi.org/10.1007/978-3-031-33418-4_77

Several years ago, I was invited to present some of my team's research about teaching high school biology through socioscientific issues. Following the presentation, one of the visit's hosts walked me across campus and asked what I thought about using socioscientific issues in elementary classrooms. At the time, I had not thought much about the question. My team had done work at the middle school, high school, and undergraduate levels, but we had not tried incorporating socioscientific issues in elementary school. We often struggled to frame the complexity and the associated science when designing socioscientific learning experiences for middle and high school learners; I worried about how even younger learners would deal with the complexity. I knew from experience with middle and high school teachers that issues-based teaching could be a challenging pedagogical approach; I wondered what struggles elementary teachers might face enacting issues-based instruction. However, my host's question pushed me to consider how issues-based teaching made great sense for elementary classrooms. Young learners are naturally curious and therefore are poised to ask important questions about complex issues. Socioscientific issues are inherently interdisciplinary—that is, they deal with numerous topics and ways of knowing from science to economics, politics to ethics. Elementary classrooms, especially as compared to secondary spaces, tend to embrace interdisciplinarity, so issues-based teaching may actually fit much better within elementary classrooms than other formal education settings.

Despite the potential of using socioscientific issues to teach elementary science, there are few examples of the approach being used, which made me interested in Laura's case. Laura and her colleagues did many things that contributed to the success of the experience. First, they framed the focal issue in a way that their third-grade students could easily relate to: they chose to focus on the declining population of butterflies, an organism that the students had surely experienced and could recognize from their everyday lives. Declining butterfly populations represents a specific instantiation of habitat degradation and loss of biodiversity, which can be tied to climate change and other global trends. However, framing the issue in terms of these much grander challenges would have been difficult for the students to appreciate and understand. Second, the unit built coherently from ideas that the students had explored in their previous grade's science explorations: ecological relationships between plants and butterflies. Third, Laura and her fellow teachers connected their unit to initiatives that were ongoing within the district, including an emphasis on place-based education and a district-wide program for removing an invasive plant species. Fourth, while the unit emphasized science ideas, it also created opportunities for students to engage with other subjects and competencies. They incorporated argumentative writing, an important language arts goal, as well as mathematics in the form of data representations (see Fig. 76.1 from Laura's case) and interpretations. As I suggested above, socioscientific issues are interdisciplinary, and in this case, student exploration of the issue created opportunities for learning and practice in multiple subject areas.

For several years, my team (Sadler et al., 2017) in addition to other research groups (Marks & Eilks, 2009) have been working toward a clear articulation of socioscientific issues-based teaching. While some specific aspects of the frameworks vary, several similarities exist, and Laura's case exemplifies many of these key features for socioscientific issues teaching. For example, Laura started the unit with a focus on the issue by engaging students with the movie about Monarch migration and a discussion of Monarchs in their local environment. This stands in stark contrast to approaches wherein an issue or application is introduced only after the science content has been fully presented. By positioning student exploration of declining butterfly populations as the initial stage of the unit, the issue truly contextualizes learning rather than just serving as an example of how ideas might be applied. Another key dimension of socioscientific issues teaching is student opportunities to work with evidence and engage with scientific practices. In Laura's work, students worked with data related to butterfly population numbers. This, in turn, created opportunities for students to engage in scientific practices, including data analysis and modeling. Exemplary socioscientific teaching also prioritizes student discourse, and Laura's example showcases multiple opportunities for students to present their ideas, express disagreements, and share feedback. Finally, most socioscientific teaching frameworks suggest concluding a unit with a culminating activity that allows students to synthesize what they have learned and their own personal perspectives and to consider actions moving forward. In this case, Laura concluded the unit with a writing exercise in which students shared their class vote with the school principal.

This culmination of the unit seems to be where Laura and her students bumped into a challenge. While some students were excited to transform their schoolyard into a butterfly garden, others were upset that they would not have a place to play soccer. Regardless of their positions, all the students felt that their decision was binding, and it sounds like Laura had just assumed that everyone understood this to be an academic exercise. By the end of the case, Laura is questioning whether she should have even tried the SSI approach in the first place. My answer to Laura's question is…absolutely. The experience seems to have been a powerful learning opportunity in which students came to better understand a challenging issue and became passionate about what they were learning. This is precisely the kind of outcome that we as educators should advocate for. At the same time, it is important to acknowledge that several students were upset by the end of the unit, and most of the class had an unrealistic idea about how their decision might lead to changes in the school grounds. However, rather than throwing out the whole approach, I think Laura could consider ways to avoid the disappointing end. The task itself could have been framed a bit differently to help students understand that they might offer a recommendation, but that they alone would not determine the fate of the schoolyard. Another option would be to introduce options that might be more realistic for the students to enact. For example, rather than taking out a soccer field, students could have considered planting milkweed in smaller areas. These are just a couple options, and I am sure there are other possibilities, but I strongly encourage Laura, as well as other elementary teachers, to think about ways to minimize some of the challenges that can emerge through socioscientific issues teaching, but also to keep trying this innovative approach, which can have lasting positive impacts for their students.

References

Marks, R., & Eilks, I. (2009). Promoting scientific literacy using a sociocritical and problem-oriented approach to chemistry teaching: Concepts, examples, experiences. *International Journal of Environmental and Science Education, 4*(3), 231–245.

Sadler, T. D., Foulk, J. A., & Friedrichsen, P. J. (2017). Evolution of a model for socio-scientific issue teaching and learning. *International Journal of Education in Mathematics, Science and Technology, 5*(1), 75–87. https://doi.org/10.18404/ijemst.55999

Troy D. Sadler is the Thomas James Distinguished Professor in Experiential Learning at the University of North Carolina at Chapel Hill. Sadler conducts research related to how students negotiate complex socioscientific issues and how these issues may be used as contexts for science learning. He is interested in how issues-based learning experiences can support student learning of science and development of practices essential for full participation in modern democratic societies.

Part VIII

Caring for Living Organisms

Deborah J. Tippins

Children of all ages need rich opportunities to engage and live with animals in appropriate ways. The natural curiosity that young children have about living organisms can be a starting point for the emergence of meaningful questions that stem from everyday experiences with animals. These questions can serve as a centerpiece for fostering children's interest and respect for animals of all kinds and encourage rich conversations.

Many children have relationships with animals through interactions with domesticated pets, service, or rescue animals or through activities such as fishing, hunting, farming, animal activism, and visits to zoos and aquariums. However, as noted in the case, *You Just Aren't Yourself These Days*, children, for the most part, have little firsthand experience with non-game animals such as reptiles and amphibians. Thus, it comes as no surprise when the fourth-grade students in this case are concerned when their classroom pets (a corn snake, box turtle, and tiger salamander) appear to be sick or less active with the approach of the fall season.

Some children have had opportunities to engage in animal activism around high-profile issues such as the decline of pollinators or the effects of plastics on marine animals. Yet other children, with little direct exposure to animals, may hold the belief that their breakfast comes from the grocery store and exotic animals only exist in zoos or on television. Still, other children may have a genuine fear of animals due to limited contact with them, fueled by images in cartoons or story-animal "villains" such as the Big Bad Wolf and similar caricatures. We witnessed this firsthand in a kindergarten classroom where children were learning about earthworms. As the kindergarten children enthusiastically observed and handled the earthworms, we cannot forget the stark fear on the face of one child who ran to the corner of the room crying and expressing genuine fear of this small animal. We are also reminded of Courtney Cazden's (2001) story of a young girl who wrote in her journal about butterflies for a classroom assignment. As she shared her story with the class, reading "I killed all the butterflies," she was met with the teacher's glare of disapproval. The teacher followed with a clear message that "butterflies are beautiful, and we should not kill them." She was unaware that

D. J. Tippins
Department of Mathematics and Science Education, Mary Frances Early College of Education, University of Georgia, Athens, GA, USA
e-mail: dtippins@uga.edu

the young child and her grandparents lived on a farm where they grew their own food—from their perspective, butterflies were pests that needed to be killed.

Increasingly, the line between humans and other animals is becoming blurred as wild animals move closer to cities, drawing attention to many thorny ethical issues. At the same time, the intellectual understanding of our relationship with animals continues to advance as we learn more about how animals such as birds and whales communicate and the important ecological roles they hold in ecosystems. Amidst this every changing present and future context, we emphasize the need for teachers to create spaces where children can have meaningful encounters that foster deep connections with other living things, along with an ethic of care for their wellbeing. As we see in the case of *And Then There Were None*, these encounters may involve opportunities to both celebrate and mourn as part of children's growing understanding of the web of life. Or, as in the case of *Baby Shark, Doo Doo Doo Doo Doo Doo*, the teacher's desire to move children toward an appreciation of biodiversity had unintended consequences when the students had unexpected reactions to shark specimens. This case highlights the complex ethical issues that surround the role of animals in science teaching and learning. Clearly, young children enter the classroom with a wide range of encounters and perspectives involving animals.

Children can explore the complexity of living organisms in spaces both within and outside the classroom. In the case, *Getting Down and Dirty*, a school pond is used to create an authentic learning environment for children's inquiries around the small animals found within it. Through this case, we come to recognize that children's personal and collective relationships with the pond require the teacher to design pluralistic ways for the students in her class to engage with the pond organisms. This case illustrates the importance of recognizing the need for children to have multiple forms of knowing and engaging with living things.

Across all of the cases in this part, there is an inherent assumption that, as teachers, we must take care to avoid using a "us versus them" perspective in communicating with children about their experiences with animals. We can avoid this dichotomous thinking by asking children to simultaneously focus on both similarities and differences between animals. We can also draw on children's literature, music, art, writing, drama, and much more to illustrate the mutual dependence of all living organisms. As children have direct, everyday interactions with animals, they have many opportunities to actively engage their imaginations, language, and creativity. The generation of stories that depict the fascinating life histories of animals is but one example of how children can learn about animals in ways which cut across traditional disciplinary boundaries. It is through numerous direct, multidisciplinary experiences with animals that we can ultimately hope to instill in children a commitment to the wellbeing of the animal life that is all around them.

Reference

Cazden, C.B. (2001). The language of teaching and learning. Heinemann.

Case: You're Just Not Yourself These Days

78

Nick Fuhrman

Abstract

Ms. Nicole Deal has been teaching fourth grade for almost three years and recently attended a teacher's conference on integrating experiential learning activities into the elementary classroom. Many of her students live in a rural area of Virginia and are often exposed to local wildlife like deer, turkeys, and squirrels but have limited knowledge about non-game animals like reptiles and amphibians. Nicole decides to acquire a permit to keep an Eastern box turtle, a corn snake, and a tiger salamander in separate aquariums in her classroom. She is so excited to engage her students with the animals, helping them to observe their behavior and even hold them while delivering presentations in class and to younger grades at her school. As autumn sets in, her students start to notice their classroom animals are not as active as they were earlier in the year, the students are deeply concerned for their well-being, and Nicole is unsure if this is normal behavior or if the animals are in need of medical attention. How should she handle this potential "teachable moment?"

It's only been a week since Sharon (the Eastern box turtle), Snowy (the corn snake), and Sanford (the tiger salamander) arrived in Nicole's classroom, but her students' excitement is still brimming. The fourth graders have been arriving in the classroom earlier than normal to check on the animals, and they are very attentive to each animals' behavior. The box turtle is most active in the mornings when the sun shines through the classroom window into the aquarium. She seems to enjoy the warmth and walks the most when she's warm. On the other hand, the corn snake and tiger salamander are both most active late in the day as the students are

N. Fuhrman (✉)
Department of Agricultural Leadership, Education and Communication, University of Georgia, Athens, GA, USA
e-mail: fuhrman@uga.edu

preparing to go home. Both move around their aquariums more then and the corn snake even pushes her nose against the lid on the top of her aquarium as it gets darker outside. Interestingly, on rainy days in particular, the box turtle and tiger salamander are most active in their aquariums.

The students have been watching the animals closely and have started an Animal Activity Record Book where they note the time of day when the animals are most active and even when they eat the most. Other classes at Fuhrman Elementary School have heard about the animals in Nicole's classroom, and her students want to share the animals with other classes through some "traveling nature shows." Nicole sees this as an ideal opportunity for her students to shine as science "experts" while drawing on communication and language arts skills as well. Initially, her students engage in mini-research projects to discover interesting facts about Eastern box turtles, corn snakes, and tiger salamanders. She then asks them to create a short presentation where they bring each animal out to other classrooms and share facts with younger students.

As her students engage in these wildlife presentations, Nicole starts noticing that her students are becoming less and less anxious about speaking in front of an audience. In fact, when they are holding the box turtle, corn snake, or tiger salamander, they seem more relaxed and conversational with their audience of younger students and even laugh as they handle the animals which often squirm and wiggle around in their hands. It was an "ex-shell-ent" learning experience for Nicole's student presenters and the younger audience of students!

The questions that the younger kids ask Nicole's students prompted them to think on their feet and relate the often jargon rich science content to an audience of younger learners. Jason, one of her students, explains, "when the young boy asked why the corn snake is called a corn snake, I knew the snake didn't eat corn, but then I realized that the corn snakes live around corn fields because of the mice they like to eat and that must be where their name comes from…it was a few steps in explaining that to the kid, but we both finally understood the reasoning behind their name."

As autumn starts to set in and the mornings get cooler, Nicole's students start noticing that Sharon, Snowy, and Sanford are not nearly as active as they were earlier in the year when the weather was warm outside. Even though their aquariums are temperature controlled with heat lamps and warming pads, the animals are not moving around as much and have even started eating less. The corn snake, Snowy, stops eating all together, and this makes Nicole's students very concerned. The box turtle, Sharon, starts digging deeper into the soil in her aquarium and moves very little. The students even start talking to the animals, saying, "you're just not yourself these days…why aren't you as active or eating as much?" The empathy they feel toward their scaly friends is touching to Nicole, and she so desperately wants to help her students understand why the animals are acting this way.

Crystal, one of Nicole's fourth graders, decides to do some research on why the animals might act more sluggish during the cooler autumn months. Nicole encourages this, and in fact, she offers an incentive to students who gather information as

Fig. 78.1 Children experiencing a presentation with a western hognose snake

part of their detective work. Students who engage in research and compile information are asked to develop a short PowerPoint presentation to share with their classmates. For doing that, Nicole awards extra credit. The small team of students who engage in the research present two different scenarios as reasons why their classroom animals might be acting sluggish, and the rest of the class is asked to be the jury for this "rep-trial" presentation in a court-case type of scenario.

Despite students' very informative presentations to peers, as Nicole listens to the jury conversations, she realizes that there are still some students who continue to be upset about the change in animal behaviors. She wonders what more she can do to alleviate her students' concerns (Fig. 78.1).

For Reflection and Discussion

1. Why would the box turtle and tiger salamander be most active on rainy days?
2. Why would the corn snake and tiger salamander be most active late in the day, as the sunshine starts to fade?
3. How would understanding the hibernation behaviors of reptiles and amphibians help Nicole's students solve the mystery of why the classroom animals move less and eat less as wintertime gets closer?
4. Why do reptiles and amphibians hibernate in the first place?
5. Why do you think as the students handle the animals and deliver presentations to younger children, they start to feel less anxious about public speaking?

Nick Fuhrman is the Josiah Meigs Distinguished Teaching Professor of Environmental Education in the Warnell School of Forestry and Natural Resources at the University of Georgia. Also known as "Ranger Nick," he hosts a monthly national television segment called "Ranger Nick" where he focuses on interesting agricultural and natural resources topics in an edutaining way. He regularly uses lives animals (such as snakes, turtles, salamanders, and birds) in his classes and his speaking engagements around the country.

Commentary: The Tremendous Benefits of Classroom Animals

Teresa Shume

Abstract

This is a commentary to the case narrative, *"You're Just Not Yourself These Days"* written by Nick Fuhrman.

Welcoming animals into your classroom can enhance the learning environment in ways that bring a wealth of benefits to students. As a former classroom teacher who now teaches science methods and supervises student teachers, I know firsthand the excitement, wonder, and joy that arise from relationships between children and animals in classroom settings. From butterflies and guppies to corn snakes and geckos, the possibilities are vast. This commentary will explore some of the benefits, challenges, and factors to consider when deciding how to approach adding scaley, furry, or feathered friends to your classroom community.

Classroom animals can offer powerful opportunities for experiential learning. Questions that arise authentically from direct experience can motivate students to seek answers to questions and formulate solutions to problems. Learning about science concepts such as brumation (hibernation of reptiles and amphibians) and body temperature regulation of ectotherms (so-called "cold blooded" animals) can occur as part of the planned science curriculum. Nicole's students, however, initiated their own learning on these topics when student-generated questions and "teachable moments" arose spontaneously through interactions with Sharon, Snowy, and Sanford.

In Nicole's classroom, the wonder and curiosity that children experienced, together with their empathy and concern, fueled discovery through research and collaboration. The students naturally engaged in a number of NGSS science

T. Shume (✉)
School of Education, North Dakota State University, Fargo, ND, USA
e-mail: teresa.shume@ndsu.edu

© The Author(s), under exclusive license to Springer Nature Switzerland AG 2023
S. Jeong et al. (eds.), *Navigating Elementary Science Teaching and Learning*,
Springer Texts in Education, https://doi.org/10.1007/978-3-031-33418-4_79

and engineering practices, such as asking questions, constructing explanations, and obtaining, evaluating, and communicating information. Moreover, students applied NGSS cross-cutting concepts when they noticed patterns in animal behaviors based on data in their Animal Activity Record Book and sought to explain cause-and-effect relationships. Similarly, Nicole's students will tap into additional components of the NGSS if they extend their investigation by interviewing experts over Zoom, inviting guest speakers to the classroom, or drawing on their own knowledge of local wildlife in the rural area where they live (e.g., chipmunks and ground squirrels hibernate). The presence of classroom animals can drive authentic inquiry learning over extended periods of time.

For this type of learning to be successful, it is necessary for the teacher to know when to intervene with a timely question or comment, and when to step back to allow the students to take the reins. This level of trust and patience can be challenging for new teachers, who frequently desire a solid sense of control in the classroom. Avoiding the temptation to guide students directly and immediately to the answer, however, can yield powerful learning outcomes. Students who are given voice and choice in their learning are more likely to develop into autonomous, self-confident, and life-long learners.

Beyond chances to develop scientific knowledge and skills, classroom animals offer important avenues to advance students' moral and ethical development. Teachable moments arising from experiences with classroom animals may relate not only to science content but also to questions and concerns about what constitutes an adequate living environment, sufficient care, or appropriate handling of animals. Though such value-laden conversations can seem daunting, they offer meaningful opportunities for students to think about moral considerations and embrace their ethical responsibilities toward animals.

In addition to moral development, classroom animals can contribute to social and emotional growth among elementary students. Caring for an animal and being responsible for its wellbeing can foster the development of empathy and compassion. For example, a third-grader I knew named Jake whose abruptness often caused conflicts with classmates showed remarkable gentleness toward Coconut, the classroom guinea pig. Emma, a girl who noticed this contrast, complimented Jake on his kindness toward Coconut and encouraged him to be kind to classmates as well. Interactions with the classroom animal provided opportunities for both Jake and Emma to practice compassion and kindness.

For many children, interacting with a classroom animal leads to bonding not only with the animal but with classmates as well. Bonds with Coconut were a shared commonality between Jake and his peers and allowed Jake's classmates to see another side of him. In Nicole's classroom, students are sure to remember their experiences with Sharon, Snowy, and Sanford for years to come. Sharing their knowledge of animals with other children would be memorable not only for Nicole's students but students in the other classrooms as well. Working together with classmates to feed animals and clean cages fosters cooperation and teamwork. A classroom animal can contribute significantly to a sense of kinship within a classroom community.

Additionally, interacting with animals is known to decrease anxiety for many children. Watching fish or insects in an aquarium provides some alone-time that can sooth a student who is agitated or upset. Holding an animal is a sensory experience that can calm a student who is encountering stress or the effects of childhood trauma. Interactions with classroom animals can provide valuable channels for learning to self-soothe and manage anxiety. Conversely, the death of a classroom animal can be a painful and even traumatic experience for children. Similarly, concerns for classroom animals can sometimes become a source of worry for some children, as described in the story of Nicole's classroom. Relationships with classroom animals can contribute to developing self-awareness and self-management of emotional regulation, key components of elementary students' emotional growth.

Though classroom animals bring a wide array of benefits for elementary learners, deciding to welcome an animal into your elementary classroom is a big decision. Many variables factor into determining whether and which animals to house in your classroom. Some questions to ponder when making an informed decision include the following:

- Am I prepared to provide for the animal's dietary, housing, social, and veterinary needs?
- Is my classroom safe and suitable for this animal in terms of noise and activity levels, and amount of attention/nurturing needed?
- What is the average lifespan of the animal (e.g., butterflies live for weeks or months, whereas iguanas can live for a decade or longer)?
- Are there concerns with allergies?
- Are there diseases that this animal could transfer to humans? (See https://www. cdc.gov/healthypets/forinformation.)
- What is my school and/or district policy regarding classroom animals?
- Are permits required? Are some species restricted in my area? What ordinances need to be followed?
- Who will take care of the animal outside of school hours and during breaks?
- How will the initial costs (e.g., cage or enclosure) and ongoing costs (e.g., food) be covered?
- Is the animal an invasive species (e.g., non-native butterflies) that must not be released?
- Will the animal be active during school hours, or is it most active at dusk or night?
- Do I want a classroom animal that children can handle (e.g., guinea pig) or one that is primarily watched (e.g., fish)?

Classroom animals can bring potent opportunities for experiential learning of science concepts and practices that resonate with the NGSS. Expanding your classroom community to include an animal can contribute to elementary children's moral, emotional, and social development. While adding an animal to your classroom involves effort and commitment, your students will experience tremendous learning benefits and gain many fond memories.

Teresa Shume is an associate professor of teacher education in the School of Education at North Dakota State University. She earned her Ph.D. in teaching and learning from the University of North Dakota and has worked as a classroom science teacher, college science instructor, science teacher educator, and science education consultant. Her teaching and research focus on preparation of science teachers committed to inclusiveness and making science accessible to all.

Case: Getting Down and Dirty

80

Sarah J. Carrier and M. Gail Jones

Abstract

This case focuses on introducing students to patterns and cycles in the natural world using authentic science practices and science process skills including observations of the microhabitat on their school grounds. Ms. Sandra Martin's 3rd grade students used dip nets to collect samples in their school's retention pond. When rainfall is abundant, the retention pond fills with water, and during dry times, the water recedes. While some students enthusiastically dug their hands into the pans of water as they tried to identify and classify the living and non-living organisms in the water, other students were reticent about getting dirty and were frightened of the living animals in the water. Ms. Martin's lesson goals were to engage her students' interest in life science concepts with a focus on: LS2.C Ecosystem Dynamics, Functioning, and Resilience, and LS4.D Biodiversity and Humans by providing students with first-hand observations of ecosystem dynamics as identified in the Next Generation Science Standards (NGSS, Lead States in Next generation science standards: for states, by states, The National Academies Press, 2013). But Sandra remained at a loss as to how she might help some of her students get over their fears of getting down and dirty with the Earth. Sandra decided that the best way to address her students' fears would be to provide them authentic outdoor experiences observing organisms in their natural habitats.

S. J. Carrier (✉)
Department of Teacher Education and Learning Sciences, North Carolina State University, Raleigh, NC, USA
e-mail: sjcarrie@ncsu.edu

M. G. Jones
Department of STEM Education, North Carolina State University, Raleigh, NC, USA
e-mail: mgjones3@ncsu.edu

Sandra's group of 3rd grade students gathered around the edge of the pond. They used long-handled dip nets to sample the pond water and shouted with excited calls when something live moved inside the net. Working in pairs, the students emptied their nets into shallow tubs of pond water so they could see what they had sampled. There were lots of "Ooohs" and "Aaahs" as dragonfly larvae and tadpoles fell from the nets into the tubs. Kiesha shouted, "it bit me!" as she picked dragonfly larvae out of the net. While Keisha just laughed and continued picking tadpoles and other materials from the net, other students within earshot stepped back from their nets nervously. As Sandra walked between groups huddled around nets and tubs, she noticed how some students exhibited high energy and enthusiasm while others attempted to physically distance themselves from the activity as their bodies and faces communicated their discomfort.

"Andrew, don't you want to match your group's organisms with our photos of macroinvertebrates?" Andrew shook his head and said, "it's just mud and dirt, and they all will either bite or sting me." "Yeah," said Kelly, "I don't want to get my dress dirty or get mud on my new shoes. It's yucky anyway." "And I don't like that smell," Jen interjected. Recognizing that the apprehensions expressed by Andrew, Kelly, and Jen could interfere with their opportunity to learn, Sandra considered ways to include Andrew and the other hesitant students in the learning process.

As a child, Ms. Martin had spent many hours of free time in nature, digging in dirt, and collecting rocks and insects. These "memorable life experiences" structured her appreciation of the natural world, providing a base of experiences that supported her desire to help students learn about the patterns and cycles in nature. Ms. Martin's schoolyard expedition goals were to provide her students with opportunities for authentic engagement in nature. She knew that students' learning is influenced by their emotions and social connections. Ms. Martin hoped to provide her students positive, authentic experiences working together as they explored concepts around ecosystems in the outdoors, rather than simply reading about them in a textbook. These activities would serve as a base for her students' observations of the biotic and abiotic parts of ecosystems. Rather than having her students identify and name each organism, Ms. Martin instead wanted her students to carefully observe and, through drawings and writing, describe the organisms and begin to organize them into classification categories. She also knew that by having students make their own collections and take inventory of their discoveries, their enthusiasm and interest to explore more invertebrates in their environment would be harnessed in productive ways.

Sandra's participation in professional development workshops on learning designs made her consider how to adapt instruction to meet the needs of her students. In her class, Sandra had several students with individual education plans (IEP), including students identified as advanced, students with behavioral challenges, and students with learning disabilities. She had learned that transitioning from a learning environment with the characteristics of a "traditional classroom" to the outdoor pond setting had potential to reach some of her non-traditional learners as well as students without documented learning challenges. Ms. Martin hoped that through these outdoor experiences, she could pique her students' interests and desire to actively participate in the learning process.

Sandra started the lesson on ecosystems and life cycles with a review of classification systems and asked her students to describe examples of animals in the categories of vertebrates and invertebrates. Her class had studied the characteristics of five classifications of vertebrates, so Sandra reviewed the needs of living things and how plants' and animals' needs are met in their environment. The school's retention pond was located by the path that students took each day to the school's cafeteria, and when Sandra explained that they were going to examine their school's retention pond, it was clear that few of her students had noticed this isolated ecosystem on their school grounds. Prior to going outside, Sandra reviewed her students' health records for allergies and parent permission slips. She reminded her students about the class expectations for the outdoor classroom, and then as she asked her students questions about the retention pond to identify their prior knowledge, she realized that her students had failed to notice the varying water levels and that they had never considered it as an ecosystem. Sandra asked the students to predict what kinds of living and non-living things they would find in the pond. Students each made their own lists, and she noticed a number of the students wrote only "fish." As they headed outside, the students took their holding tubs and dip nets and headed to the retention pond. As some students collected and described the samples, other students documented the living and non-living findings through written descriptions and drawings, but Ms. Martin still pondered how to include Andrew, Jen, and Kelly who stood in the back and clearly felt squeamish and put off by the mud and organisms.

Sandra remembered Andrew's interest in gaming. She then asked Andrew to return to the classroom and bring back four Chromebooks. When Andrew returned, she put Andrew and the other reticent students in charge of searching for macroinvertebrate photos, habitat descriptions, and life cycles. As one group sampled, another recorded, and a third developed a classification system. Soon, Andrew's group's new role as information providers drew student teams to them to ask for more information.

In addition, as the pond activity continued, Ms. Martin noticed how Andrew, Jen, and Kelly demonstrated a sense of engagement and pride as information sources. She also noticed that Andrew moved closer and closer to the pans of water samples. When she saw him ask another student, Jennie, to hold his Chromebook so he could use a magnifying glass to observe a dragonfly nymph he was holding in his hand, she encouraged the entire class to listen to Andrew's careful description of the segmented abdomen. "Andrew, do you think this animal has a backbone?" All the students chorused "no, it's an invertebrate!" As the sampling activity progressed, students developed T-charts to document "biotic" for all the living things in their sample water and "abiotic" for all the non-living things.

When they returned to the classroom, the class used a cart to transport their samples with them. Ms. Martin poured each sample into a large classroom aquarium so students' observations and investigations would extend beyond the outdoor activity. Subsequent classroom investigations were guided by focus questions asking about the interconnections of biotic and abiotic parts of ecosystems, life cycles, and patterns in nature that were situated in the context of the retention pond. One of Ms. Martin's goals was to facilitate her students working together as they engaged in science practices such as making observations (including using magnifying glasses to extend their sense of sight), communicating (drawing, words, or audio recordings), classifying (using their self-selected groups), and measuring (i.e., length, volume, and time). In addition, students' authentic investigations connected science content with literacy practices. Shelly asked, "Ms. Martin, where is the book about algae?" Shawn wanted to read more about dragon flies and damsel flies, and Ellen used the classroom books about geology to write about how sedimentary rocks are formed.

As the year went by, the students' interest and enthusiasm about all the hidden life in the retention pond persisted, particularly in the early spring when the frogs could be heard calling. But when the seasonal rains waned and the pond's water level dropped, Andrew expressed a sense of worry about the water creatures and their survival. Sandra was able to channel Andrew's concerns and share them with other students as the class continued the documentation of patterns and life cycles. While her students' comfort levels with sampling and touching the organisms varied, this set of activities provided students with shared experiences and a base for further learning.

For Reflection and Discussion

1. What are some other strategies Ms. Martin could use to include all the students in this learning process?
2. How can Ms. Martin address the loss of habitat when the retention pond dries up?
3. How can Ms. Martin further support connections of science and other subjects: literacy, social studies, and mathematics?
4. What are some ways to build the reticent students' identities as science learners while also respecting their fears about the aquatic environment?
5. How can Ms. Martin reduce the fear that some of the students felt while sampling the pond water while also building their confidence in investigating science outdoors?
6. How can you re-imagine Ms. Martin's lesson to include a student who might be in a wheelchair in the investigations? A student with visual impairment?

Reference

NGSS Lead States. (2013). *Next generation science standards: For states, by states*. The National Academies Press.

Sarah J. Carrier is a former elementary school teacher who has a Ph.D. in science education from the University of Florida. She is a professor of elementary science education at North Carolina State University and teaches preservice and in-service teachers. Dr. Carrier's research focuses on environmental and outdoor science instruction, including supporting cognitive and affective connections building on awe in the natural world.

M. Gail Jones has a Ph.D. in Science Education from North Carolina State University. Dr. Jones currently serves as professor of science education at NC State University teaching preservice and in-service teachers and conducting research on virtual reality, nanotechnology, and teacher education. Dr. Jones' research is currently investigating the role of awe as an epistemic emotion to promote cognitive growth in science.

Commentary: Diversity in the Pond and Instruction

Michael Svec

Abstract

This is a commentary to the case narrative, *"Getting Down and Dirty"* written by Sarah J. Carrier and M. Gail Jones.

In the book *Last Child in the Woods: Saving our Children from Nature-Deficit Disorder* (2005), Richard Louv made the argument that children are experiencing an increasing disconnect from the natural world. The causes are many, and some of the consequences he identified include exaggerated fears of nature as well as an increase in the frequency of emotional and cognitive ailments. He then issued a call for more thoughtful exposure of children to nature as well as increased opportunity for unstructured outdoor play.

In this case, Sandra is serving her students well by providing an authentic outdoor experience at the school's retention pond. The cited standards are focused on ecosystems and biodiversity, and the associated science practices include analyzing and interpreting data as well as using evidence to construct explanations. Sandra engaged the students by providing a rich experience creating a new opportunity that went beyond the curriculum. It was a thoughtful exposure of the natural world surrounding the students' school. However, the new experience of engaging with mud and critters seems to have intimidated some students, a reasonable response to a new experience. Fortunately, Sandra sought instructional ways to ease these students' fear and engage them with the content and their peers.

Sandra's instincts and learning through professional development are serving her and the students well. Her desire to engage all of the students and adapting instruction to their needs is harmonious and consistent with *Universal Designs for*

M. Svec (✉)
Department of Education, Furman University, Greenville, SC, USA
e-mail: michael.svec@furman.edu

Learning (UDL). UDL is a framework for improving learning for all students that helps customize instruction, assessment, and goals to meet individual student needs (CAST, 2021). UDL seeks to minimized barriers for student learning. For example, a universally designed video will include closed captioning. Closed caption is an appropriate accommodation for students with hearing impairments, but it also helps language learners, helps with recognition of new or complicated vocabulary, and benefits those who want to watch a video without disturbing others near them. What was originally an accommodation turns out to have many benefits and is just good design. Just because we have a set of standards to address does not mean we have to have one "standard way" to achieve that science standard.

The UDL guidelines provide concrete methods and strategies to ensure all students can access and participate in science (Waitoller & King Thorius, 2016). Within the UDL guidelines are specific strategies for engaging learners. Among the engagement strategies, Sandra is demonstrating minimizing threats and distractions as a means to recruit student interest. Her willingness to use technology as a scaffold aligns with varying the sources of information so that the resource, in this case information accessed on Chromebooks, is more authentic to the children, providing the hesitant learners with a new pathway to engagement.

Sandra demonstrated awareness of the students' interest and found a meaningful way to use those interests. By varying the demands of the lesson and providing a new method for the students to contribute, she turned their hesitancy into a positive contribution that benefited all the learners. As information providers, the students started from a strength, the technology resource they were comfortable with, and slowly move toward experience they were uncomfortable with, interacting with the mud and animals. The simple act of allowing the students to use the Chromebooks changed the task so that all the learners could actively participate, explore, and experiment consistent with the goals of the lesson.

Ms. Sandra Martin could build on this success in future lessons. She could supplement the observations with the addition of a digital microscope or digital camera. Students could then use the digital microscope to capture images of the critters to further supplement their drawings and observations. These additional observations could be done in the field or back inside with the samples in the classroom aquarium. Digital images of the ecosystem animals could also be gathered and analyzed with the students identifying biotic and abiotic parts of the ecosystem. The digital images could provide another way to facilitate the goals of helping students make observations and communicate results through another medium.

Louv favors engaging children and adults in immersive experiences with nature in which the children's senses become fully engaged, much like Ms. Martin's childhood experiences. Although the use of technology can work against this immersion, the approach of using digital microscopes and cameras extends the senses and provides a new avenue to engage the students who would normally avoid nature. Louv acknowledges that reading is another means to revive a sense of natural wonder. Sandra's efforts to build on students' comfort with technology is, like reading, another means to engage the learners and open a door to an appreciation of the natural world that surrounds the children.

References

CAST. (2021). About Universal Design for Learning. https://www.cast.org/impact/universal-design-for-learning-udl
Louv, R. (2005). *Last child in the woods: Saving our children from Nature-Deficit Disorder.* Algonquin Books.
Waitoller, F., & King Thorius, K. (2016). Cross-Pollinating Culturally Sustaining Pedagogy and UDL. *Harvard Educational Review, 86*(3), 366–390.

Michael Svec is a professor of Science Education at Furman University with over two decades of experience preparing elementary and secondary science teachers. His scholarship and teaching include history of astronomy, science fiction, and international comparisons.

Case: And Then There Were None...

Chelsea M. Sexton

Abstract

Chelsea is a fifth-year teacher in a second-grade classroom in the suburbs of Atlanta. About half of her students have been identified as gifted and talented by the state. She planned an interactive unit on animal life cycles which included opportunities for her students to be able to watch frogs complete their metamorphosis from eggs, to tadpoles, to froglets, and finally into adult frogs. It was the circle of life. The stability and change of plants and animals were a big idea for second-grade science, and her unit incorporated 3D practices she had seen presented during preplanning development. This open case explores a situation where Chelsea had to confront a life cycle of a whole different kind.

Three minutes after the kids energetically rushed through the door after a long weekend, I heard it.

"Ms. Sexton, why are they all gone?"

No. They couldn't all be gone. I racked my brain. Their numbers had been steadily decreasing since the gelatinous eggs had hatched into tiny tadpoles last month. We started with twenty little swimmers, but how could they all be gone now?

Over winter break, I was daydreaming about my classroom, instead of actually taking a vacation from work. I wanted to plan a different anchoring phenomenon in addition to the state-suggested "animal bodies collect and transfer pollen from one flower to another." During preplanning, I attended the elementary science sessions, and the whole focus was on 3D learning. Along with the content standards, we were supposed to include cross-cutting concepts and science practices, the other

C. M. Sexton (✉)
Department of Mathematics, Science, and Social Studies Education, Mary Frances Early College of Education, University of Georgia, Athens, GA, USA
e-mail: cmsexton@uga.edu

© The Author(s), under exclusive license to Springer Nature Switzerland AG 2023   419
S. Jeong et al. (eds.), *Navigating Elementary Science Teaching and Learning*,
Springer Texts in Education, https://doi.org/10.1007/978-3-031-33418-4_82

dimensions of science instruction. We also were "encouraged" by our principal to teach life skills to our students along with the standard content.

Since our last big idea for the year was stability and change in plants and animals, I wanted the kids to be able to observe frogs through their life changes. While we observed the cross-cutting concepts of stability and change and patterns, I thought that we could also focus on asking questions and obtaining, evaluating, and communicating information as practices. I also thought it might be a good way to incorporate life skills about responsibility by taking care of a classroom pet. I saw that I could order live frog eggs from our teacher catalog, so I ordered a set along with a 10-gallon aquarium tank for the classroom. I set the delivery for February, so my students and I could have time to plan for our new classmates as a group.

During our first week back to school in January, I pitched the idea. My room was BUZZING with excitement. Students began volunteering their past experiences. Jimmy exclaimed, "I have two turtles at home that I take care of!", and Kaitlyn, Caitlin, Sahir, and Susan mentioned that they all had fish. About half the class also had a cat or dog at home and shared that they helped with their care and daily feedings. I explained that it would be a group effort, and these would be OUR tadpoles, not just Ms. Sexton's. They quickly agreed, and we diverted the day's science lesson from the force of kicking a soccer ball to brainstorming for the new habitat with a last-minute media center reservation to conduct online research.

Later that day in the media center, one group looked up the types of food tadpoles would need and what a feeding schedule should look like. Another searched for the kind of decorations and habitat items we might want to add to the aquarium. A third group worked with the librarian to find books about frogs that they could share with the class during our unit. Finally, the fourth group took on the task of determining what was important to know about the water. Susan and Caitlin, in this fourth group, called me over to their computer pod because they weren't sure what they were reading. "What are nitrates and ammonia?" Susan timidly queried.

Wow. I hadn't thought about those words since high school chemistry and college biology, and I was expecting 7-year-olds to understand water quality and chemistry? Quickly scanning the computer page they were on, I recovered from my initial shock and replied, "Well, those are just some of the things that get left in the water after the tadpoles poop. Everybody poops!"

"Ewwwww," grimaced Caitlyn and Marcus, while Susan laughed a giggly, "That's cool."

"But don't worry," I continued. "We can get a water test kit with strips of paper. All we have to do is dunk the paper in the water and check to make sure it turns the right color. If it doesn't look right, we might have to try something else to fix the water."

"Ohh. Ok." Marcus sighed with relief.

When February came, the students were ready. The habitat was set up and, to my relief, the water test strip showed all the right colors. The students did not have much to draw in their journal for the first week—really they were just drawing circles representing the clusters of eggs. But when the eggs hatched, the real fun began. For many students, counting the tadpoles was a challenge at first because they were small and moved like lightning wriggling through the water. Nevertheless, the kids had so much fun scooping them into small dishes and looking at them through magnifying glasses.

One of our class jobs became feeding the tadpoles each day. It was a coveted task and engaged the whole classroom. The pollywogs would swarm the anchovy and soybean food pellets that the food group chose in January, and the pellets would spin as the tadpoles thrashed their tails back and forth, eager for a bite. Apparently, a diet with fish meat and vegetables was exactly what the tadpoles wanted.

And the tiny pollywogs started to grow! They seemed to enjoy exploring the habitat that the students picked out. There would always be two or three hiding under the half log at the bottom, and another few would scurry between the leaves of the fake aquatic plants. None of the tadpoles seemed to enjoy the dragon figurine placed near the water filter, but he was added because it is our school mascot. On Tuesdays and Fridays, Caitlin, Marcus, or Susan would dip a water test strip to make sure the colors matched the right numbers on the box.

One Wednesday morning, Sammy tugged on my sleeve.

"Ms. Sexton, two of our baby frogs are missing."

"Are you sure you counted right, Sammy? And did you check if anyone was drawing in their journal right now? They may have scooped them out to check out their legs."

"Yes, Ms. Sexton. I double checked both."

"Humm, that is strange. I will look into it while you all are in music class today."

I forgot to investigate the missing class pets. But we had already moved on from the conversation about the absent amphibians. After all, we had reading and math and social studies to finish today too.

Late the next week, Sammy and Marcus called me over to our frog habitat.

"Ms. Sexton, Ms. Sexton! We KNOW there are missing frogs. We only count eleven now!"

Remembering with a jolt that I was supposed to have looked into the reasoning for our missing friends, I joined Sammy and Marcus at the tank. We were certainly missing half of our tadpoles. The ones we had looked heathy and were still going after the food with vigor. But I could not find any dead bodies of the missing nine tadpoles.

"They probably just got caught in the filter. It's ok. We still have plenty of frogs for our journals," I reassured them, though I was not confident in my explanation.

While the students were journaling on Friday, we discovered individual toes starting to appear on our pollywog pets' legs, but by now, there were only six left, limiting the number of students who could observe at the same time. The students observed that some of the tadpoles were looking healthier and bigger than others, so we brainstormed how to help out the smaller ones. Maria shouted, "We have to feed our dogs in different rooms because Shadow will try to eat Missy's food. Maybe we can feed them separately?".

"That's a great idea, Maria," I responded calmly, "but next time, let's remember to raise our hand and not shout with our outdoor voice."

Caitlin also volunteered to check the water chemistry and found that the ammonia was a little bit high. "Ms. Sexton, do you think you could do a water change after school?" I agreed and felt secure in the new procedure for feeding that the students had agreed upon.

The next week, we were down to three large tadpoles and at a loss as to why they were all disappearing. I was ready to ask my grade-level team if someone was playing a prank on me. My teacher peers seemed as confused as I was. Deborah, who shared an extra classroom door with me, emphatically assured me, "No, Chelsea, I have not done anything with your students' aquarium. I like pets with fur more than the slippery, slimy ones."

Tuesday morning, after a long weekend, I opened my door and encouraged the students to begin morning work. That was when I heard Kaitlyn's voice from across the room.

"Ms. Sexton, why are they all gone?"

Oh no. I could not, for the life of me, see a single pollywog. There were two left yesterday during our teacher workday, and they seemed strong as if they were growing well. They had fully formed legs by this point. I continued investigating the habitat and checked, yet again, in the filter system. This time I found one dead pollywog.

This large tadpole was caught in the bottom of the filter. I could barely see the silhouette of the leg and tail of another pollywog hanging out of its mouth.

My mouth dropped open. I was prepared to talk about the transformation of frogs through their life cycle. I wanted them to ask questions and communicate what they found. Since the tadpoles started disappearing, I could even discuss the idea of an animal coming to the end of its life cycle and what that means. But this, no. I was not prepared to have a conversation about tadpole siblings eating each other.

About cannibalism.

For Reflection and Discussion

1. How could Chelsea explain all of the tadpoles disappearing to her students?
2. How can Chelsea help her second-grade students appreciate birth and death as a natural part of life cycles?
3. What other experiences might Chelsea design to help extend her students' understanding of animal life cycles?
4. What ethical dilemmas may be embedded in this situation that Chelsea did not notice?

Chelsea M. Sexton is a former high school environmental science and research methods teacher with a background in marine ecology. She currently teaches elementary preservice teachers while working on her doctoral degree in science education at the University of Georgia. Her research interests center on education for sustainability, justice-centered and case-based pedagogies, and preservice science teacher preparation.

Commentary: I Should Have Stuck to a Video

Michael J. Reiss

Abstract

This is a commentary to the case narrative, '*And Then There Were None...*' written by Chelsea M. Sexton.

When I was a schoolteacher, I took pride in (almost) never arriving late for my lessons. On one of the handful of occasions that I failed to meet my own standards, I entered the science classroom to see a number of visibly upset students. The classroom had a pair of gerbils that lived in a large cage and were favorites with some of the students. Within seconds, it was clear that one of the two gerbils had died and, far worse, the other one had started to eat his (or her) dead companion—something I now know, but didn't then, is not uncommon among gerbils.

I remember shepherding the students away from the cage, 'phoning our wonderful lab technician,' and dealing with the immediate situation (upset students) as best as I could. But what I can't remember—I think I was as shocked as most of the students were—is whether I tried to extract any lessons for science or life in general from the situation. I rather think that all I did was apologize profusely.

When, years earlier, I was a pupil in elementary school, I remember a scorching hot day when our teacher decided to abandon the stiflingly hot classroom and take us out into the modest school grounds, where we sat on a small patch of grass under a tree and continued with the lesson. Without wanting to overstate the significance of her actions, I think the experience was something of an epiphany for me. Previously, I had simply assumed that learning took place indoors and that the outdoors was for other things. Now, the boundary had been transgressed, rather

M. J. Reiss (✉)
Institute of Education, University College London, London, UK
e-mail: m.reiss@ucl.ac.uk

as Chelsea's tadpoles and my gerbil transgressed the rules of what 'nice' animals should do.

Biology is like that. We humans anthropomorphize—we almost automatically assume that animals, deep down, are pretty much like simplified versions of ourselves, whereas it may be that many of them are simply very distinct from us and have very different ways of being that we may find difficult to imagine (Nagel, 1976). This can be unsettling for us. We don't like to think of rabbits routinely eating their own soft feces (coprophagy) or tadpoles or gerbils eating one another. We prefer a Disneyfied version of reality in which animal societies mostly resemble good human societies.

Media such as films and videos have a considerable effect on how we see animals. Think of how these well-known films (not just ones produced by Disney) portray animals from *Bambi* (released in 1942) onwards: *101 Dalmatians* (1961), *Charlotte's Web* (1973), *The Lion King* (1994), *A Bug's Life* (1998), *Ice Age* (2002), *Finding Nemo* (2003), *Happy Feet* (2006), *Kung Fu Panda* (2008), *Penguins of Madagascar* (2014), and *Zootopia* (2016). It might be supposed that viewers of all ages would be able to draw a line between the portrayal of animals in such films and their behavior in 'reality' but this is not entirely the case. In one study, most of what children aged 4, 8, 11, or 14 'knew' about animals came from what they had learned out of school, and from a wide variety of sources, fictional as well as factual (Tunnicliffe & Reiss, 1999a).

My point is not to complain about the portrayal of animals in animated films—the genre is entertainment not documentary—but rather to note that children will (do) learn from anywhere. Now, I don't particularly want elementary children to have to learn about 'Nature, red in tooth and claw' through their pet tadpoles being eaten alive; there are less traumatic ways of learning about the natural world. In another study I undertook with Sue Dale Tunnicliffe, we analyzed the conversations that year 2 and year 5 pupils had while observing meal worms (*Tenebrio molitar*) or brine shrimps (*Artemia salina*) during science activities in an elementary school (Tunnicliffe & Reiss, 1999b). Both these species have quite short life cycles (and are a lot easier to keep in a classroom than are frogs or gerbils). We found that of the 583 comments that the children made, 18% were to do with sex or reproduction, 27% with life history, 19% with emotions, and 15% with death or violence.

So, what do I conclude from Chelsea's tadpoles, my gerbils, and all this? For a start, teachers sometimes get it wrong. I like the way Chelsea admitted that she had forgotten to investigate the missing class pets: 'After all, we had reading and math and social studies to finish today too.' Children need to know that however good and kind their teachers are, their teachers are human, and that part of being human is to get things wrong and sometimes let others down. A Kleinian perspective on early human development has the new-born dividing the mother's breast into the 'good breast'—warm, milk-producing and comforting—and the 'bad breast'—which fails to produce milk and is taken away or absent. One of the great achievements (for most of us) as we grow up is to go through what is called the 'depressive position' and realize that our parents combine goodness and badness (Hinshelwood, 1991).

Then there is the fact that tempted as we might be to put on a video rather than spending hours tending frogs, gerbils or any other living organisms in science classrooms, there is much of value, both for science education (growth, reproduction, life cycles, etc.) and for education about life more generally (including bereavement education), that can be got from children having first-hand experience of animals, both in the classroom and beyond.

A final point is about what teachers might do when children are upset, angry, or hurt. In the context of teaching evolution to children who come from creationist or similar backgrounds, I have suggested that a useful rule of thumb is for teachers to teach with sensitivity (Reiss, 2019). Teachers, particularly (I suspect) elementary teachers, are used to teaching sensitively. Good teachers react with empathy and care when they see others, particularly their pupils, who are upset. Just as the large majority of us eventually come to accept that our parents are not entirely at our beck and call, so we want pupils who come to learn that the natural world is a source (to us) of both joy and pain but that part of being human is to come to understand this and to value the natural world for its own sake.

References

Hinshelwood, R. D. (1991). *A dictionary of Kleinian thought* (2nd ed.). Free Association Books.

Nagel, T. (1976). What is it like to be a bat? *The Philosophical Review, 83*(4), 435–450.

Reiss, M. J. (2019). Evolution education: Treating evolution as a sensitive rather than a controversial issue. *Ethics and Education, 14*(3), 351–366.

Tunnicliffe, S. D., & Reiss, M. J. (1999a). Building a model of the environment: How do children see animals? *Journal of Biological Education, 33*, 142–148.

Tunnicliffe, S. D., & Reiss, M. J. (1999b). Opportunities for sex education and personal and social education (PSE) through science lessons: The comments of primary pupils when observing meal worms and brine shrimps. *International Journal of Science Education, 21*, 1007–1020.

Michael J. Reiss is a professor of science education at University College London, a fellow of the Academy of Social Sciences, and visiting professor at the Royal Veterinary College. The former director of Education at the Royal Society, he is president of the Association for Science Education, president of the International Society for Science and Religion, and a member of the Nuffield Council on Bioethics. He has written extensively about curricula, pedagogy, and assessment in science education and has directed a very large number of research, evaluation, and consultancy projects over the past twenty-five years funded by UK research councils, government departments, charities, and international agencies.

Case: "Baby Shark, Doo Doo Doo Doo Doo Doo"

84

Austin David Heil and Aarum Youn-Heil

Abstract

Austin is an informal science educator working at a summer camp at a local elementary school inland from the "Forgotten Coast" of Florida. This summer camp was centered around marine science and was open to all ages of elementary school students. Austin surprised the campers at the end of the week with a "Sharks at Dark!" day with ten ethanol-preserved deep-sea sharks laid out on tables. This lesson was designed to show the unique diversity of deep-sea sharks that researchers study. This open case explores a situation where Austin had to confront a group of third graders' concerns about the ethics of sacrificing animals for the purpose of science.

"Ewwww!" The group of third-grade summer campers walked in smelling the sharks before they could see them. I observed some of the campers pulling their t-shirt collars over their noses trying to mask what the kids called, "the stink of death." With the lights turned off, I heard the kids pretending to scare one another with ghost noises and camp counselors shushing the students while attempting to keep each group of ten campers herded together.

As students' eyes adjusted to the darkness, one camper, Will, waved over to me and yelled, "Austin! Austin!" His group all turned their heads toward me as I waved back and gestured excitedly for them to come to our table. Evan began to notice our deep-sea sharks and grabbed Will's shirt to point at our table. "Look! Look! Shark Week!! RAWR!!" Evan yelled to Will while mimicking a

A. D. Heil (✉)
Department of Mathematics and Science Education, University of Georgia, Athens, GA, USA
e-mail: Austin.Heil@uga.edu

A. Youn-Heil
Grady School of Mass Communication and Journalism, University of Georgia, Athens, GA, USA
e-mail: younf@uga.edu

shark mouth. Another camper, Ava, shook her head at Evan and Will and said in disagreement, "No, no! He's cute!! Like a little baby!" Ava then turned to me and pretended to cradle a shark in her arms.

It was day five of a weeklong summer camp at our local elementary school. Our summer camp was built around marine science. We had teamed up with the local college to provide opportunities for students to explore different aspects of the ocean. Over the course of the week, students explored marine habitats at the coast, touched common shore animals, and met with real marine scientists. Every day, we would hear from the campers, "I wish we got to see a shark!" "Will we get to see sharks today?" "I want to see a shark!" The campers had no idea what was in store for them on their last day. For their last day, we planned a "Sharks at Dark!" event to bring preserved deep-sea sharks to the elementary school.

I was excited to see this group of third graders interested in learning more about deep-sea sharks. After the campers met some marine scientists, the opportunity to see and touch sharks was uniquely afforded to them when the scientists offered to provide animals from their research labs. I knew I wanted them to learn about the diversity of these sharks. Thus, my goal was for campers to think about adaptations for deep-sea sharks compared to other organisms like starfish, crabs, and sea urchins that they had observed earlier in the week. When I heard how much the campers wanted to see a real shark, I thought this was my chance as an educator to bring some deep-sea life to them! On such short notice, we were lucky to get ten adult deep-sea sharks, albeit small in size. The sharks ranged from the Cuban dogfish (< 2 ft.) to the smallest known shark species, the dwarf lanternshark (about ½ ft.). My goal as an educator was to shake up the campers' expectations of what they previously thought about sharks. What were they hoping to learn from seeing and touching a shark? Where did their ideas about sharks come from? What made deep-sea sharks so different from the animals they interacted with from the shallow waters? Despite all of these questions and expectations I had about this event, I could not have planned for what happened next.

I am a marine science educator who floats around the local elementary schools and teaches students about marine life not too far from our inland city. Every summer, I get the chance to help out with Wells Elementary's marine science camp. I love engaging with elementary schoolers about marine science because that is the very time in my life when I started to dream about becoming a marine scientist. I grew up in this same inland city wanting to learn more about the coast that was not too far away, which was one of the main reasons why I became a marine science educator. I want students to reflect on how much the ocean and marine life are part of their lives even though they do not see it every day.

On this final day of camp, I started my lesson with a simple question just to gauge what students might already know about sharks. I pointed at the Cuban dogfish, a deep-sea shark with bulging eyes and a dorsal spine next to its fin, and asked the group of campers, "What is special about this shark that is different from what you have seen on TV and movies?" All the campers were hesitant to answer as they stared at the different deep-sea sharks on the table in front of them. After a moment, Max's eyes moved away from the shark and he said, "Like Nemo?" In an upbeat voice, I exclaimed, "Yeah! Kind of like the shark in the Nemo movie, yeah!" I continued to nod and looked around at the campers who whispered back and forth to one another. Willow giggled and said softly, "But that shark talked a lot!" Max nudged and elbowed Willow in agreement, "Yeah! He moved a lot too!".

I was waiting to allow the group to look at the sharks for a couple more seconds before I asked them another question when Willow pointed to one of the sharks, the dwarf lanternshark. "Why doesn't he move? Is he... alive?" she said curiously. Taken aback, I paused to form my words but fumbled in the process. "Weh-weh well, this is our only way to know things and learn things about them! When we catch them from the deep sea, we can't just throw them back into the water because they're rare," I said matter of factly.

Jerking his head up, Evan chimed in, "Why not? My daddy throws fish back from the boat *alllll the time!*" Will and Evan made eye contact, and Will looked up with his arms crossed and said, "Yeah! That's not fair."

Frustrated and dejected, Ava lifted her head up. "Bu-but-but... why do you... why do you have to make them die? Austin, why do you have to kill the little sharks?" Ava asked as she brought her arms close again to her body, pretending to again cradle the shark in her arms.

I opened my mouth to respond but found myself thankful for the lights off because I was left speechless. I had never questioned the mortality of deep-sea sharks or any marine life for the purpose of research, but here stood an 8-year-old and her nine friends confronting me about the ethics of science. I thought they would ask me why the shark's bone was sticking out or why one had more fins than the other but not this. I took a deep breath and thought to myself, Austin, what did you just do?

For Reflection and Discussion

1. What are your thoughts about using preserved marine specimens to help young children learn about ocean biodiversity?
2. How could Austin explain to the campers the moral issues behind sacrificing animals for scientific purposes?
3. What could be Austin's role as a summer camp teacher in helping students learn about death and scientific sampling practices?
4. How could Austin respond to Ava and the other campers in this situation?

Austin David Heil is teaching faculty in the Department of Marine Science at the University of Georgia. His research explores instructional decisions in introductory biology courses and the teaching and learning of academic science writing. Austin received his PhD in Science Education and M.S in Biological Oceanography. He has worked as a marine scientist and marine science educator.

Aarum Youn-Heil is a doctoral student in the Grady College of Journalism and Mass Communication at the University of Georgia. Her research interest is in interracial communication and how it relates to entertainment media as a form of critical pedagogy. Specifically, she focuses on interracial communication *in* and *about* media. Through a critical lens, she investigates how mediated racial representations impact racial identities and interracial communication.

Commentary: Is the Stink of Death Worth It?

85

Donna GovernorGovernor

Abstract

This is a commentary to the case narrative, *"Baby Shark, Doo Doo Doo Doo Doo Doo"* written by Austin David Heil and Aarum Youn-Heil.

When my daughter was four, we took her to the beach where we would take daily walks up and down the coastline. As with any trip to the beach, waves would occasionally wash up and deposit a lifeless fish or other sea creature on the shore. I remember one day while walking with her we found a dead crab and she was fascinated! "Why did it die?" she asked. Then she began to poke at it to see what was "inside." She was hooked. When she enrolled in college some 14 years later, she became a biology major.

Opportunities to observe animals, living, dead, or virtually, have the potential to engage and excite students in unparalleled ways. Curious students, like my daughter, can explore the structure and function of organisms and discover an appreciation for the complexity of life. While the National Science Teaching Association's Position Statement on the Responsible Use of Live Animals and Dissection in the Classroom (NSTA, 2008) doesn't specifically address the use of preserved specimens for observation, it does recognize that providing students the opportunity to interact with organisms gives them the opportunity to build observation skills, appreciate the complexity of life, and explore the structures and processes of various organisms. A systems approach to how the body works to carry out life's functions can be explored as students observe the arrangement and organization of an organism's external anatomy. In the case presented here,

D. Governor (✉)
Department of Middle, Secondary, and Science Education, University of North Georgia, Dahlonega, GA, USA
e-mail: donna.governor@ung.edu

comparative anatomy can inspire curiosity and engage students in the practice of asking questions based on scientific observations.

Both NSTA (2008) and the National Association of Biology Teachers (NABT, 2019) have position statements that address live animals in the classroom and dissection but neither makes recommendations for the use of preserved specimens for other activities, such as the one highlighted in this case. However, both do offer some guidelines that apply to this case. Among these guidelines are making sure that activities are well prepared and appropriate for the age and maturity level of students. Additionally, it is recommended that preserved specimens should ALWAYS be obtained from reputable resources; either reliable scientific supply companies or FDA-inspected facilities. Finally, appropriate safety guidelines must be in place such as goggles, gloves, adequate ventilation, and proper supervision. Specific learning outcomes should be clear, and there should be an emphasis on fostering respect for living organisms and all we can learn from them, whether in their natural habitat or in the lab.

Because Austin was an informal science educator working at a research facility, the sharks would have been collected using processes approved by the institution's Animal Care and Use Committee for scientific research. Therefore, Austin has taken care to use appropriately obtained specimens. Further reading of the case shows that he has specific learning goals related to comparative anatomy, with an instructional focus on the crosscutting concept of structure and function guiding the lesson. Austin followed all safety guidelines for his students and returned the sharks to the researchers after the activity.

However, in activities involving the use of live animals or preserved specimens with young children, there are some additional considerations. Austin was caught off-guard by the reactions of some students when they were presented with the opportunity to observe preserved sharks. The age and maturity of the students may have accounted for their response, catching Austin by surprise. In this regard, it is important to prepare children for experiences with living or dead organisms in advance. Finally, as is recommended for dissection activities, it is always a good idea to offer alternatives for those students who object to or are not mature enough for studying preserved animals.

Were the third graders in this case mature enough to handle preserved speci-mens and issues related to life and death? While students' maturity level can be difficult to judge, it is important for teachers to tailor activities to the maturity level of the students when planning instructional activities involving preserved animals. The reaction from Willow and Max indicates that they associate sharks with a lovable character from one of their favorite movies, and they appear to be upset by seeing the dead sharks lined up for investigation on the lab table. Evan's response, equating the lifeless sharks to the fish his father throws back into the water, indicates he understands that living things die, but his comments may be counterproductive to fostering the respect for living organisms necessary to the activity. This wide range of student responses can result in confusion for students at this age or work against the learning goals of the activity. As a summer camp educator, Austin probably has experience with students of all ages and likely understands child development; however, he may not have enough history with this group of students to assume that they will all engage equally in the experience or have the maturity to be appropriately respectful throughout the activity.

If teachers determine that their students have the necessary maturity for a lesson involving preserved organisms, they should prepare them for "the stink of death" before the activity. Preserved specimens may literally stink from formalin, and students of any age may have an aversion to the chemical smell. Perhaps this could have been an outside activity. In this case, Austin had intentions of excitement with the element of surprise, but that may have caught some students off-guard and unprepared. Considering the young age of children in this case, it is especially important to provide them with an orientation to the activity prior to being shown the shark specimens. A discussion with students could center around the purpose of the activity and the real-world applications for what scientists can learn from observing structural anatomy. Students need opportunities to express their concerns or work through potential repulsion in advance. Since this is likely the first experience these children have with preserved specimens, Austin may want to explain the purpose of the safety precautions he has put in place. If time permits, children should have opportunities to ask questions and discuss any concerns in advance. More than likely, the third graders in this case would be interested in learning about how these sharks were obtained and when using them is (and is not) appropriate. This kind of discussion can foster consideration and respect for living organisms.

Such a conversation could also address the final and most important concern for this experience: the option for students to participate in an alternative activity. Ava's response suggests that she sees these animals as cuddly, living organisms that have been unjustly slaughtered. Ava is expressing an objection to using dead animals for learning with her age-appropriate language. Additionally, some students may even be from vegetarian or vegan families who believe that killing animals for their own benefit is cruel and harmful. It is likely that, for many children, this was their first encounter with preserved specimens of any kind. Thus, the reactions of some students when presented with shark specimens are understandable.

Since Austin used the sharks in the context of a (voluntary) summer camp, parents likely chose to sign their students up for a program that involved deep-sea shark specimen observation. However, in a traditional classroom setting, a teacher should provide an alternative activity for students who are not ready or willing to investigate with dead animals. Although the position statements of both NSTA and NABT recommend alternatives be offered for dissection only, teachers should consider the appropriateness of an alternative for this activity as well, given the age of the students. While many deep-sea sharks are not able to be housed in typical aquariums and are seldom caught on film, there are other opportunities for observing most animals. Observing such beautiful creatures while living (in the aquarium or on video) can yield an understanding of differences in anatomy and behavior. Are the adults all the same size? Do they all move the same way when swimming? Do they have the same fins in the same places? What does the eye placement tell you about them as predators? What else can students learn about various sharks from watching them in their natural habitat?

Austin's enthusiasm and dedication is admirable. It is important for informal educators like Austin, as well as classroom teachers, to consider a wide range of developmentally appropriate factors before presenting students with preserved animals to study in any type of investigation. Student age and maturity is definitely an important consideration in this case, and students must be adequately prepared for genuine learning to occur. With the considerations discussed above in mind, teachers can design the best learning experience for their young students to meet learning goals, with or without using preserved specimens.

References

National Science Teaching Association. (2008). *Responsible use of live animals and dissection in the science classroom.* https://static.nsta.org/pdfs/PositionStatement_LiveAnimalsAndDissect ion.pdf
National Association of Biology Teachers. (2019). *Position statements.* https://nabt.org/Position-Statements-The-Use-of-Animals-in-Biology-Education

Donna Governor   is an associate professor in the College of Education at the University of North Georgia. She has earned degrees in education from the University of West Florida and a PhD in science education from the University of Georgia. She is currently the Preservice Division Director for the National Science Teachers Association and author of two NSTA press books. Dr. Governor has 32 years of K-12 classroom experience and has earned several teaching awards, including the 2007 Presidential Award for Excellence in Science Teaching from Georgia.

Part IX
Conclusion

Conclusion: Cases in Elementary Science Teaching—Now and in the Future

Natasha Hillsman Johnson and Sophia Jeong

Abstract

Teaching and learning science in elementary grades has always included a unique set of hurdles for early childhood educators. For some educators, challenges arise from their self-perception of having limited science content knowledge and a lack of self-efficacy. We contend that case-based pedagogy in science education provides a unique opportunity to meet these challenges. It has the potential to meet the professional learning needs of elementary science teachers in a research-based, convenient, and context-specific manner. In this book, we share diverse case narratives that incorporate both the expected and unforeseen teaching challenges to equip classroom teachers with the necessary skills to tackle the increasing demands of a rapidly changing educational landscape. As we look to the future, case-based pedagogy will continue to evolve and fill a critical gap in the content- and context-specific professional needs of educators in a variety of settings.

Teaching and learning science in elementary grades has always included a unique set of hurdles for early childhood educators. For some educators, challenges arise from their self-perception of having limited science content knowledge and a lack of self-efficacy (Johnson & Dabney, 2018). However, Johnson and Dabney (2018)

N. H. Johnson
Department of Teacher Education, University of Toledo, Toledo, OH, USA
e-mail: natasha.johnson@utoledo.edu

S. Jeong (✉)
Department of Teaching and Learning, The Ohio State University, Columbus, OH, USA
e-mail: jeong.387@osu.edu

have identified additional "constraints that enthusiastic and well-prepared first year elementary teachers encounter while trying to deliver high-quality science instruction" (p. 245). The factors that can impede elementary teachers' efforts to teach science include time; classroom management; access to materials and resources; scheduling; and prioritizing elementary science instruction (Johnson & Dabney, 2018). In recent years, elementary educators have faced increased pressure to prepare students for the future, including two specific challenges: shifting to an integrated science, technology, engineering, and mathematics curricula; and teaching lifelong learning skills such as creativity, communication, critical thinking, and collaboration (Holdren et al., 2010). The responsibility has been placed on the primary grades because this time in students' lives is when initial STEM interest often sparks.

86.1 How Can the Science Education Community Improve the Preparation of Elementary Teachers to Deliver High-Quality Science Instruction?

We contend that case-based pedagogy in science education provides a unique opportunity to meet these challenges. Case-based pedagogy, which originated at Harvard Law School, has proven an effective method in the education of law, medical, and business students and holds equal promise for the field of education (Merseth, 1991; Puri, 2022). It has the potential to meet the professional learning needs of elementary science teachers in a research-based, convenient, and context-specific manner. It is imperative that we equip classroom teachers with the necessary skills to tackle the increasing demands of a rapidly changing educational landscape. Failure to properly contextualize elementary science teaching and learning may continue to have detrimental effects on teacher morale, teacher retention, and learning outcomes for "vulnerable" populations, including underrepresented "minority" students, economically disadvantaged students, and students with disabilities. Case-based pedagogy can be used by teacher preparation programs, professional learning communities, and local schools to explore complex issues included in this book, such as science standards; science and engineering practices; technology integration; diversity, equity, and inclusion; socioscientific issues; assessing student learning; using living organisms; and designing of science instruction.

86.2 How Does Case-Based Pedagogy Benefit Students and Teachers in an Educational Environment?

Standardized assessments, such as the educative Teacher Performance Assessment (edTPA) and the National Board for Professional Teaching Standards, used in the training, licensing, and certification of preservice and in-service teachers, have reinforced the need for competent teachers to effectively plan content, deliver instruction, and assess student learning. One of the most important components of these assessments is the ability for the candidate or teacher to engage in reflection about one's pedagogical practices. According to Rosen (2008), the use of case-based pedagogy in teacher education can "convey a strong representation of a complex teaching and learning context, its participants, and the realistic challenges in the situation" (p. 28). Research has also demonstrated that case-based pedagogy can result in "improved reflection on educational theories and instructional practices that facilitate children's learning" (Rosen, 2008, p. 33).

86.3   What is the Future of Case-Based Pedagogy?

Case-based pedagogy will continue to evolve and fill a critical gap in the content- and context-specific professional needs of educators in a variety of settings, including urban, suburban, and rural classrooms; traditional and alternative certification programs; and undergraduate and graduate programs. This innovative approach will usefully provide education to elementary science teachers without the hindrances of budget constraints, limited planning time, multiple course preparations, and teacher isolation. New cases will be developed to incorporate the expected and unforeseen teaching challenges associated with current and future global pandemics, legislation related to divisive issues, increasing incidents of school violence, social justice in science education, technological advancements, and the social and emotional needs of students.

As we look to the future, the science education community will continue to work collaboratively with teachers to navigate the challenges of elementary science teaching and to better understand the dilemmas of practice in order to meet the needs of all learners.

References

Holldren, J. P., Lander, E. S., & Varmus, H. (2010). Prepare and inspire: K-12 education in science, technology, engineering, and math (STEM) for America's future. *Executive Report Washington, DC: President's Council of Advisors on Science and Technology, 2*, 68–73.

Johnson, T. N., & Dabney, K. P. (2018). Voices from the field: Constraints encountered by early career elementary science teachers. *School Science and Mathematics, 118*(6), 244–256.

Merseth, K. K. (1991). The early history of case-based instruction: Insights for teacher education today. *Journal of Teacher Education, 42*(4), 243–249.

Puri, S. (2022). Effective learning through the case method. *Innovations in Education and Teaching International, 59*(2), 161–171.

Rosen, D. (2008). Impact of case-based instruction on student teachers' reflection on facilitating children's learning. *Action in Teacher Education, 30*(1), 28–36.

Natasha Hillsman Johnson is an assistant professor of Science Education in the Department of Teacher Education at the University of Toledo. The overarching goal of her scholarship is to increase interest, access, and achievement in the sciences for all students. Issues related to equity, social justice, and the amplification of marginalized voices continue to be the focus of her past, current, and future research interests in the area of science education.

Sophia Jeong is an assistant professor of Science Education in the Department of Teaching and Learning at The Ohio State University. Her work draws on theories of new materialisms to explore ontological complexities of subjectivities by examining socio-material relations in the science classrooms. Her research interests focus on equity issues through the lens of rhizomatic analysis of K-16 science classrooms. She is passionate about fostering creativity, encouraging inquisitive minds, and developing sociopolitical consciousness through science education.

Index

Printed in the United States
by Baker & Taylor Publisher Services